GENETICS, GENOMICS AND BREEDING OF BANANAS

Genetics, Genomics and Breeding of Crop Plants

Series Editor

Chittaranjan Kole
Department of Genetics and Biochemistry
Clemson University
Clemson, SC
USA

Books in this Series:

Published or in Press:

- Jinguo Hu, Gerald Seiler & Chittaranjan Kole: *Sunflower*
- Kristin D. Bilyeu, Milind B. Ratnaparkhe & Chittaranjan Kole: *Soybean*
- Robert Henry & Chittaranjan Kole: *Sugarcane*
- Kevin Folta & Chittaranjan Kole: *Berries*
- Jan Sadowsky & Chittaranjan Kole: *Vegetable Brassicas*
- James M. Bradeen & Chittaranjan Kole: *Potato*
- C.P. Joshi, Stephen DiFazio & Chittaranjan Kole: *Poplar*
- Anne-Françoise Adam-Blondon, José M. Martínez-Zapater & Chittaranjan Kole: *Grapes*
- Christophe Plomion, Jean Bousquet & Chittaranjan Kole: *Conifers*
- Dave Edwards, Jacqueline Batley, Isobel Parkin & Chittaranjan Kole: *Oilseed Brassicas*
- Marcelino Pérez de la Vega, Ana María Torres, José Ignacio Cubero & Chittaranjan Kole: *Cool Season Grain Legumes*
- Yi-Hong Wang, Tusar Kanti Behera & Chittaranjan Kole: *Cucurbits*
- Michael Pillay, George Ude & Chittaranjan Kole: *Bananas*

GENETICS, GENOMICS AND BREEDING OF BANANAS

Editors

Michael Pillay
Department of Biosciences
Vaal University of Technology
Vanderbijlpark
South Africa

George Ude
Department of Natural Sciences
Bowie State University (BSU)
Bowie, MD
USA

Chittaranjan Kole
Department of Genetics and Biochemistry
Clemson University
Clemson, SC
USA

CRC Press
Taylor & Francis Group
an informa business
www.taylorandfrancisgroup.com

6000 Broken Sound Parkway, NW
Suite 300, Boca Raton, FL 33487
711 Third Avenue
New York, NY 10017
2 Park Square, Milton Park
Abingdon, Oxon OX14 4RN, UK

Science Publishers
Jersey, British Isles
Enfield, New Hampshire

Published by Science Publishers, an imprint of Edenbridge Ltd.

- St. Helier, Jersey, British Channel Islands
- P.O. Box 699, Enfield, NH 03748, USA

E-mail: info@scipub.net Website: www.scipub.net

Marketed and distributed by:

CRC Press
Taylor & Francis Group
an informa business
www.taylorandfrancisgroup.com

6000 Broken Sound Parkway, NW
Suite 300, Boca Raton, FL 33487
711 Third Avenue
New York, NY 10017
2 Park Square, Milton Park
Abingdon, Oxon OX14 4RN, UK

ISBN 978-1-57808-788-4

Cover photograph reproduced by kind courtesy of Moses Nyine of IITA, Uganda.

Library of Congress Cataloging-in-Publication Data

Genetics, genomics and breeding of bananas / editors, Michael Pillay, George Ude, Chittaranjan Kole. -- 1st ed.
p. cm. -- (Genetics, genomics and breeding of crop plants)
Includes bibliographical references and index.
ISBN 978-1-57808-788-4 (hardcover : alk. paper)
1. Bananas--Genetics. 2. Bananas--Genome mapping. 3. Bananas--Breeding. I. Pillay, Michael, 1950- II. Ude, George (George Nkem) III. Kole, Chittaranjan. IV. Series: Genetics, genomics and breeding of crop plants.
SB379.B2G46 2011
634.772--dc23

2011038880

The views expressed in this book are those of the author(s) and the publisher does not assume responsibility for the authenticity of the findings/conclusions drawn by the author(s). Also no responsibility is assumed by the publishers for any damage to the property or persons as a result of operation or use of this publication and/or the information contained herein.

Printed in the United States of America

Foreword

Banana and plantain (*Musa* spp.) are very important fruits worldwide. Export of banana is probably only surpassed by citrus in trade volumes. *Musa* cultivars evolved by natural hybridization between two species: *M. acuminata* (contributing genome A) and *M. balbisiana* (contributing genome B), which originated in Southeast Asia.

The production of dessert bananas (AAA, AAB), cooking bananas (ABB) and plantains (AAB) relies on the widespread dissemination of a few superior landraces and cultivars that are well known by vernacular names in each location of the tropics and subtropics where they are grown.

Genetic improvement by cross-breeding is achieved by using triploid landraces and cultivars as maternal parent and diploids as male parents. Advanced ploidy manipulations may lead to secondary triploid hybrids resulting from crosses between selected tetraploid hybrids with elite diploid stocks, both producing *n* gametes. Triploid *Musa* hybrids may also occur due to unilateral sexual polyploidization among selected diploid stocks, if one of the parents produces 2*n* gametes. The late Dirk Vuylsteke, Norman Simmonds, Ken Shepherd, Phil Rowe, and Ren Golsalves are among the most respected *Musa* breeders worldwide. Their pioneering work in the genetic enhancement of bananas and plantains during the 20th century in tropical America and sub-Saharan Africa paved the way for improving these crops for smallholders and for understanding the *Musa* genome.

Molecular and cell biology research and ensuing biotechnology tools could help breeders to genetically enhance the crop and meet this century's demand for banana and plantain. Available techniques for selecting and transferring genes could significantly shorten the breeding process and overcome some of the agronomic and environmental problems, which may not be possible through conventional methods. Likewise, molecular marker-assisted breeding has the potential to dramatically enhance the pace and efficiency of genetic improvement in *Musa*.

As pointed out in the last paragraph of the PhD Thesis of Dirk Vuylsteke "A broad-based, improved *Musa* germplasm with pest/disease resistance will be a major component to achieve sustainable production of this vegetatively propagated, perennial crop. Such germplasm can be produced through conventional cross-breeding, enhanced by the utilization of innovative methods for the introduction of additional genetic variation. Also, the increased use of molecular markers will accelerate the process of recurrent selection of improved *Musa* germplasm and, hence, facilitate the development of new hybrids. The prospects of banana and plantain breeding are unlimited and increased efforts will at once initiate a new phase of *Musa* evolution". I fully concur with his vision, and am confident that new breakthroughs in *Musa* genomics and transgenics will allow their use in breeding new cultivars, which are urgently needed by banana and plantain farmers worldwide to sustainably produce this crop, especially at a time of changing climate and high food prices.

Rodomiro Ortiz, PhD (Univ. Wisconsin-Madison 1991)
Chaclacayo, February 2011

Preface to the Series

Genetics, genomics and breeding has emerged as three overlapping and complementary disciplines for comprehensive and fine-scale analysis of plant genomes and their precise and rapid improvement. While genetics and plant breeding have contributed enormously towards several new concepts and strategies for elucidation of plant genes and genomes as well as development of a huge number of crop varieties with desirable traits, genomics has depicted the chemical nature of genes, gene products and genomes and also provided additional resources for crop improvement.

In today's world, teaching, research, funding, regulation and utilization of plant genetics, genomics and breeding essentially require thorough understanding of their components including classical, biochemical, cytological and molecular genetics; and traditional, molecular, transgenic and genomics-assisted breeding. There are several book volumes and reviews available that cover individually or in combination of a few of these components for the major plants or plant groups; and also on the concepts and strategies for these individual components with examples drawn mainly from the major plants. Therefore, we planned to fill an existing gap with individual book volumes dedicated to the leading crop and model plants with comprehensive deliberations on all the classical, advanced and modern concepts of depiction and improvement of genomes. The success stories and limitations in the different plant species, crop or model, must vary; however, we have tried to include a more or less general outline of the contents of the chapters of the volumes to maintain uniformity as far as possible.

Often genetics, genomics and plant breeding and particularly their complementary and supplementary disciplines are studied and practiced by people who do not have, and reasonably so, the basic understanding of biology of the plants for which they are contributing. A general description of the plants and their botany would surely instill more interest among them on the plant species they are working for and therefore we presented lucid details on the economic and/or academic importance of the plant(s); historical information on geographical origin and distribution; botanical origin and evolution; available germplasms and gene pools, and genetic and cytogenetic stocks as genetic, genomic and breeding resources; and basic information on taxonomy, habit, habitat, morphology, karyotype, ploidy level and genome size, etc.

Classical genetics and traditional breeding have contributed enormously even by employing the phenotype-to-genotype approach. We included detailed descriptions on these classical efforts such as genetic mapping using morphological, cytological and isozyme markers; and achievements of conventional breeding for desirable and against undesirable traits. Employment of the in vitro culture techniques such as micro- and megaspore culture, and somatic mutation and hybridization, has also been enumerated. In addition, an assessment of the achievements and limitations of the basic genetics and conventional breeding efforts has been presented.

It is a hard truth that in many instances we depend too much on a few advanced technologies, we are trained in, for creating and using novel or alien genes but forget the infinite wealth of desirable genes in the indigenous cultivars and wild allied species besides the available germplasms in national and international institutes or centers. Exploring as broad as possible natural genetic diversity not only provides information on availability of target donor genes but also on genetically divergent genotypes, botanical varieties, subspecies, species and even genera to be used as potential parents in crosses to realize optimum genetic polymorphism required for mapping and breeding. Genetic divergence has been evaluated using the available tools at a particular point of time. We included discussions on phenotype-based strategies employing morphological markers, genotype-based strategies employing molecular markers; the statistical procedures utilized; their utilities for evaluation of genetic divergence among genotypes, local landraces, species and genera; and also on the effects of breeding pedigrees and geographical locations on the degree of genetic diversity.

Association mapping using molecular markers is a recent strategy to utilize the natural genetic variability to detect marker-trait association and to validate the genomic locations of genes, particularly those controlling the quantitative traits. Association mapping has been employed effectively in genetic studies in human and other animal models and those have inspired the plant scientists to take advantage of this tool. We included examples of its use and implication in some of the volumes that devote to the plants for which this technique has been successfully employed for assessment of the degree of linkage disequilibrium related to a particular gene or genome, and for germplasm enhancement.

Genetic linkage mapping using molecular markers have been discussed in many books, reviews and book series. However, in this series, genetic mapping has been discussed at length with more elaborations and examples on diverse markers including the anonymous type 2 markers such as RFLPs, RAPDs, AFLPs, etc. and the gene-specific type 1 markers such as EST-SSRs, SNPs, etc.; various mapping populations including F_2, backcross, recombinant inbred, doubled haploid, near-isogenic and pseudotestcross; computer software including MapMaker, JoinMap, etc. used; and different types of genetic maps including preliminary, high-resolution, high-density, saturated, reference, consensus and integrated developed so far.

Mapping of simply inherited traits and quantitative traits controlled by oligogenes and polygenes, respectively has been deliberated in the earlier literature crop-wise or crop group-wise. However, more detailed information on mapping or tagging oligogenes by linkage mapping or bulked segregant analysis, mapping polygenes by QTL analysis, and different computer software employed such as MapMaker, JoinMap, QTL Cartographer, Map Manager, etc. for these purposes have been discussed at more depth in the present volumes.

The strategies and achievements of marker-assisted or molecular breeding have been discussed in a few books and reviews earlier. However, those mostly deliberated on the general aspects with examples drawn mainly from major plants. In this series, we included comprehensive descriptions on the use of molecular markers for germplasm characterization, detection and maintenance of distinctiveness, uniformity and stability of genotypes, introgression and pyramiding of genes. We have also included elucidations on the strategies and achievements of transgenic breeding for developing genotypes particularly with resistance to herbicide, biotic and abiotic stresses; for biofuel production, biopharming, phytoremediation; and also for producing resources for functional genomics.

A number of desirable genes and QTLs have been cloned in plants since 1992 and 2000, respectively using different strategies, mainly positional cloning and transposon tagging. We included enumeration of these and other strategies for isolation of genes and QTLs, testing of their expression and their effective utilization in the relevant volumes.

Physical maps and integrated physical-genetic maps are now available in most of the leading crop and model plants owing mainly to the BAC, YAC, EST and cDNA libraries. Similar libraries and other required genomic resources have also been developed for the remaining crops. We have devoted a section on the library development and sequencing of these resources; detection, validation and utilization of gene-based molecular markers; and impact of new generation sequencing technologies on structural genomics.

As mentioned earlier, whole genome sequencing has been completed in one model plant (Arabidopsis) and seven economic plants (rice, poplar, peach, papaya, grapes, soybean and sorghum) and is progressing in an array of model and economic plants. Advent of massively parallel DNA sequencing using 454-pyrosequencing, Solexa Genome Analyzer, SOLiD system, Heliscope and SMRT have facilitated whole genome sequencing in many other plants more rapidly, cheaply and precisely. We have included extensive coverage on the level (national or international) of collaboration and the strategies and status of whole genome sequencing in plants for which sequencing efforts have been completed or are progressing currently. We have also included critical assessment of the impact of these genome initiatives in the respective volumes.

Comparative genome mapping based on molecular markers and map positions of genes and QTLs practiced during the last two decades of the last century provided answers to many basic questions related to evolution, origin and phylogenetic relationship of close plant taxa. Enrichment of genomic resources has reinforced the study of genome homology and synteny of genes among plants not only in the same family but also of taxonomically distant families. Comparative genomics is not only delivering answers to the questions of academic interest but also providing many candidate genes for plant genetic improvement.

The 'central dogma' enunciated in 1958 provided a simple picture of gene function—gene to mRNA to transcripts to proteins (enzymes) to metabolites. The enormous amount of information generated on characterization of transcripts, proteins and metabolites now have led to the emergence of individual disciplines including functional genomics, transcriptomics, proteomics and metabolomics. Although all of them ultimately strengthen the analysis and improvement of a genome, they deserve individual deliberations for each plant species. For example, microarrays, SAGE, MPSS for transcriptome analysis; and 2D gel electrophoresis, MALDI, NMR, MS for proteomics and metabolomics studies require elaboration. Besides transcriptome, proteome or metabolome QTL mapping and application of transcriptomics, proteomics and metabolomics in genomics-assisted breeding are frontier fields now. We included discussions on them in the relevant volumes.

The databases for storage, search and utilization on the genomes, genes, gene products and their sequences are growing enormously in each second and they require robust bioinformatics tools plant-wise and purpose-wise. We included a section on databases on the gene and genomes, gene expression, comparative genomes, molecular marker and genetic maps, protein and metabolomes, and their integration.

Notwithstanding the progress made so far, each crop or model plant species requires more pragmatic retrospect. For the model plants we need to answer how much they have been utilized to answer the basic questions of genetics and genomics as compared to other wild and domesticated species. For the economic plants we need to answer as to whether they have been genetically tailored perfectly for expanded geographical regions and current requirements for green fuel, plant-based bioproducts and for improvements of ecology and environment. These futuristic explanations have been addressed finally in the volumes.

We are aware of exclusions of some plants for which we have comprehensive compilations on genetics, genomics and breeding in hard copy or digital format and also some other plants which will have enough achievements to claim for individual book volume only in distant future. However, we feel satisfied that we could present comprehensive deliberations on genetics, genomics and breeding of 30 model and economic plants, and their groups in a few cases, in this series. I personally feel also happy that I could work with many internationally celebrated scientists who edited the book volumes on the leading plants and plant groups and included chapters authored by many scientists reputed globally for their contributions on the concerned plant or plant group.

We paid serious attention to reviewing, revising and updating of the manuscripts of all the chapters of this book series, but some technical and formatting mistakes will remain for sure. As the series editor, I take complete responsibility for all these mistakes and will look forward to the readers for corrections of these mistakes and also for their suggestions for further improvement of the volumes and the series so that future editions can serve better the purposes of the students, scientists, industries, and the society of this and future generations.

Science publishers, Inc. has been serving the requirements of science and society for a long time with publications of books devoted to advanced concepts, strategies, tools, methodologies and achievements of various science disciplines. Myself as the editor and also on behalf of the volume editors, chapter authors and the ultimate beneficiaries of the volumes take this opportunity to acknowledge the publisher for presenting these books that could be useful for teaching, research and extension of genetics, genomics and breeding.

Chittaranjan Kole

Preface to the Volume

This book entitled *"Genetics, Genomics and Breeding of Banana and Plantain"* is a collection of chapters written by eminent authors who have been researching on various aspects of banana and especially in the field of genomics. Over 100 million metric tons of bananas and plantains are produced annually in over 140 countries in the tropical and subtropical regions of the world. It is a staple food and a source of income for millions of people. The diseases and pests that affect banana are responsible for huge crop losses that are threatening the livelihoods of millions. Conventional breeding alone may not be able to generate elite hybrids to prevent crop losses due to the complex biology of the species that includes polyploidy, heterozygosity, multiple genomes, and vegetative propagation. Molecular breeding including the tools of genomics, proteomics, metabolomics and bioinformatics could play a greater role in the development of enhanced *Musa* germplasm. The purpose of this book is to gather in one volume the latest information on the genomic researches of *Musa*. A group of leading experts in *Musa* genetics, genomics and breeding contributed to this volume in a style that will be understood by students and researchers. This book provides basic as well as advanced information for those interested in learning more about the banana genome and its improvement as well for those pursuing further research in the crop.

Chapter 1 written by Michael Pillay, Abdou Tenkouano and Rodomiro Ortiz, individuals who spent many years working with the crop, provide a detailed botanical description of the plant, history of the crop, its academic importance and a description of the germplasm and genepools.

In **Chapter 2** Michael Pillay provides an overview of efforts in classical mapping, its limitations and the utility of molecular mapping in *Musa*. In addition, breeding objectives and achievements in conventional breeding are outlined. The limitations of traditional breeding and a case for molecular breeding in *Musa* is provided. **Chapter 3** by Kornel Burg and Ratri Boonruangrod outlines the usefulness of molecular studies in unraveling the origin of extant bananas. They highlight the value of using organellar DNA from the chloroplast and mitochondrial in tracing the ancestry of our modern triploid bananas from the ancestral diploid species, *Musa* *acuminata* and *M. balbisiana*. In **chapter 4,** Michael Pillay, Kaliyaperumal Ashokkumar, Andrew James, Silvas Jabekumar Prince Kirubakaran, Robert Miller, Rodomiro Ortiz and Elayabalan Sivalingam describe the various molecular techniques that have been used in *Musa* to address many research questions. The authors advocate the use of the latest techniques such as SNPs, TILLING and high resolution melting to enhance the knowledge on genomes in *Musa*.

Chapter 5 by Cory Johnson and Christopher Cullis address the strategies, resources and achievements of molecular mapping in *Musa* with examples. Various types of populations used in mapping are described. Mapping software and the development of genetic and molecular maps are discussed.

In **Chapter 6** Rodomiro Ortiz draws from his personal experience and not only discusses the importance of simply inherited traits but describes their transmission genetics- research in which he was directly involved. The chapter also describes the tagging of genes with molecular markers.

Chapter 7 written by Rodomiro Ortiz addresses the problems associated with mapping of quantitative traits in *Musa*. Framework maps and markers used to characterize breeding populations are discussed. He suggests that complex traits that are not controlled by many genes may involve simpler pathways that can more easily be measured phenotypically and characterized at the DNA sequence level.

In **Chapter 8** Andrew James, Rodomiro Ortiz and Robert Miller use their vast combined experience in molecular genetics of *Musa* to highlight the current knowledge of map-based cloning in the crops.

The chapter addresses genomic resources available in *Musa* for map-based cloning, cDNA and EST libraries, available segregating populations, genomic libraries and future prospects for map-based cloning.

In **Chapter 9** Michael Pillay, Sivalingam Elayabalan, Kaliyaperumal Ashokkumar and Silvas Jabekumar Prince review some of the techniques for gene functional analysis with special emphasis on insertional mutagenesis, targeted gene disruption by homologous recombination, RNA interference, VIGS technology, and TILLING. In addition, progress in some areas of transcriptomics research in *Musa* is reviewed. In **Chapter 10,** Hoang Lan Chi Dinh and Peer M. Schenk show how proteomics and metabolomics have been utilized in crop development. This chapter reviews the current advances in different areas of *Musa* metabolome and proteome studies. Experimental and analytical tools such as NMR, MS, 2D-gel electrophoresis, GC-MS, MALDI-TOF have been described as important requisites for the generation of *Musa* metabolome and proteome data. In **Chapter 11,** Mathieu

Rouard, Sebastien Christian Carpentier, Stephanie Bocs, Gaëtan Droc, Xavier Argout, Nicolas Roux, and Manuel Ruiz deliberate on the role of bioinformatics as a tool in *Musa*. The chapter discusses the various databases available for gene and genomes, comparative genomics, gene expression, molecular marker and genetic maps, protein or metabolome databases. The chapter also addresses the integration of different data.

In **Chapter 12** Cláudia Fortes Ferreira, Hermínio Souza Rocha, Ricardo Franco Cunha Moreira and Lahyre Izaete Silveira Gomes have poooled their experiences to address the development of molecular markers, sequencing initiatives and other biotechnological tools to examine *Mycosphaerella fijiensis*. In **Chapter 13,** Raman Thangavelu, P. Ganga Devi, Mohammed M. Mustaffa, Subramaniam Sreeramanan, and Xavier Rathinam present a chapter on the genomics of Fusarium wilt, one of the major diseases of banana. The causal agent, its classification into races and VCGs is backed with new evidence from molecular classification. Markers to identify the disease and strategies to generate banana tolerance to fungal pathogens are discussed. Finally transgenic technology to develop resistant cultivars is described.

Chapter 14 by Santy Peraza-Echeverria, Virginia Aurora Herrera-Valencia and Andrew James provide an overview about the cloning and molecular characterization of two classes of disease resistance (R) gene homologues in banana, the *NBS* (nucleotide-binding site) and the *Pto* (serine/threonine kinase) classes of R genes. Furthermore, they provide insights on the recent cloning and characterization of three banana genes showing homology to the non-expresssor of pathogenesis related gene 1 (*NPR1*), which is a master transcriptional regulator of systemic acquired resistance (SAR). The potential applications of these genes to develop disease resistance in this crop are discussed. **Chapter 15,** by Michael Pillay, Kaliyaperumal Ashokkumar, Arun Shiva Kumar Shunmugam and Sivalingam Elayabalan provide a case for molecular breeding in *Musa*. The breeding challenges, production constraints, and main objectives are described. A case is built for molecular breeding in *Musa*. Progress in identifying molecular markers for important traits is described. The chapter also addresses marker-assisted gene introgression and gene pyramiding. Finally the achievements and prospects of molecular breeding in *Musa* are discussed. In the concluding **Chapter 16** Michael Pillay looks at prospects in *Musa* research and assesses some of the latest molecular tools such as TILLING, proteomics, RNA interference and metabolomics for their use in *Musa* research. The chapter also discusses the importance of breeding for stress tolerance in view of global climate changes.

We trust that those interested in banana and plantain will gain valuable insights and inspiration as they expand their scientific knowledge with information from this book. This book is intended for students, teachers and banana researchers. We hope that academics throughout the world interested in developing tropical crops will also find use for this book. We thank all the contributors for their time and patience in sharing their knowledge to make this book a success. We are grateful to Dr. Rodomiro Ortiz for writing the Foreword for this volume and for his timely editing of many of the chapters.

Michael Pillay
George Ude
Chittaranjan Kole

Contents

List of Contributors

Xavier Argout
CIRAD, Biological Systems Department, UMR AGAP-TA40/03, 34398 Montpellier, France.
Tel: +33 4 67 61 58 00
Fax: +33 4 67 61 56 05
Email: *xavier.argout@cirad.fr*

Kaliyaperumal Ashokkumar
Department of Plant Sciences, University of Saskatchewan, 51 Campus Drive, Saskatoon, SK, Canada, S7N 5A8.
Tel: 001-306-966-5859
Fax: 001- 306 966-5015
Email: *ashok.kumar@usask.ca*

Stephanie Bocs
CIRAD, Biological Systems Department, UMR AGAP-TA40/03, 34398 Montpellier, France.
Tel: +33 4 67 61 56 31
Fax: +33 4 67 61 56 05
Email: *stephanie.sidibe-bocs@cirad.fr*

Ratri Boonruangrod
Department of Horticulture, Faculty of Agriculture at Kamphaengsaen, Kasetsart University, Kamphaengsaen, Nakhon Pathom, 73140 Thailand.
Tel: +66 34 281 053-6 ext. 114
Email: *ratri.b@ku.ac.th*

Kornel Burg
Health & Environment Department Bioresources, AIT Austrian Institute of Technology GmbH, 2444 Seibersdorf, Austria.
Tel: +43(0) 50550-3524
Fax: +43(0) 50550-3666
Email: *kornel.burg@ait.ac.at*

Sebastien Christian Carpentier
Department of Biosystems, Division of Crop Biotechnics, Katholic University of Leuven, Leuven, Belgium.
Tel: +32 16 321633
Fax: +32 16 321993
Email: *sebastien.carpentier@biw.kuleuven.be*

Hoang Lan Chi Dinh
School of Biological Sciences, The University of Queensland, St. Lucia, Queensland 4072, Australia.
Tel: +61-7-3365 2815
Fax: +61-7-3365 1699
Email: *dinhhoanglanchi@gmail.com*

Christopher Cullis
Department of Biology, Case Western Reserve University, 10900 Euclid Avenue, Cleveland, Ohio 44106-7080.
Tel: +1 216 368 3557
Fax: +1 216 368 4672
Email: *cac5@case.edu*

Gaëtan Droc
CIRAD, Biological Systems Department, UMR AGAP-TA40/03, 34398 Montpellier, France.
Tel: +33 4 67 61 58 00
Fax: +33 4 67 61 56 05
Email: *gaetan.droc@cirad.fr*

Cláudia Fortes Ferreira
Molecular Biology Department, Rua Embrapa, C.P. 007, Cruz das Almas, Bahia, Brazil.
Tel: +55 75 3312-8061
Fax: +55 75 3312-8097
Email: *claudiaf@cnpmf.embrapa.br*

Lahyre Izaete Silveira Gomes
Federal University of Viçosa–UFV, Department of Phytopathology, Rua PH Rolfs 36570-000 Viçosa –MG, Brazil.
Tel: +55 31 3899 2026
Email: *lahyreizaete@yahoo.com.br*

Virginia Aurora Herrera-Valencia
Unidad de Biotecnología, Centro de Investigación Científica de Yucatán, Calle 43 No. 130 Colonia Chuburná de Hidalgo, Mérida, Yucatán, México. C.P. 97200.
Tel: +52 999 9428330
Fax: +52 999 9813900
Email: *vicky@cicy.mx*

Andrew James
Centro de Investigacion Cientifica de Yucatan, Calle 43 No. 130, Col. Churburna de Hidalgo, Merida, C.P. 97200, Yucatan, Mexico.
Tel: +52-999-942 83 30;
Fax: +52 -999-981-3900
Email: *andyj007@cicy.mx*

Cory Johnson
Department of Biology, Case Western Reserve University, 10900 Euclid Avenue, Cleveland, Ohio 44106-7080.
Tel: 216-368-5301
Fax: 216-368-4672
Email: *clj18@case.edu*

Silvas Jabekumar Prince Kirubakaran
Department of Plant Molecular Biology and Biotechnology, Center for Plant Molecular Biology, Tamil Nadu Agricultural University, Coimbatore-641 003-India.
Email: *symsil07@gmail.com*

Robert Miller
Laboratório de Microbiologia: Interação Planta-Praga, Instituto de Ciências Biológicas, Departamento de Biologia Celular, Universidade de Brasília, Campus Universitário Darcy Ribeiro, Asa Norte, 70910-900, Brasília, D.F., Brasil.
and
Universidade Católica de Brasília, SGAN Qd. 916 Módulo B, Av. W5 Norte CEP: 70.790-160, Brasília/DF, Brazil.
Email: *rmiller@pos.ucb.br or bruna@tba.com.br*

Ricardo Franco Cunha Moreira
Federal University of Bahia Recôncavo, Center for Agricultural Sciences, Environmental and Biological, Rui Barbosa, 710-University Campus, CEP 44380-000, Cruz das Almas, Bahia, Brazil.
Tel: + 55 75 3621 9417
Fax: + 55 75 3621 6389
Email: *ricardofcm@ufrb.edu.br*

Mohammed M. Musataffa
National Research Centre for Banana, Thiruchirapalli 620102, Tamil Nadu, India.
Tel: +914312618106
Fax: +919443589882
Email: *directornrcb@gmail.com*

Rodomiro Ortiz
Department of Plant Breeding and Biotechnology, Swedish University of Agricultural Sciences, Box 101, SE-230 53, Alnarp, Sweden.
Tel: + 51 1 3585530
Email: *rodomiroortiz@gmail.com*

Santy Peraza-Echeverria
Unidad de Biotecnología, Centro de Investigación Científica de Yucatán, Calle 43 No. 130 Colonia Chuburná de Hidalgo, Mérida, Yucatán, México. C.P. 97200.
Tel: +52 999 9428330
Fax: +52 999 9813900
Email: *santype@cicy.mx*

Ganga Devi Perumal
National Research Centre for Banana, Thiruchirapalli 620102, Tamil Nadu, India.
Tel: +914312618106
Fax: +919443589882
Email: *ganga_gns@yahoo.com*

Michael Pillay
Vaal University of Technology, Private Bag X021, Vanderbijlpark 1900, Gauteng, South Africa.
Tel: 016 950 9614
Fax: 016 950 9794
Email: *mpillay@vut.ac.za*

Thangavelu Raman
National Research Centre for Banana (ICAR), Thiruchirapalli 620102 Tamil Nadu, India.
Email: *bananathanga@rediffmail.com*

Xavier Rathinam
Department of Biotechnology, AIMST University (AIMST), Semelling, 08000, Kedah, Malaysia.
Tel: +6016-4591514
Fax: +604-4298009
Email: *rxavier77@yahoo.com*

Hermínio Souza Rocha
Rua Embrapa, C.P. 007, 44380-000, Cruz das Almas, Bahia, Brazil.
Tel: +55 75 3312-8061
Fax: +55 75 3312-8097
Email: *herminio@cnpmf.embrapa.br*

Mathieu Rouard
Bioversity International, Commodities for Livelihood Program, 34397 Montpellier, France.
Tel: +33 4 67 61 29 08
Fax: +33 4 67 61 03 34
Email: *m.rouard@cgiar.org*

Nicolas Roux
Bioversity International, Commodities for Livelihood Program, 34397 Montpellier, France.
Tel: +33 4 67 61 99 46
Fax: +33 4 67 61 03 34
Email: *n.roux@cgiar.org*

Manuel Ruiz
CIRAD, Biological Systems Department, UMR AGAP - TA40/03, 34398 Montpellier, France.
Tel: +33 4 67 61 65 29
Fax: +33 4 67 61 56 05
Email: *manuel.ruiz@cirad.fr*

Peer M. Schenk
School of Biological Sciences, The University of Queensland, St. Lucia, Queensland 4072, Australia.
Tel: +61-7-33658817
Fax: +61-7-33651699
Email: *p.schenk@uq.edu.au*

Arun Siva Kumar Shunmugam
Department of Plant Sciences, University of Saskatchewan, 51 Campus Drive, Saskatoon, SK, Canada, S7N 5A8.
Tel: 001-306-966-5859
Fax: 001-306 966-5015
Email: *anbeshivam481@gmail.com*

Elayabalan Sivalingum
Crop Tech Ltd., P.O.Box 1367, Arusha, Tanzania.
Tel: +255-658-381983
Email: *balabiotech@gmail.com*

Sreeramanan Subramaniam
School of Biological Sciences, Universiti Sains Malaysia (USM), Georgetown, 11800, Penang, Malaysia.
Tel: +6016-4141109
Fax: +604-6565125
Email: *sreeramanan@usm.my*

Abdou Tenkouano
Regional Center for Africa, AVRDC-The World Vegetable Center, P.O. Box 10, Duluti, Arusha, Tanzania.
Tel: +255-27-255-3093/255-3102
Fax: +255-27-255-3125
Email: *Abdou.Tenkouano@worldveg.org*

Raman Thangavelu
National Research Centre for Banana, Thiruchirapalli 620102, Tamil Nadu, India.
Tel: +914312618106
Fax: +919443589882
Email: *bananathanga@rediffmail.com*

1

Introduction

Michael Pillay,[1,*] *Abdou Tenkouano*[2] and *Rodomiro Ortiz*[3]

ABSTRACT

Bananas originated in Southeast Asia and the western Pacific region and spread to the rest of the world. Although the origin and evolution of banana is still speculative, new ideas obtained from phytoliths and molecular research are presented. The chapter describes the taxonomic position of *Musa* and highlights the questionable nature of sections in view of new molecular data. The general morphology of the plant is presented and the different genomes, their sizes and ploidy levels in *Musa* are described. The role of *Musa* as a food crop, the main banana producing countries and the number of hectares that the crop occupies in the world is highlighted. Nutritional information on banana is summarized. The academic importance of the crop with respect to genomics studies, polyploidy, as a non-model crop for proteomics studies, breeding and cytogenetics are outlined. Although banana breeding is hampered by many factors the crop offers many new opportunities for genetic improvement by the discerning scientist. Banana breeders have largely relied on the introgession of genes from closely related wild plants to increase genetic variation in the cultivated varieties. The value of wild plants as sources of resistance for major diseases and pests of banana are discussed. While writing this chapter it became evident that research in *Musa* remains dynamic and new research groups are constantly adding information about the crop.

Keywords: *Musa*, origin, economic importance, academic importance, germplasm, breeding

[1]Vaal University of Technology, Private Bag X021, Vanderbijlpark 1900 South Africa; e-mail: *mpillay@vut.ac.za*
[2]Regional Center for Africa, AVRDC-The World Vegetable Center, P.O. Box 10, Duluti, Arusha, Tanzania; e-mail: *Abdou.Tenkouano@worldveg.org*
[3]Department of Plant Breeding and Biotechnology, Swedish University of Agricultural Sciences, Box 101, SE-230 53, Alnarp, Sweden; e-mail: *rodomiroortiz@gmail.com*
*Corresponding author

List of abbreviations after the text.

1.1 Brief History of the Crop

Despite its long history, the origin and evolution of banana is still regarded as speculative. Perhaps, the main reason for this is that bananas do not fossilize. Some of the thoughts on the origin of banana are changing as new data, especially from phytoliths, which are highly durable pieces of silica that form in plant cells, and molecular research, accumulate. Some new ideas on the early history and origin of bananas are described by Lejju et al. (2006) and De Langhe et al. (2010), respectively and references therein. However, it is accepted that bananas originated in the natural forests of Southeast Asia and the western Pacific region where wild inedible, seeded, diploid species (*Musa acuminata*) are still extant (Robinson 1996). This region is considered as the primary center of *Musa* diversity on the basis of Vavilov's (1935) assumptions that important crop plants originated in those regions in which the largest number of varieties can be found in the wild. Recent genetic studies have confirmed that the wild seeded *Musa acuminata* ssp. *banksii* F. Muell., was domesticated in New Guinea and then dispersed to Southeast Asia (Lebot et al. 1993; Lebot 1999). All edible bananas, except for a small group called "Fei" bananas, are thought to be derived from inter- and or intra- specific hybridization from two wild diploid ($2n = 2x = 22$) species, *M. acuminata* Colla and *M. balbisiana* Colla whose genomes are designated AA and BB, respectively (Simmonds 1962). The morphology of many banana varieties considered to have originated from crosses between the A and B-genome donors shows a bias towards the A or B phenotype and does not correspond to the simple genome formulas proposed by Simmonds and Shepherd (1955). This may indicate that the origin of cultivated banana varieties was not a single-step affair and that the domestication involved a set of backcrosses and human selection leading to a modern-day crop (De Langhe et al. 2010). The idea that selected plants grown by early humans produced seed progeny after backcrossing is supported by the observation of residual fertility in most of clonally propagated banana varieties (De Langhe et al. 2009). New ideas on the origin of present day hybrid bananas are expressed in Boonruangrod et al. (2008, 2009). There are a number of subspecies in the *Musa acuminata* complex.

The following account of the early history of banana is provided by Robinson (1996). After many years of natural crossing among the inedible diploid subspecies of *M. acuminata* produced numerous interspecific hybrids. Some of these hybrids were probably parthenocarpic and female sterile and were discovered by local inhabitants to have edible fruits. Such varieties were selected, cultivated and distributed locally as food. Some of the diploid x diploid crosses produced triploid varieties, perhaps resulting from the production of $2n$ gametes in the diploids. This is perhaps how AAA and possibly AAAA varieties originated. This is plausible since $2n$

gametes have been reported in *Musa* (Dodds 1943; Dodds and Simmonds 1946; Sathiamoorthy and Balamohan 1993; Ortiz 1997). Further selection of the edible triploid varieties on the basis of vigor, fruit size and adaptability were promoted while the inferior diploids were neglected. *Musa balbisiana* is thought to have originated in drier monsoon regions of India and the Philippines where it occurred naturally. *Musa balbisiana* still occurs wild in parts of India, Sri Lanka, Myanmar and Southwest China (Simmonds and Shepherd 1995). It is believed that wild banana seeds identified from an early Holocene site in Sri Lanka are probably those of *M. balbisiana* (Kajale 1989). When diploid and triploid selections of *M. acuminata* were transported by man to the drier areas, interspecific hybridization took place and produced *M. acuminata* x *M. balbisiana* diploid and triploid accessions. The earliest archeological evidence for a domesticated AB hybrid is probably that of *Musa* phytoliths from the Harappan site of Kot Diji in Sindh, which dates back to the second half of the third millennium (Kajale 1989; Fuller and Madella 2001). Another species that was probably involved in the origin of the cultivated species, especially some of the diploids in Papua New Guinea, is *M. schizocarpa* (Sharrock 1989). Molecular studies showed a close affinity between *M. schizocarpa* and the *M. acuminata* complex (Ude et al. 2002a). It has also been reported that hybrids between *M. acuminata* and *M. schizocarpa* do exist in nature (Argent 1976; Shepherd and Ferreira 1984; Tezenas du Montcel et al. 1995). The relationship between *M. schizocarpa* and the *M. acuminata* accessions suggests that the S-genome, characteristic of *M. schizocarpa*, could be exploited in breeding programs.

The dissemination of banana from its center of origin to other parts of the world is still speculative, but was probably due to the sailing nations. Such early voyaging across the Indian Ocean has been tentatively credited to Proto-Malayo-Polynesian- speaking populations (Lejju et al. 2006). It has been suggested that bananas from Indonesia were taken across the Indian Ocean to Madagascar about 500 BC and from there to East Africa, Zaire (now D.R. Congo) and West Africa (Robinson 1996). Plantains arrived much later. When the Portuguese arrived in West Africa in the 15th century, they found both bananas and plantains already in the region. It is thought that they introduced bananas from West Africa to the Canary Islands from where it was taken to Haiti in the 16th century. Bananas were then introduced to the remaining of the Carribean and other tropical American locations. It is also assumed that bananas may have been introduced to Africa several times. AA and AAA cultivars may have been introduced directly from Southeast Asia, whereas AAB and ABB hybrids are more likely to have reached Africa from India or Sri Lanka (Lejju et al. 2006). The recent discovery of banana phytoliths at Munsa in Uganda provides evidence for the existence of bananas in Africa during the fourth millennium BC (Lejju et al. 2006). This paper also reports that bananas were probably being cultivated in Africa as long as 5,000 years ago (Lejju et al. 2006), suggesting that the dissemination of bananas cannot be dissociated from their cultivation history. The cultivation of bananas was first documented in ancient scriptures of India dating back to 500–600 BC (Reynolds 1927), but it is now accepted that cultivation in Papua New Guinea occurred earlier, 7,000–10,000 years ago (Denham et al. 2003). Exactly when and by what routes bananas were introduced to tropical Africa and South America remains unresolved (Harris 2006). However, it appears that Islamic conquerors and Arab merchants were key players in the process. The word banana is of West African origin, but it is derived from the Arabic *banaan* (finger), and passed into English via Spanish or Portuguese.

Secondary centers of diversity are recognized for some of the subgroups of *Musa*. The humid lowlands of West and Central Africa are considered to be the secondary center of diversification for plantains where at least 115 known cultivars exist (Rossel 1991; Swennen and Rosales 1994) while the Highlands of East Africa are recognized as the secondary center of diversity for the East African Highland cooking and beer bananas (De Langhe 1961; Stover and Simmonds 1987). In both cases, an accumulation of somatic mutations that was enhanced by the long history of cultivation and experimentation increased the diversity of these banana subgroups (De Langhe 1964).

1.2 Botanical Descriptions

Banana belongs to the order Zingiberales and the family Musaceae. There are only two genera in the family Musaceae, *Musa* and *Ensete* (Simmonds 1966). It is estimated that *Musa* comprises about 70 species (Hakkinen and De Langhe 2001) and 500 cultivars (Simmonds 1966). These numbers are constantly changing since many new species have been recently identified (Hakkinen 2004; Hakkinen and Meekiong 2004; Valmayor et al. 2004; Hakkinen and Hong 2007; Hakkinen 2009; Argent 2010; Hakkinen et al. 2010). It is believed that many more *Musa* species may be identified if new explorations are made within its center of diversification (Hakkinen and Wong 2007). *Musa* is divided into five sections on the basis of chromosomes numbers and inflorescence morphology. The sections are *Eumusa, Rhodochlamys, Australimusa, Callimusa* and *Ingentimusa*. Sections *Eumusa* and *Rhodochlamys* have $2n = 22$ chromosomes while those in *Australimusa* and *Callimusa* have $2n = 20$ chromosomes. A single species, *M. ingens,* with $2n = 14$ chromosomes makes up section *Ingentimusa*. Section *Eumusa* has about 13 to 25 species and is considered the most ancient and diversified (Purseglove 1972). The section includes dessert bananas, cooking bananas and plantains that are widespread in tropical and subtropical countries. Sections *Callimusa* and *Rhodochlamys* consist of about nine non-parthenocapic species with no nutritional value and are important only as ornamentals. *Australimusa* consists of parthenocarpic edible types, known as Fei' cultivars. The distinguishing characteristics of the Fei' bananas are the erect fruit bunches and red sap, which sets it apart from other cultivated bananas. Bananas in this section are important for food and fiber and a valuable red dye obtained from the pseudo-stems. There is no consensus on the origin of Fei' bananas. Cheesman (1950) suggested that Fei' is closely related to *M. lolodensis* while Simmonds (1956) postulated that *M. maclayi* is the most likely ancestor. Restriction fragment length polymorphism (RFLP) analysis by Jarret et al. (1992) revealed that Fei' was indeed closest to *M. lolodensis.*

The validity and usefulness of sections in *Musa* has been questioned by some authors who examined sectional relationships in *Musa* using molecular tools (Wong et al. 2002; Nwakanma et al. 2003a). Amplified fragment length polymorphism (AFLP) and chloroplast analysis provided evidence for the merger of sections *Rhodochlamys* and *Eumusa* (Wong et al. 2002; Nwakanma et al. 2003a). This view has been already supported with morphological data and cross hybridization experiments (Simmonds 1954), RFLP of the nuclear genome (Gawel et al. 1992) and AFLP data (Ude et al. 2002b). Wong et al. (2002) also suggested that section *Callimusa* and section *Australimusa* should be combined into a single section. Chloroplast DNA (Nwakanma et al. 2003a) and AFLP data (Ude et al. 2002b) also showed a close affinity between the two sections.

Most of the cultivated banana accessions are diploid or triploid and it is believed that they originated from intra-and interspecific hybridizations between seed bearing subspecies of *M. acuminata* (A-genome donor) and *M. balbisiana* (B-genome donor) (Cheesman 1948; Simmonds and Shepherd 1955). The subspecies in *M. acuminata* are recognized on morphological characters and hybridization studies and include: *banksii*, *burmannica*, *burmannicoides*, *errans*, *malaccensis*, *microcarpa*, *siamea*, *truncata* and *zebrine.* However, their classification remains questionable based on molecular data (Carreel et al. 2002; Ude et al. 2002b). Molecular data based on nuclear as well as on cytoplasmic genomes will be appropriate for elucidation of genetic relationship of the present day subspecies; however, to reach precise and conclusive classification the analysis of more genotypes will be necessary (Boonruangrod et al. 2008, 2009).

Musa balbisiana is widely distributed in tropical and subtropical Asia, including China, India, Indonesia, Malaysia, Myanmar, Nepal, Papua New Guinea, the Philippines, Sri Lanka, Thailand and Vietnam (Ge et al. 2005). No subspecies of *M. balbisiana* have been described although there is substantial molecular and morphological differentiation within the species (Ude et al. 2002a; Ge et al. 2005). *Musa balbisiana* is resistant to the major pests and diseases of banana and is able to thrive in dry and cold environments. There is no natural edible parthenocarpic diploid or triploid from of *M. balbisiana* (Simmonds 1995).

1.2.1 General Morphology

A banana plant is a giant monocotyledonous, perennial herb ranging in size from 1.5 to 3 m height with some species attaining a height of 9 m. It consists of a true stem that is partly or wholly underground called "corm" or rhizome with roots. Robinson (1996) suggests that the banana stem is a rhizome and not a corm. Roots develop from the rhizome from the region between the inner zone (central cylinder) and the outer zone (cortex). A healthy rhizome may produce from 200 to 500 primary roots. Each primary root gives rise to a system of secondary and tertiary roots. The rhizome is an important storage organ for sustaining growth of the bunch and for the developing suckers. The apical meristem lies at the tip of the rhizome. The meristem remains below the soil level until flowering when it develops into the inflorescence axis that bears the bunch. Leaves develop from the meristem of the rhizome and consist of sheaths, a petiole and lamina or blade. The leaf sheaths of successive leaves overlap and closely encircles each other forming the pseudostem or the false stem. As new leaves develop at the meristem, the older leaves are pushed outwards, turn yellow and dry out (Simmonds 1962). Most banana varieties produce approximately 30–40 leaves in their life time. After a fixed number of leaves are produced, the meristem ceases to produce young leaves and starts to develop an inflorescence, which grows upwards through the middle of the pseudostem. The flowering stem emerges in the middle of the leaf crown and a complex inflorescence of flower clusters develops. The female flowers appear first and have large ovaries that develop into fruits. As the inflorescence develops, a bulb-shaped male bud containing small flowers develops at the end. However, in most cultivated bananas, the fruit develops by parthenocarpy preventing formation of seeds that would otherwise make the fruit unsuitable for human consumption. Three types of flowers are produced on the banana inflorescence. The female (pistillate) flowers develop into fruit, while the male (staminate) flowers found in the male bud may produce pollen that may or may not be fertile. The third type of flowers, called hermaphrodite or neuters, are found on the inflorescence axis or rachis between the female flowers and the male bud. They are usually sterile. The female flowers of most cultivated bananas are almost always sterile and the fruits develop by parthenocarpy. In all bananas the growing shoot dies after fruiting once (Simmonds 1962) and its life is perpetuated by means of suckers, which develop from adventitious buds produced on the rhizome. The suckers are the major form of vegetative planting material and form the subsequent vegetative generation. When the first plant fruits and dies, the maiden sucker (large but non-fruiting ratoon with foliage leaves) continues the growth cycle. Bananas are propagated vegetatively through suckers, although wild species can also be propagated by seed (Stover and Simmonds 1987). Sucker development consists of three distinct stages; peeper (young sucker bearing scale leaves only), sword sucker (sucker bearing narrow sword leaves) and maiden sucker (large but non-fruiting ratoon with foliage leaves) (Simmonds 1966; Swennen et al. 1984). The cluster formed by the mother plant and the surrounding suckers is referred to as "mat".

1.2.2 Genomes and Genome Size

Cultivated bananas comprise four genomes A, B, S and T (Simmonds and Shepherd 1955; Tezenas du Montcel 1988; Sharrock 1989). Three of the four genomes (A, B and S) are found only in section *Eumusa.* A majority of cultivated bananas are composed of the A and B genomes, while the S-genome is present in a single species, *M. schizocarpa.* The T-genome is characteristic of species in section *Australimusa.* The composition of the genomes in *Musa* has played a key role in its classification. The major genomic groups in *Musa* include diploids (AA, AB, BB), triploids (AAA, AAB, ABB) and tetraploids (AAAA, AAAB, AABB, ABBB). Genome groups with the S-genome include AS, AAS, ABBS while the T-genome is found in combination with the A and B genomes represented by AAT, AAAT and ABBT. Molecular markers have been identified specific for the A and B genomes (Howell et al. 1994; Pillay et al. 2000; Nwakanma et al. 2003b) while none have been identified for the other genomes as yet.

The nuclear genome size of *Musa* is approximately 552 to 697 Mbp with the A-genome being larger than the B-genome (Dolezel 2004). The 2C values of diploid *Musa* species range from 1.16 pg for B-genome species to 1.33 pg for the A-genomes species (Kamate et al. 2001). The B-genome was found to be smaller than the A-genome by 12% (Lysak et al. 1999). The *Musa* genome is about 25% larger than that of rice (490 Mbp) but relatively small at approximately 500 to 600 Mbp, which is dispersed over 11 chromosomes. Consequently the *Musa* genome is amenable to complete sequencing and *Musa* could be a model polyploid species for genomics as described later in this chapter. The sequences will provide an opportunity for extensive functional characterization of the genes.

1.2.3 Ploidy Levels

There are four basic chromosome numbers and three main ploidy levels in the genus *Musa*. Basic chromosome numbers were useful in dividing the genus into different sections (Cheesman 1947). Most edible bananas are triploids ($2n = 3x = 33$). The other major ploidy levels are diploids ($2n = 2x = 22$) and tetraploids $2n = 4x = 44$). Determination of chromosome numbers and ploidy in banana was done largely by conventional counting methods until the mid-1990s. Other phenotypic traits including stomata size, stomata density and pollen size were also used to estimate ploidy in *Musa* (Simmonds 1948; Hamill et al. 1992; Tenkouano et al. 1998). These methods were found to be inconsistent mainly due to strong influence of the genotype (Vandenhout et al. 1995; Van Duren et al. 1996). Flow cytometry, which is rapid and reliable, has virtually replaced chromosome counting for ploidy determination in *Musa* (Dolezel et al. 1994; Dolezel 2004). Flow cytometry is especially useful in banana improvement programs where large populations are regenerated from crosses involving the same or different ploidy levels. Knowledge of the chromosome number of each progeny is vital in a breeding program. Ploidy and genome segregation routinely occurs in the *Musa* breeding process (Bakry and Horry 1992; Osuji et al. 1997a; Pillay et al. 2004) and constitutes a major hindrance for selection because such features are not easily detectable at the morphological level (Oselebe et al. 2006). Flow cytometry can also be used as a rapid screen to detect *Musa* genotypes producing $2n$ gametes. A preponderance of $3x$ individuals from a $2x$ x $2x$ cross would suggest that at least one parent produces $2n$ gametes.

1.3 Economic Importance

For most people not living in tropical regions of the world, the word "banana" conjures up the image of the sweet or dessert banana, which is one of the most popular fruits on the world market. This fact may be due to its flavor and texture and because it is easy to peel and eat (Robinson 1996). But there are many other types of bananas. The name banana also includes the large starchier variety known as plantain. In general, banana (*Musa* spp.) is divided in two main categories: dessert and cooking bananas (Jones 2000). Dessert bananas make up 43% of the world production, are consumed raw when ripe (yellow), and are distinguished by the sweet flavor of the fruit. Dessert bananas are cooked in some tropical countries, especially where cooking types are unknown or unavailable. Cooking bananas make up 57% of the world's banana production and because of their high starch content, even in the ripened stage. They are boiled, fried or roasted to make them palatable. The most well known cooking bananas are plantains that constitute 23% of the world's total banana production. Other cooking banana types are steamed or cooked when the fruits are green and provides a starchy staple similar to potato in nutrition (Simmonds 1966). These include the unique East African Highland bananas that are endemic to the mid- and high-altitudes of East Africa.

Banana and plantain are grown in over 130 tropical and subtropical countries where climatic conditions appear to be ideal for their production. In terms of gross value of production, banana ranks among the most important tropical food sources after cereals staples such as rice, wheat and maize. In terms of volume they rank number one in exports and are second after citrus in regard to value. The total world production of banana is currently over 100 million metric tons and only 10% of this makes up the export trade (FAOSTAT 2005). In 2004 the world total exports of banana was approximately 15.9 million tonnes. The major banana producing countries in the world are listed in Table 1-1.

Table 1-1 The major *Musa* fruit producing countries worldwide in 2004.

Country	Production (metric tons)	Country	Production (metric tons)
India	16,820,000	Nigeria	2,103,000
Uganda	10,515,000	Mexico	2,026,610
Brazil	6,602,750	Thailand	1,900,000
Ecuador	6,552,000	Cameroon	1,830,000
China	6,420,000	Peru	1,660,310
Philippines	5,638,060	Côte d'Ivoire	1,602,423
Colombia	4,400,000	Burundi	1,600,000
Indonesia	4,393,685	Democratic Republic of Congo	1,412,000
Rwanda	2,469,741	Vietnam	1,353,800
Ghana	2,390,858	Guatemala	1,268,000
Costa Rica	2,230,000	Honduras	1,225,066

Source: FAOSTAT 2004

The 10 major banana producing countries accounted for about 75% of total banana production in 2004. About 98% of the world's banana is produced in the developing countries. Over 50% of the world's banana is produced in India, Uganda, Brazil and Ecuador. The largest exporters of banana are Ecuador, the Philippines, Costa Rica, Colombia and Guatamala. Although India, Uganda and Brazil rank as the top three producers of banana, very little banana or plantain is exported from these countries. It is estimated that over 90% of the bananas produced in some countries such as India, Uganda, Brazil and China are consumed locally in many different forms with each country having its own traditional dish and method of processing (Frison and Sharrock 1999). Banana consumption rates are quite high averaging between 200 to 250 kg per capita annually in New Guinea and countries surrounding the Great Lakes region of East Africa. In addition to being a major staple, banana and plantain provide a valuable source of income through local trade for millions of people in the world. The banana industry is also a very important source of income, employment and export earnings for the developing countries in Latin America, the Caribbean, Asia and Africa since world banana exports are valued at over US$4.7 billion per year (UNCTAD Secretariat from FAO statistics. www.unctad.org accessed 28/1/11). In some of these countries banana export is the only significant source of foreign income. According to the FAO "it is at the local and regional level where a strong bond is established between banana-generated income and household food security" (FAO 1999).

The area of banana production harvested in 2004 for some countries is listed in Table 1-2.

Table 1-2 Hectares of banana production in some countries.

Country	Hectares
India	680,000
Brazil	491.042
Philippines	415,427
Indonesia	314,708
Burundi	300,000
China	269,150
Ecuador	226,521

Source: FAO 2004

1.4. Nutritional Value

Nutritional information is being used by public agencies and agricultural industries to promote fresh produce. Foods that are rich in antioxidant vitamins (vitamins A, C and E), calcium, magnesium and potassium are being sought by health conscious consumers (Wall 2006). Greater consumption of fruits and vegetables is associated with reduced risk of cardiovascular disease, stroke and cancers of the mouth, pharynx, oesophagus, lungs, stomach and colon (cited in Wall 2006). In this context bananas are rich in carbohydrate, fiber, protein, fat, vitamins A, C and B_6 (Marriott and Lancaster 1983; Robinson 1996). The antioxidant gallocatechin was identified in Cavendish banana (Someya et al. 2002). A greater quantity (158 mg/100 g dry wt) was found in the peel than in the pulp (29.6/mg/100 g dry wt) suggesting that the banana peel, which is usually discarded, may also be assessed for potential use as a food source.

A range of mineral concentrations has been reported for bananas (Wenkam 1990; Hardisson et al. 2001; USDA-ARS 2004). The nutrition content of banana as per USDA-ARS is listed in the Table 1-3.

The exact cultivar used to obtain this nutrition information is not known, but it most likely represents those for the Cavendish subgroup (AAA). A recent comprehensive study found that dwarf Brazilian bananas (AAB) are more nutritious than Williams (AAA) for having higher content of vitamins A and C, and the minerals P, Ca, Mg, Mn and Zn. Vitamin C content was almost three-times higher in Dwarf Brazilian bananas (Wall 2006). The former study also showed that location can influence the mineral

Table 1-3 Nutritive value of raw banana per 100 g.

		US recommendations for adults
Carbohydrates	22.84 g	
- sugars 12,23g		
- dietary fiber 2.6g		
Fat	0.33 g	
Protein	1.09 g	
Thiamin (Vit. B1)	0.031 mg	2%
Riboflavin (Vit. B2)	0.073 mg	5%
Niacin (Vit. B3)	0.665 mg	4%
Pantothenic acid (B5)	0.334 mg	7%
Vitamin B6	0.367 mg	28%
Folate (Vit. B9)	20µg	5%
Vitamin C	8.7 mg	15%
Calcium	5 mg	1%
Iron	0.26 mg	2%
Magnesium	27 mg	7%
Phosphorus	22 mg	3%
Potassium	358 mg	8%
Zinc	0.15 mg	1%

Source: USDA nutrient database (www.nal.usda.gov) accessed 21/4/11

and vitamin content of bananas. Nutritional analyses of fruits often do not consider the effects of cultivar, maturity and/or environment on vitamin and mineral content. The nutritional composition of a fruit type at harvest can vary widely depending on cultivar, maturity, climate, soil type and fertility (Lee and Kader 2000). Carotenoid content can increase in banana with maturation and ripeness but ascorbic acid content can decrease during ripening (Lee and Kader 2000).

The advantages of banana as a food source are dealt with adequately by Robinson (1996). Dessert bananas are low in fats, cholesterol and salt and have a therapeutic value in many special diets. Ripe bananas are an excellent food for babies as well as the elderly since the fruit is easily digestible. The low sodium and fat content with no cholesterol makes it ideal for managing patients with high blood pressure and heart disease. Due to the low sodium and protein content, banana is used in special diets for patients with kidney problems. Bananas are also useful for people with peptic ulcers and other stomach problems such as gastritis and gastro-enteritis. The low lipid/high palatability combination is ideal for the diet of obese people.

Bananas can play a major role in alleviating micronutrient deficiencies in many countries. Micronutrient deficiencies in foods have received current attention by the World Health Organization and the World Bank (World Bank 1994). Micronutrient malnutrition is primarily evident in women, infants and children from impoverished families in developing countries (WHO 1992; Mason and Garcia 1993). Women and children in

Sub-Saharan Africa, South and Southeast Asia, Latin America and the Caribbean are at risk of disease, premature death, and impaired cognitive abilities because of diets poor in crucial nutrients, particularly iron, vitamin A, iodine, and zinc. Yet, these are the main banana producing countries of the world. Engelberger (2003) demonstrated that banana cultivars from the Federated States of Micronesia are rich in provitamin A carotenoids. Our unpublished research also showed that banana varieties from Papua New Guinea are rich in vitamin A, iron and zinc (Fungo 2007). In view of these data, it appears that efforts should be made to identify and distribute bananas with high nutrient values to countries where serious micronutrient deficiencies exist. However, banana consumption should be seen as a component of diet diversification as banana products alone only contribute to a fraction of the dietary requirements (Honfo et al. 2007).

1.5 Academic Importance

The two model species currently used in plant genomics are *Arabidopsis* and *Oryza sativa* (rice). The advantages of these plants as model species for genomics include their comparatively small genome sizes (*Arabidopsis* 125 Mbp and rice 490 Mbp), ease of handling and short life cycles. They are also easily transformed and have wide natural distributions. Model species have been proposed by other interest groups with sound reasoning for their choices. For example, the tree species *Populus* has been proposed as a model for tree genetics (Taylor 2002) while legumes such as *Medicago truncatula* and soybean have also been proposed as model plants for genomic studies of plants (Barker et al. 1990; Gepts et al. 2005). *Musa* has also been proposed as a model polyploid species for genomic studies and for providing insights into the genomes of other polyploidy species (Frison et al. 2004). These authors have provided several reasons why *Musa* should also be considered a model plant for plant genomics. In addition to its agricultural importance, *Musa* is considered to be a good model species for several fundamental aspects of structural and functional genomics, which cannot be done with *Arabidopsis* and rice. There is a wealth of knowledge on the structure of *Musa* species from morphology, organelle and nuclear DNA markers. There is great diversity in *Musa* providing a good model for examining gene regulation. *Musa* is a good model for genetic studies since it is one of the few plant species with biparental cytoplasmic inheritance in which mitochondria are paternally inherited while chloroplasts show maternal inheritance. As a vegetatively propagated crop, *Musa* offers good potential to study the role of somaclonal variation and phenomena such as "imprinting". Genomic imprinting is an epigenetic phenomenon which, in most cases, is believed to occur in gametogenesis. Genomic imprinting occurs when both maternal and paternal alleles are present, but one allele will be expressed while the other remains inactive (Albertson et al. 1996). The genome size of *Musa* (500–600 Mbp) is about 25% larger than that of rice and sequencing the entire genome could provide information on gene characterization and function. *Musa* is also regarded as a perfect model for studying plant evolution at the genome level. *Musa* has evolved over thousands of years in the same environment and the genomes of many sterile clones have been fixed by vegetative propagation. The presence of partially fertile and highly fertile diploids that have evolved over the same period could provide vital clues on genome evolution. A model plant should be readily transformable. A number of transformation systems have been developed for *Musa*. Protocols for electroporation of protoplasts derived from embryogenic cell suspensions (Sagi et al. 1994), particle bombardment of embryogenic cells (Sagi et al. 1995) and co-cultivation of meristems (May et al. 1995; Tripathi et al. 2005) or cell suspension cultures with *Agrobacterium tumefaciens* (Ganapathi et al. 2001; Khanna et al. 2004) are available for bananas and plantains.

1.5.1 Polyploidy

An organism may be defined as a polyploid if it has more than two haploid sets of chromosomes. Polyploidy has long been recognized as playing an important role in plant evolution. There has been a resurgence of interest in polyploidy since advances in technology have now made it possible to study genome evolution and the effects of genome duplication (Mable 2003). In addition to a revision of polyploidy in plants by Tate et al. (2005), a number of articles on polyploidy covering a range of topics including gene duplication and evolution (Lynch 2002), genome evolution (Wendel 2000), mechanisms of novel gene expression (Osborn et al. 2003), the effect of duplicated genes on the transcriptome (Adams et al. 2003), photosynthetic rates (Wullschleger et al. 1996), quantitative trait loci (QTL) affecting plant height and flowering (Ming et al. 2002), structural and physiological characteristics of polyploids relative to water deficit (Li et al. 1996) and the advantages and disadvantages of polyploidy (Comai 2005) have appeared recently. Since banana is regarded both as an auto- and allo-polyploid, it could be used as a model plant to study a number of aspects such as the functional genetics of polyploidy, gene silencing and tissue- or developmental-specific expression of duplicate genes, the alternative fitness of diploid and tetraploid cytotypes in different habitats as described by Mable (2003), plant evolution and the formation and role of polyploidy in plants. Interesting ideas for new research areas in banana are contained in a paper addressing new perspectives in polyploid research by Mable (2003). Some of these ideas are described below. Although the production of *2n* gametes has been considered to be the primary mechanism of polyploidy, somatic doubling has also been considered to play a major role. The cytological mechanisms and the number and location of genes involved in the production of 2*n* gametes are current goals of plant research. Many advances in cytogenetic techniques such as chromosome painting, chromosome sorting, high resolution banding, micro-dissection and *in situ* hybridization should allow more precise dissection of factors that control homologous and homeologous pairing during meiosis in polyploids that may lead to factors affecting fertility in triploids. It is believed that the extra genome copies in polyploids enable them to tolerate and invade harsher environments than their diploids. Polyploids might be more tolerant to water deficit (Wullschleger et al. 1996) and or have higher levels of photosynthesis (Li et al. 1996). Ecological adaptations of plants are now being studied at the molecular level by looking at changes in the regulatory gene (Lawton-Rauh et al. 2003) and QTL (Ming et al. 2002).

Evidence of extensive gene duplication in diploid plants suggests that polyploidy contributed to evolutionary diversification. Understanding the mechanisms and consequences of such changes is an important part of polyploidy research in plants.

Duplicate genes may have many fates—silencing, development of novel functions, sub-division of functions among tissue types or developmental stages (Mable 2003).

In plants an intermediate triploid stage has been proposed as the most likely route to tetraploid formation. This idea has been questioned because of the frequent sterility of triploids. The low frequency of viable gametes in triploids might be able to overcome this barrier. Majority of edible bananas are triploids and it is now known that fertility does occur at the triploid level (Pillay et al. 2002).

Polyploidy in *Musa* is accompanied by an increase in plant size and ecological and reproductive differences compared to diploids. The causes or consequences of such developmental differences for the adaptive potential of polyploids in new environments are not yet explored (Mable 2003).

Different types of polyploids have been described adequately by Tate et al. (2005). In a strict sense, autoploypoids are formed by intraspecific hybridization within a species either from genome doubling in a single individual or fusion of 2*n* gametes from genetically similar individuals. Inter-racial autopolyploids are formed by hybridization from genetically distinct individuals within a species but with structurally similar chromosomes. Both types of autopolyploids are characterized by having multivalents and polysomic inheritance. True or genomic allopolyloidy are formed by interspecific hybridization between distantly related species. A distinction is made between allopolyploidy and autoallo-polyploidy. In the latter case, alloploidy is followed by genome doubling. Allopolyploidy is characterized by having bivalents and disomic inheritance whereas in autoallo-polyploids chromosome pairing results in multivalents or bivalents and inheritance is disomic for some loci and polysomic for others. Segmental allopolyploidy are regarded as plants within a species but having genomes that differ from each other in a large number of genes or chromosomal segments. They are unstable polyploids that evolve towards auto-or true-allopolyploidy. They are characterized by having bivalents and disomic inheritance or multivalents and polysomic inheritance. There is little information on inheritance patterns of genes in *Musa*. However, various chromosomal configurations from univalents to multivalents were observed in cytological analyses of banana genotypes (Adeleke 2001). *In situ* hybridization studies reported cross hybridization between the different genomes of banana (Osuji et al. 1997b; D'Hont et al. 2000). Greater cross hybridization was found between the A and B genomes than between the A and S genomes. The least cross hybridization was observed with the T-genome and the A and B genomes. The intensity of cross hybridization is a reflection of the sequence homologies and affinities between the genomes. The above information may suggest that banana may be more than a true auto- or allo-polyploid. Therefore banana could be used a model plant in polyploid research.

1.5.2 Proteomics

"Proteomics" was introduced in scientific literature for the first time in 1994. Since then tremendous progress has been made in this field especially in plants. Proteomics is described as the systematic analysis of proteins expressed by a genome. It includes the identification of their primary amino acid sequence, determination of their relative amounts, specific activity, state of modification and association with other proteins or molecules of different types, subcellular localization and three dimensional structures (Barbier-Brygoo and Joyard 2004; Kwon et al. 2006).

Proteomics research is likely to make greater progress in species whose genomes are completely sequenced. Proteomics research in banana is still in its infancy and at present has only a few publications (Carpentier et al. 2007; Samyn et al. 2007). Despite this banana has been suggested to be an excellent model to study the meristem proteome (Carpentier et al. 2007). Over 130 proteins were identified from the top layer of 1–2 mm of meristematic tissue of banana. Banana has multiple shoot meristems that can be propagated by a well-developed shoot meristem culture technique (Strosse et al. 2006). This makes it possible to do multiple experiments from a single meristem. For example, Carpentier et al. (2007) investigated the influence of sucrose mediated osmotic stress in banana. In this manner proteins that were significantly up- or down-regulated were identified. In a similar manner, the meristem proteome in banana could be used to study the quantitative differences in proteins by subjecting it to other stresses.

Interest in plant proteomics is on the increase and is not restricted to the species that are current models in genomics. Functional proteomics is being used as a tool to analyze plant –specific processes and the responses to various biotic and abiotc stresses. Thousands of proteins remain to be discovered in plants after sequencing of model genomes. It is believed that functional proteomics and the mining of protein expression profiles will be the next challenge in plant proteomics (Rossignol et al. 2006).

1.5.3 Breeding

Although banana breeding is hampered by triploidy and low male and female fertility, the crop offers many new opportunities for genetic improvement by the discerning scientist. Most edible triploid varieties do not carry genes for disease and pest resistance. Consequently these genes must be introgressed from fertile diploid accessions. Various schemes have been devised to breed bananas (Rowe and Rosales 1996; Vuylsteke et al. 1997; Tenkouano 2005; Pillay and Tripathi 2007). One of these schemes involves crossing the triploid varieties with diploids as male parents. In this scheme, tetraploids are selected for characters of interest. But it was observed that the tetraploids are fertile and could set seeds if exposed to fertile pollen. Therefore tetraploids are crossed with improved diploids to produce secondary triploids. It is also possible to cross tetraploids with tetraploids. The predominant pollen class in 4*x* individuals is hypoploid with half the chromosome content of the haploid class, resulting from double reduction (Tenkouano 2005). Therefore, the progeny from 4*x* x 4*x* crosses are essentially triploids, which accumulate genes originating from their triploid maternal grandparents. Not surprisingly therefore, progeny from these 4*x* x 4*x* crosses show a greater resemblance to the maternal grandparents, with regard to morphology and bunch characteristics (M Pillay, unpubl.).

The acceptability of such hybrids by consumers is expected to be higher since they resemble the original varieties that farmers generally prefer. The 4*x* x 4*x* crosses can also act as a genetic bridge between triploid varieties that otherwise cannot be crossed with each other due to their sterility. In this way, the 4*x* x 4*x* breeding scheme could enhance new recombination events that were previously not likely. The 4*x* x 4*x* breeding scheme for banana could act as a model for other crop plants.

1.5.4 Cytogenetics

Early cytogenetic studies in *Musa* were focused on determination of chromosome numbers (Cheesman 1932a, b; Cheesman and Larter 1935).

Thereafter, research was focused on chromosomes pairing studies in artificially produced hybrids (Dodds 1943; Wilson 1946). A comprehensive study of meiotic studies in *Musa* was carried out by Shepherd (1996). Agarwal (1987, 1988a, b) published a series of papers on meiotic studies in *Musa*. With the advent of molecular biology, it appears that there was a general decline in interest for cytogenetic analyses of organisms since it was now possible to examine their DNA. The small size and poor staining ability of the banana chromosomes was a further deterrent for research in *Musa* cytogenetics.

The individual chromosomes in bananas have not yet been identified and numbered partly because of the small size of the chromosomes and the absence of cytological markers. Therefore, any published research on karyotypes in *Musa* must be regarded as preliminary. The only chromosomes that are easily identifiable in *Musa* are the satellited ones that represent the longest chromosomes in both *M. acuminata* and *M. balbisiana*. In many plant species with small chromosomes, the pachytene stage of meiosis has provided an alternative method of developing karyotypes. Better staining techniques made it possible to observe pachytene chromosomes for the first time in *Musa* (Adeleke et al. 2002). A few studies applied molecular cytogenetic techniques in *Musa* that provided interesting results for genome identification, genome composition of hybrids, the location of the NOR regions (Osuji et al. 1998; Dolezelova et al. 1998; D'Hont et al. 2000) the presence of mobile genetic elements (Baurens et al. 1997) and transposons (Balint-Kurti et al. 2000). These aspects of the genome are important for a banana breeding program especially in a crop that has multiple genomes and with a complicated genetic system.

1.5.5 *Aneuploidy*

Aneuploidy has been reported in root tip meristems of a range of triploid *Musa* clones (Shepherd 1996; Shepherd and Da Silva 1996). In field trials these plants displayed various characteristics: some were fairly normal and vigorous; others were weak and died without flowering; some were inviable and many did not grow to maturity (Shepherd 1996). An aneuploid series of banana mutants may be a valuable genetic tool to identify the genes found on the different banana chromosomes. Rapid detection of aneuploidy in *Musa* is now possible via flow cytometry (Roux et al. 2003). Despite its value there appears to be no interest for this type of basic research in *Musa*.

1.6 Germplasm, Gene Pools and Utilization of Wild Species

Genetic resources refers to the diversity of genetic material found in farmers' varieties, modern cultivars, wild relatives of crops and other wild species that provide farmers and plant breeders the ability to develop new and more productive crops. Wild species and primitive cultivars of many crop plants harbor useful genes for resistance traits, landraces for their quality characteristic and hybrids for superior agronomic characteristics. Banana breeders have largely relied on the introgession of genes from closely related wild plants to increase genetic variation in the cultivated varieties. Landraces, crossable wild relatives and hybrids constitute the primary genetic pool in banana breeding. In the section that follows we examine some of the major diseases and pests of banana and discuss the value of wild plants as sources of resistance.

1.6.1 Diseases and Pests

1.6.1.1 Black Sigatoka

Black Sigatoka caused by *Mycospharella fijiensis* is the most economically important disease of bananas and plantain throughout the world (Carlier et al. 2000). The most important source of resistance for black Sigatoka is a wild diploid banana *Musa acuminata* ssp. *burmannicoides,* widely known as "Calcutta 4" (Rowe and Rosales 1990). Calcutta 4 has been used in many breeding programs as a source of resistance to black Sigatoka. "Pisang lilin", a parthenocarpic diploid (AA) accession in the *M. acuminata* ssp. *malaccensis* complex and "Tuu Gia" have also been used as sources of resistance for black Sigatoka. By crossing "Calcutta 4" with other bred diploids, the Fundacion Hondureña de Investigación Agrícola (FHIA) program was able to develop several diploids with exceptional bunch sizes and resistance to diseases and pest. Notable diploids with resistance to black Sigatoka include SH 2829, SH 3437, SH 3142, and SH 3217. It appears that the success of the FHIA breeding program is largely linked to the initial development of these resistant diploids. Similarly, the International Institute of Tropical Agriculture (IITA) used "Calcutta 4" and "Pisang lilin" as the main sources for introducing host plant resistance to African plantains (Vuylsteke et al. 1997).

1.6.1.2 Fusarium Wilt

Fusarium wilt caused by a soil borne fungus (*Fusarium oxysporum* Schlect. F. sp. *cubense* (E.F. Smith) Snyder and Hansen is regarded as one of the most destructive diseases of banana. There are four races (race 1, 2, 3 and 4) of *F. oxysporum* (Ploetz 1994). Somatic compatibility has also been used to identify genetically isolated populations of *Fusarium*. The groups called vegetative compatibility groups (VCGs) are identified by heterokaryon formation between nitrate-non-utilizing auxotrophic mutants. There are about 16 VCGs or VCG complexes in *F. oxysporum* f. sp. *cubense* (Ploetz 1990; Ploetz et al. 1997).

The wild diploids *M. acuminata* ssp. *malaccensis* and "Calcutta 4", and the edible diploid "Pisang lilin" were reported to be good sources of resistance to races 1 and 2 of *Fusarium* (Shepherd et al. 1994; Rowe and Rosales 1996). "Pisang lilin" was the first plant to be used as the source of resistance to develop a "Gros Michel" like cultivar with resistance to *Fusarium* wilt. Two resistant cultivars, "Bodles Altafort" or "1847" from a "Gros Michel" x "Pisang lilin" cross and "2390-2" from the cross "Highgate" x "Pisang lilin" were released (Jones 2000). Later, male parents were selected from crosses between *M. acuminata* ssp. *banksii* and "Paka", an AA diploid from Zanzibar. In the 1960s a male parent from a cross between *M. acuminata* ssp. *malaccensis* and *M. acuminata* ssp. *banksii* was used for breeding. Later, improved diploids were obtained by crossing diploids obtained from *M. acuminata* ssp. *malaccensis* x ssp. *banksii* with others including those derived from *M. acuminata* ssp. *banksii* x "Paka".

The FHIA diploids SH3142, SH3362 and SH3437 were reported to be resistant to races 1 and 2, and SH3362 is also resistant to race 4 (Rowe and Rosales 1996). Using these diploids, FHIA has developed two tetraploid hybrids FHIA 17 and FHIA 23 that are resistant to race 1. In addition, FHIA 01 has been show to be resistant to race 1 and race 4. The Brazilian breeding program at EMBRAPA has bred many clones with resistance to *Fusarium* (Shepherd et al. 1994). These include PA 03-22 (AAAB), PV 03-44 (AAAB), PC12 -05 (AAAB) that are derived from "Prata Ana", "Pacovan" and "Prata", respectively. These clones are resistant to race 1 and are being tested in many countries. *Fusarium* wilt is the one of major constraints affecting "Pisang Awak" (ABB) and "Sukali Ndizi" (AAB) in eastern and southern Africa. Nine genotypes including "Pisang lilin", "Fougamou", "Pisang Awak", "SH3217", "SH3142", "SH3362", "Yangambi Km5", and "Kikundi" were tested for resistance in a *Fusarium*-infected field in Uganda. Of the nine genotypes, "Pisang Awak" and "Fougamou" showed typical symptoms of Fusarium wilt including leaf discoloration and wilting, pseudostem splitting and browning of the vascular tissue within six months. These clones are known to be susceptible to race 1 of *Fusarium*. The genotypes "Pisang lilin", "SH3217", "SH3142", "SH3362", "Yangambi Km5" , and "Kikundi" have not displayed any symptoms of Fusarium wilt after being in the infected field for 60 months. The latter genotypes should be considered as being resistant (M. Pillay, unpubl.).

1.6.1.3 Banana Xanthomonas Wilt

Xanthomonas wilt is a relatively new disease affecting bananas in East Africa. The causal bacterium, originally known as *Xanthomonas campestris* p. *musacearum,* is now known as *X. vasicola* (Carter et al. 2009). The disease causes total yield loss because the fruits are inedible in the diseased plants.

The disease is now prevalent in Uganda, Rwanda, Eastern Democratic Republic of Congo, Tanzania and has recently been identified in Kenya (Mbaka et al. 2007, 2009). Typical symptoms of the disease include wilting, pre-mature yellowing and rotting of fruits (Biruma et al. 2007). It is estimated that the disease is causing an annual loss of US$ 360 million to the Ugandan economy. If unchecked, the disease could cause massive losses in the main banana producing districts of Uganda and in the neighboring countries where intensive banana cultivation is practiced. Cultural control measures are useful in curbing the spread of the disease but needs to be adopted by all farmers for its effective containment (Eden-Green 2004, 2007; Blomme et al. 2005a, b).

To date no banana genotype has been found to be resistant to Xanthomonas wilt.

An attempt was made at screening a range of genotypes to identify sources of resistance to *Xanthomonas* in Uganda. Initial observations from this trial indicated that the genotypes "Yangambi km5", "Kikundi", "Kisubi", TMBx660K1,TMBx201K, "Yalim", "Nakasabira", TMBx1438K-1, TMBx7197-2, SH3217, SH3362, SH3640-9 and "Nakitengu" did not appear to get infected naturally even if were surrounded by infected plants. Although current literature is not clear on how the disease is transmitted, (M.Pillay pers. observ.) indicated that transmission occurs through vectors, contaminated farm tools, infected planting material and planting in existing infested fields. Vector transmission is perhaps the most important means of disease spread. Disease transmission is believed to occur when the vector (insects, birds or bats) transfers inoculum (bacteria) from diseased to healthy plants. Entry points for the disease appear to be the "cushions" of the floral bracts where disease symptoms are initially observed. Plants that do not shed their floral bracts do not provide an entry point for the disease and appear to "escape" the disease. Among the East African Highland Bananas, genotypes in the "Nakitembe" clone set do not shed their floral bracts. In an experiment conducted in Uganda it was shown that banana varieties such as "Nakitembe" and "Mbwazirume" that do not shed their floral bracts did not get infected (M. Pillay, unpubl.).

1.6.1.4 Viral Diseases

Four well-characterized viruses are known to affect bananas and other *Musa* species. These are banana bunchy top virus (BBTV), banana streak virus (BSV), cucumber mosaic virus (CMV), and banana bract mosaic virus (Pietersen and Thomas 2001).

As techniques to observe viruses become more sophisticated, virologists are able to discover new viruses in banana. The presence of virus-like particles observed in bananas from Africa, the Americas, Southeast Asia, and Australia led to the discovery of a new virus called banana mild mosaic virus. Another new virus, banana die back virus, has been described from Nigeria (Hughes et al. 1998). Of all the viral diseases, banana bunchy top disease is the most serious and devastating disease (Rishi 2009). Early identification and eradication of infected plants is the best means of controlling virus diseases.

Screening procedures for viruses are not well developed, although virus diagnosis techniques have been improved recently. Large scale screening of germplasm for virus resistance is limited.

Consequently, sources of resistance for the different viruses affecting banana are not known. Breeding for virus resistance cannot make any progress in the absence of suitable parental material. In this regard, Dahal et al. (2000) assessed natural incidence of BSV based on symptoms and virus indexing, as well as relative concentration of BSV antigens in leaf tissues —measured by ELISA, and some growth and yield traits. Virus occurrence and symptom expression, and the relative concentration of BSV antigens fluctuated significantly across seasons during the cropping cycle: the rainy season being the highest and negligible in the hot dry season. Relative concentration of BSV antigens in leaf samples was high in most polyploid hybrids but low in some landraces. Despite the high natural incidence and relative high antigen concentration in the leaf tissue, some IITA and FHIA hybrids suffered low or no bunch weight loss and could be regarded as "field tolerant to BSV".

1.6.1.6 Weevils

The banana weevil, *Cosmopolites sordidus,* is the most important pest of banana and plantain. Adult weevils lay eggs at the base of the plant and upon hatching the larvae bore into the corm to feed and develop. Extensive tunneling of the corm by the larvae reduces water and mineral uptake and results in weakening of the plant and reduced yield (bunch weight). In severe cases of weevil infestation, toppling and premature death of the plant are common scenarios among infested plants. Pseudostem splitting at the base and distortion of young leaves are the other observed symptoms of extreme weevil damage.

Few studies have evaluated banana germplasm for resistance to weevils (Fogain and Price 1994; Kiggundu et al. 2006). Plantains (AAB) are considered to be more susceptible than other banana genotypes (Ittyeipe 1986; Fogain and Price 1994). Pavis (1991) demonstrated that cultivars from the "Pisang awak" group exhibits high levels of tolerance despite heavy tunneling while "Yangambi Km5" was almost free from attack. "Pisang awak" and "Bluggoe" (AABs) are generally considered more resistant than plantains (AAB) and the East African Highland bananas (Abera et al. 1997). Crosses between plantains and "Calcutta 4" demonstrated that most of the diploid hybrids were resistant to weevils while majority of the polyploids were susceptible (Ortiz et al. 1995). On this basis, "Calcutta 4" was considered to be a source of resistance for weevils. Other genotypes with weevil resistance include "Sannachenkadali" (AA), "Sakkaki" (ABB), "Senkadali" (AAA), "Elacazha" (BB), "Njalipovan" (AB), FHIA03, TMBx612-74, TMB2x6142-1,TMB2x8075-7, TMB2x7197-1, "Long Tavoy", "Njeru", "Muraru", "Bluggoe" and *M. balbisiana* (Kiggundu et al. 2003). A new method for resistance screening of banana genotypes for the banana weevil has been developed (Kassim et al. 2010). With this method it is possible to screen for resistance or susceptibility to weevils within 7 months in a screen house compared to 2 to 3 years in the field.

1.6.1.7 Nematodes

Nematodes are recognized as important pests of bananas in most regions of the world. The average annual yield loss due to nematodes is estimated at about 20% worldwide (Sasser and Freckman 1987). The most important and widespread banana nematodes are *Radopholus similis* (burrowing nematode), *Helicotylenchus multicinctus* (spiral nematode) *Pratylenchus coffeae* and *P. goodeyi* (root lesion nematodes) and *Meliodogyne incognita* (root knot nematode). In addition, other minor nematodes such as *Heterodera oryzicola* (India), *Radopholus reniformis*, *Helicotylenchus dihystera* (Malaysia), *Rotylenchus* (Indonesia) also occur in bananas in some countries. The most damaging nematodes are the migratory endoparasitic lesion-nematodes *R. similis* and *O. coffeae* (Sarah et al. 1996).

Resistance to nematodes has been identified in wild species, landraces and new synthetic plants developed by breeding programs. The wild species *M. acuminata* "Calcutta 4" shows field resistance to nematodes, especially *R. similis*. Its resistance to *R. similis* was confirmed in pot trails (Viaene et al. 2003). Other subspecies of *M. acuminata* (*malaccensis*, *microcarpa*, and *zebrina*) have been found to have moderate to good resistance to *R. similis* (Wehunt et al. 1978; Fogain 1996). The bananas "Marau", "Pora Pora", "Kokopo", "Pisang mas", "Saba, Gia Hiu" and "Vudu papua" are resistant to *R. similes*, whereas "Pitu", "Yalim", "Yanun yefan" and *M. balbisiana* showed partial resistance to *R. similis* (Dochez et al. 2006).

Musa acuminata ssp. *malaccensis* has been used by FHIA as a source of nematode resistance in the breeding program. Some diploid cultivars such as "Pisang Mas", "Pisang Lidi" have moderate resistance to *R. similis* (Wehunt et al. 1978; Fogain 1996). The "Pisang jari buaya" (AA) complex of diploids and "Pisang Batuau" are recognized to be resistant to *R. similis* (Wehunt et al. 1978; Fogain 1996; Elsen et al. 2002). "Pisang jari buaya" was utilized as a female parent by the FHIA to produce SH 3142, a synthetic diploid with burrowing nematode resistance. The diploid AB clone "Kunnan" has consistently shown high levels of resistance to *R. similis* and *P. coffeae* (Collingborn and Gowen 1997). "Yangambi Km 5" (AAA) has been reported to be resistant to *R. similis* (Pinochet et al.1998). Except for cv. Muvubo, East African highland bananas are as susceptible to *R. similis* as the dessert banana "Valery" (Dochez et al. 2005). Four out of 13 tetraploid hybrids were identified with resistance to *R. similis*, as well as 13 out of 19 diploids and five out of 18 secondary triploids. In this regard, IITA's banana breeding program has developed a number of diploid and tetraploid hybrids that are resistant to *R. similis*. These include TMB2x 9128-3, TMB2x 5265S-1, TMB2x 3107S-4, TMB2x 2582S-1, TMB2x 4443S-1, TMB2x 2569S-1 and TMB2x 5105-1. Among the Fe'i varieties, Stoffelen et al. (1999) identified two cultivars "Rimina" and "Menei" as being resistant to *R. similis*. Most accessions of *M. balbisiana* have been found to be resistant to *R. similis* (Fogain 1996) while the ABB cultivar "Pelipita" is reported to be moderately resistant (Price and McLaren 1996).

1.7 Mutants

Edible banana was presumed to be selected in prehistoric times from spontaneous mutations (Buddenhagen 1987). The rate of occurrence of spontaneous mutations is too low to make mutations a significant factor in the genetic improvement of *Musa*. Mutagenic agents such as radiation and certain chemicals can be used to induce mutations at a higher frequency and generate genetic variation from which desired mutants can be selected (Roux 2004). Early mutation studies were found to be ineffective in inducing mutants in *Musa* because seeds and suckers were used as the starting material. Thereafter, greater success in obtaining mutants occurred when *in vitro* plant material, especially the shoot tips that contain the meristemic tissue was used with gamma radiation. A detailed protocol for mutant induction in *Musa* is provided by Roux (2004). However, one of the main difficulties associated with mutation induction in *Musa* is the high degree of chimerism observed in the mutants (Hautea et al. 2004; Roux 2004). The problem of chimerism could be reduced when embryogenic cell suspensions are used for inducing mutations in *Musa* (Roux et al. 2004). Mutation breeding is also time consuming, with high labor and field costs since large areas of field space is required to grow the plants to maturity. To date, two banana accessions, "Novaria" and "Klue Hom Thong KU1", are the only registered improved mutant varieties. "Novaria" is an early flowering mutant derived from "Grande Nain" that flowers about 10 weeks earlier than the parental clone. "Klue Hom Thong KU1" was obtained by treating tissue cultures with gamma rays. It has a larger cylindrical bunch than its parent. Roux (2004) provides a list of variants obtained from somaclonal variation and putative mutants from gamma irradiation. The selected traits of these mutants include dwarfism, larger fruit size, earliness, tolerance to toxins from *Mycosphaerella fijiensis* and tolerance to *Fusarium oxysporum*.

1.8 Conclusion

The importance of banana and plantains as a food crop especially in the tropical regions is well established. Research trends in the crop seem to reflect similar directions that occurred in other crop plants. The initial emphasis was on morphology and cytogenetics both of which played a major role in the classification of the crop.

The development of molecular techniques witnessed a surge of papers in the molecular aspects of the crop and raised the level of its academic importance in genomics. Concurrently, the development of better tools in cytogenetics and molecular cytogenetics provided new insights into the genomes of *Musa*. The development of flow cytometric techniques for *Musa* opened new avenues for breeding of the crop. While writing this chapter it became evident that research in *Musa* remains dynamic. New species are being collected from remote areas. The value of these species in deciphering the complex taxonomy of *Musa* and for breeding holds tremendous potential for research. The carotenoid content of banana and its value in health and nutrition is being actively investigated. It was also interesting to note that *Musa* has been suggested as a model for "proteomics" research. Despite the difficulty in obtaining research funding for this "orphan" crop, the future of *Musa* research holds plenty of promise.

Abbreviations

AFLP	:	amplified fragment length polymorphism
DNA	:	Deoxyribonucleic acid
EMBRAPA	:	Empresa Brasileira de Pesquisas Agropecuarias
FHIA	:	Fundación Hondureña de Investigación Agrícola
RFLPs	:	restriction fragment length polymorphisms

References

Abera A, Gold CS, Kyamanywa S (1997) Banana weevil oviposition and damage in Uganda. African Crop Science Conference Proceedings 3: 1199–1205.

Adams KL, Cronn R, Percifield R, Wendel JF (2003) Genes duplicated by polyploidy show unequal contributions to the transcriptome and organ-specific reciprocal silencing. Proc Natl Acad Sci USA, 100: 4649–4654.

Adeleke MTV (2001) Meiotic studies in the genus *Musa* L. Ph.D Thesis, University of Port Harcourt, Port Harcourt, Nigeria.

Adeleke MTV, Pillay M, Okoli BE (2002) An improved procedure for examining meiotic chromosomes in *Musa*. HortScience 37: 959–961.

Agarwal PK (1987) Cytogenetical investigations in *Musaceae*. II. Meiotic studies in eight male sterile triploid banana varieties of India. Cytologia 52: 451–454.
Agarwal PK (1988a) Cytogenetical investigations in *Musaceae*. III. Meiotic studies in diploid *Musa* species and banana varieties of India. Cytologia 53: 359–363.
Agarwal PK (1988b) Cytogenetical investigations in *Musaceae*. IV. Cytomorphology of an interspecific triploid hybrid of *Musa acuminata* Colla. X *M. rubra* Wall. Cytologia 53: 717–721.
Albertson K, Frederickson L, Palmer S (1996) DNA Methylation and Genomic Imprinting. In: LW Browder (ed) Developmental Biology, http://www.ucalgary.ca/~browder.
Argent GCG (1976) The wild bananas of Papua New-Guinea. Notes Roy Bot Gard Edinb 35: 77–114.
Argent G (2010) A new species of wild banana *Musa arfakiana* (Musaceae) from Papua (formerly Irian Jaya) of Indonesia. Gardens' Bulletin 62(2): 243–248.
Bakry F, Horry JP (1992) Tetraploid hybrids from interploid 3x/2x crosses in cooking banana. Fruits 47: 641–647.
Balint-Kurti PJ, Clendennen SK, Dolezelovca M, Valarik M, Dolezel J, Beetham PR, May GD (2000) Identification and chromosomal localization of the monkey retrotransposon in *Musa* sp. Mol Gen Genet 263: 908–915.
Barbier-Brygoo H, Joyard J (2004) Focus on plant proteomics. Plant Physiology and Biochemistry 42: 913–917.
Barker DG, Bianchi S, Blondon F, Dattée Y, Duc G, Essad P, Flament P, Galuxi P, Génier G, Guy P, Muel X, Tourneur J, Dénarié J, Huet T (1990) *Medicago truncatula,* a model plant for studying the molecular genetics of *Rhizobium*-legume symbiosis. Plant Mol Biol Rep 8: 40–49.
Baurens FC, Noyer JL, Lanaud C, Lagoda PJL (1997) Sequence tagged site markets to draft the genomic structure of the banana chloroplast. Fruits 52: 247–259.
Biruma M, Pillay M, Tripathi L, Blomme G, Abele S, Mwangi M, Bandyopadhyay R, Muchunguzi P, Kassim S, Nyine M, Eden-Green S (2007) Banana *Xanthomonas* wilt: a review of the disease, management strategies and future research directions. Afri J Biotechnol 6: 953–962.
Blomme G, Mpiira S, Ssemakadde R, Mukasa H (2005a) Controlling banana Xanthomonas wilt through de-budding. InfoMusa 14(1): 46.
Blomme G, Mukasa H, Ssekiwoko F, Eden-Green S (2005b) On-farm assessment of banana bacterial wilt control options, 317–320. In: African Crop Science Society Conference Proceedings, Entebbe, Uganda.
Boonruangrod R, Desai D, Fluch F, Berenyi M, Burg K (2008) Identification of cytoplasmic ancestor gene-pools of *Musa acuminata* Colla and *Musa balbisiana* Colla and their hybrids by chloroplast and mitochondrial haplotyping. Theor Appl Genet 118: 43–55.
Boonruangrod R, Fluch S, Burg K (2009) Elucidation of origin of the present day hybrid banana cultivars using the 5'ETS rDNA sequence information. Molecular Breeding 24: 77–91.
Buddenhagen IW (1987) Disease susceptibility and genetics in relation to breeding bananas and plantains, In: GJ Persley, EA De Langhe (eds) Banana and Plantain Breeding Strategies. Proceedings No.21, Australian Centre for International Agricultural Research, Australia, pp 95–109.
Carlier J, Foure E, Gauhl F, Jones DR, Lepoivre P, Mourichon X, Pasberg-Gauhl C (2000) Fungal diseases of the foliage. In: DR Jones (ed) Diseases of Banana. Abaca and Enset. CABI Publishing Wallingford, UK, pp 37–141.
Carpentier SC, Witters E, Laukens K, Van Onckelen H, Swennen R, Panis B (2007) Banana (*Musa* spp.) as a model to study the meristem proteome: acclimation to osmotic stress. Proteomics 7: 92–105.
Carreel F, de Leon Gonzalez D, Lagoda PJL, Lanaud C, Jenny C, Horry JP, Tezenas du Montcel H (2002) Ascertaining maternal and paternal lineage within Musa by chloroplast and mitochondrial DNA RFLP analyses. Genome 45: 679–692.

Carter BA, Reeder R, Mgenzi SR, Kinyua ZM, Mbaka JN, Doyle K, Nakato V, Mwangi M, Beed F, Aritua V, Lewis IML, Miller SA, Smith JJ (2009) Identification of *Xanthomonas vasicola* (formerly *X. campestris* pv. *musacearum*), causative organism of banana xanthomonas wilt, in Tanzania, Kenya and Burundi. New Disease Reports 19: 25.
Cheesman EE (1932a) Genetic and cytological studies of *Musa*. 1. Certain hybrids of the Gros Michel banana. J Genetics 26: 291–312.
Cheesman EE (1932b) Genetic and cytological studies of *Musa*. II.Hybrids of the Mysore banana. J Genetics 26: 313–316.
Cheesman EE (1947) Classification of the bananas. Kew Bull 2: 97–117.
Cheesman EE (1948) The classification of the bananas. Kew Bull 3: 11–28, 145–157, 323–328.
Cheesman EE (1950) The classification of the bananas. Kew Bull 5: 27–31, 151–155.
Cheesman EE, Larter LNH (1935) Genetic and cytological studies of *Musa*. III. Chromosome numbers in the Musaceae. J Genetics 30: 31–50.
Collingborn FMB,Gowen SR (1997) Screening of banana cultivars for resistance to *Radopholus similis* and *Pratylenchus coffeae*. Infomusa 6: 3.
Comai L (2005) The advantages and disadvantages of being polyploidy. Nature Reviews 6: 836–845.
D'Hont A, Paget-Goy A, Escoute J, Carreel F (2000) The interspecific genome structure of cultivated bananas, *Musa* spp. revealed by genomic DNA *in situ* hybridization. Theor Appl Genet 100: 177–183.
Dahal G, Ortiz R, d'A. Hughes J, Tenkouano A, Thottappilly G, Vuylsteke D, Lockhart B (2000) Relationship between natural occurrence, symptom expression, relative concentration of banana streak virus antigens, growth and yield characteristics of some micropropagated *Musa* accessions. Plant Pathology 49: 68–79.
De Langhe E (1961) La taxonomic du bananier plantain en Afrique Equatoriale. J Agri Trop Bot Appl (Brussels) 8: 419–449.
De Langhe E (1964) The origin of variation in the plantain banana. Mededelingen van de Landbouwhogeschool en de Opzoekingsstations van de Staat te Gent. 29: 45–80.
De Langhe E, Vrydaghs L, de Maret P, Perrier X, Denham T (2009) Why bananas matter: an introduction to the history of banana domestication. Ethnobot Res Appl 7: 165–177.
De Langhe E, Hřibová E, Carpentier S, Doležel J, Swennen R (2010) Did backcrossing contribute to the origin of hybrid edible bananas? Ann Bot 106: 849–857.
Denham TP, Haberle SG, Lentfer C, Fullagar R, Field J, Porch N, Therin M, Winsborough B, Golson J (2003) Multi-disciplinary Evidence for the Origins of Agriculture from 6950–6440 Cal BP at Kuk Swamp in the Highlands of New Guinea. Science 301: 189–193.
Dochez C, Whyte J, Tenkouano A, Ortiz R, De Waele D (2005) Response of East African highland bananas and hybrids to *Radopholus similis*. Nematology 7: 655–666.
Dochez C, Tenkouano A, Ortiz R,. Whyte JBA, De Waele D (2006) New sources of resistance to *Radopholus similis* in *Musa* germplasm from Asia. Aust J Plant Pathol 35: 481–485.
Dodds KS (1943) Genetical and cytological studies of *Musa*. V. Certain edible diploids. J Genet 45: 113–138.
Dodds KS, Simmonds NW (1946) Genetical and cytological studies of *Musa*. VIII. The formation of polyploid spores. J Genet 47: 223–241.
Dolezel J (2004) Cytogenetic and cytometric analysis of nuclear genome in *Musa*. In: SM Jain, R Swennen (eds) Banana Improvement: Cellular, Molecular Biology, and Induced Mutations. Science Publ Enfield, NH, USA, pp 245–249.
Dolezel J, Dolezelova M, Novak FJ (1994) Flow cytometric estimation of nuclear DNA amount in diploid bananas (*Musa acuminata* and *M. balbisiana*). Biol Plant 36: 351–357.
Dolezelova M, Valarik M, Swennen R, Horry JP, Dolezel J (1998) Physical mapping of the 18S–25S and 5S ribosomal RNA genes in diploid bananas. Biol Plant 41: 479–505.
Eden-Green S (2004) How can the advance of banana Xanthomonas wilt be halted? InfoMusa 13(2): 38–41.

Eden-Green S (2007) Banana *Xanthomonas* wilt: a review of the disease, management strategies and future research directions. African Journal of Biotechnology 6: 953–962.
Elsen A, Stoffelen R, Tuyet R, Nguyen T, Baimey H, de Boulois HD, De Waele D (2002) *In vitro* screening for resistance to *Radopholus similis* in *Musa* spp. Plant Science 163: 407–416.
Engelberger L (2003) Carotenoid-rich bananas in Micronesia. InfoMusa 12: 2–5
FAO (1999) The impact of banana supply and demand changes on income, employment and food security. Gold Coast, Australia, 4–8 May 1999. (www.fao.org)
FAO (2004) Food and Agricultural Organization. Rome, Italy (www.fao.org)
FAO (2005) Food and Agricultural Organization. Rome, Italy (www.fao.org)
Fogain R (1996) Screenhouse evaluation of *Musa* for susceptibility to *Radopholus similis*: evaluation of plantains AAB and diploids AA, AB and BB.. In: EA Frison, JP Horry, D De Waele (eds) Proceedings of the Workshop, New Frontiers in Resistance Breeding for Nematodes, Fusarium and Sigatoka, Kuala Lumpur, Malaysia. INIBAP, Montpellier, France, pp 79–88.
Fogain R, Price NS (1994) Varietal screening of some *Musa* genotypes for susceptibility to the banana weevil, *Cosmopolites sordidus* (Coleoptera: Curculionidae). Fruits 49: 247–251.
Frison E, Sharrock S (1999) The economic, social and nutritional importance of banana in the world. In: C Picq, E Foure, EA Frison (eds) Proceedings of the International Symposium, Bananas and Food Security, Douala, Cameroon, INIBAP, Montpellier, France, pp 21–35.
Frison EA, Escalant JV, Sharrock S (2004) The Global *Musa* genomic consortium: A boost for banana improvement. In: S Mohan Jain, R Swennen (eds) Banana Improvement: Cellular, Molecular Biology, and Induced Mutations. Science Publishers, Inc., NH, USA, pp 341- 349.
Fuller DQ, Madella M (2001) Issues in Harappan archaeobotany: retrospect and prospect. In: S Settar, R Korisettar (eds) Indian Archaeology in Retrospect, vol II: Protohistory, Archaeology of the Harappan Civilization, Manohar, New Delhi, India, pp 317–390.
Fungo R (2007) An assessment of B-carotene, iron and zinc in a range of banana cultivars. MSc Thesis, Makerere Univ, Uganda.
Ganapathi TR, Higgs NS, Balint-Kurti PJ, Arntzen CJ, May GD, Van Eck JM (2001) *Agrobacterium* -mediated transformation of the embryogenic cell suspensions of the banana cultivars Rasthali (AAB). Plant Cell Rep 20: 157–162.
Gawel NJ, Jarret RL, Whittemore AP (1992) Restriction fragment length polymorphism (RFLP)-based phylogenetic analysis of *Musa*. Theor Appl Genet 84: 286–290.
Ge XJ, Liu MH, Wang WK, Schaa BA, Chiang TY (2005). Population structure of wild bananas, *Musa balbisiana*, in China determined by SSR fingerprinting and cpDNA PCR-RFLP. Mol Ecol 14: 933–944.
Gepts P, Beavis WD, Brummer EC, Shoemaker RC, Stalker HT, Weeden NF, Young ND (2005) Legumes as a model plant family. Genomics for food and feed. Rep of the Cross-legume Advances through Genomics conf. Plant Physiol 137: 1228–1235.
Hakkinen M (2004) *Musa voonii*. A new *Musa* species from Northern Borneo and discussion of the section *Callimusa* in Borneo. Acta Phytotaxon Geobot 55: 79–88.
Häkkinen M (2009) *Musa chunii* Häkkinen, a new species (Musaceae) from Yunnan, China and taxonomic identity of *Musa rubra*. J Syst Evol 47:87–91.
Hakkinen M, De Langhe E (2001) *Musa acuminata* in Northern Borneo. INIBAP Publ, Montpellier, France.
Hakkinen M, Meekiong K (2004) A new species of the wild banana genus, *Musa* (*Musaceae*), from Borneo. Syst Biodiver 2: 169–173.
Hakkinen M, Wong W (2007) New species and variety of *Musa* (Musaceae) from Yunnan, China. Novon 17: 440–446.
Häkkinen M, Yeh CL , Ge XJ (2010) A New combinaison and a new variety of *Musa itinerans* (Musaceae). Acta Phytotaxon Geobot (Jpn) 61: 41–48.

Hamill SD, Smith MK, Dodd WA (1992) *In vitro* induction of banana autotetraploids by colchicine treatment of micropropagated diploids. Aust J Bot 40: 887–896.

Hardisson A, Rubio C, Baez A, Martin M, Alvarez R, Diaz E (2001) Mineral composition of the banana (*Musa acuminata*) from the island of Tenerife. Food Chem 73: 153–161.

Harris DR (2006) The interplay of ethnographic and archaeological knowledge in the study of past human subsistence in the tropics. J Roy Anthropol Inst (NS) 12: S63–S78.

Hautea DM, Molina GC, Balatero CH, Coronado NB, Perez EB, Alvarez MTH, Canama AO, Akuba RH, Quilloy RB, Frankie RB, Caspillo CS (2004) Analysis of induced mutants of Philippine bananas with molecular markers. In: S Mohan Jain, R Swennen (eds) Banana Improvement: Cellular, Molecular Biology, and Induced Mutations. Science Publishers, Inc., NH, USA, pp 45–57.

Honfo FG, Kayodé PAP, Coulibaly O, Tenkouano A (2007) Relative contribution of banana and plantain products to the nutritional requirements for iron, zinc and vitamin A of infants and mothers in Cameroon. Fruits 62: 1–11.

Howell EC, Newbury HJ, Swennen R, Withers LA, Ford-Lloyd BV (1994) The use of RAPDS for identifying and classifying *Musa* germplasm. Genome 37: 328–332.

Hughes Jd'A, Speijer PR, Olatunde O (1998) Banana Die-Back Virus—a New Virus Infecting Banana in Nigeria. Plant Dis 82: 129.

Ittyeipe K (1986) Studies on host preference of banana weevil borer, *Cosmopolites sordidus* Germ. (*Curculionidae*: *Coleoptera*). Fruits 46: 375–379.

Jarret RL, Gawel N, Whittemore A, Sharrock S (1992) RFLP-based phylogeny of *Musa* species in Papua New Guinea. Theor Appl Genet 84: 579–584.

Jones DR (2000) Introduction to banana, abaca and enset. In: DR Jones (ed) Diseases of Banana, Abaca and Enset. CABI Publishing, CAB International, Wallingford, UK.

Kajale MD (1989) Mesolithic exploitation of wild plants in Sri Lanka: archaeobotanical study at the cave site of Beli-Lena. In: DR Harris, GC Hillman (eds) Foraging and Farming: The Evolution of Plant Exploitation, Routledge, London, UK, pp. 269–281.

Kamate K, Brown S, Durand P, Bueau JM, De Nay D, Trinh TH (2001) Nuclear DNA content and base composition in 28 taxa of *Musa*. Genome 44: 622–627.

Kassim S, Nyine M, Pillay M (2010) A screening method for banana weevil (*Cosmopolites sordidus* Germar) resistance using reference genotypes. African Journal of Biotechnology 9: 4725–4730.

Khanna H, Becker D, Kleidon J, Dale J (2004) Centrifugation Assisted *Agrobacterium tumefaciens*-mediated transformation (CAAT) of embryogenic cell suspensions of banana (*Musa* spp. Cavendish AAA and Lady finger AAB). Mol Breed 14: 239–252.

Kiggundu A, Gold CS, Labuschagne MT, Vuylsteke D, Schalk L (2003) Levels of host plant resistance to banana weevil, *Cosmopolites sordidus* (Germar) (Coleoptera : Curculionidae), in Uganda *Musa* germplasm. Euphytica 133: 267–277.

Kiggundu A, Gold CS, Labuschagne MT, Vuylsteke D, Luow S (2006) Components of resistance to banana weevil (*Cosmopolites sordidus*) in *Musa* in Uganda. Entomol Exp Appl 122: 27–35.

Kwon SJ, Choi EY, Choi YJ, Ahn JH, Park OK (2006) Proteomics studies of post-translational modifications in plants. J Exp Bot 57: 1547–1551.

Lawton-Rauh A, Robichaux RH, Purugganan MD (2003) Patterns of nucleotide variation in homoeologous regulatory genes in the allotetraploid Hawaiian silversword alliance (Asteraceae). Mol Ecol 12: 1301–1313.

Lebot V (1999) Biomolecular evidence for plant domestication in Sahul. Genet Resour Crop Evol 46: 619–628.

Lebot V, Aradhya KM, Manshardt R, Meilleur B (1993) Genetic relationships among cultivated bananas and plantains from Asia and the Pacific. Euphytica 67: 163–175.

Lee SK, Kader A (2000) Preharvest and postharvest factors influencing vitamin C content of horticultural crops. Postharvest Biol Technol 20: 207–220.

Lejju JB, Robertshaw P, Taylor D (2006) Africa's earliest bananas? J Archaeol Sci 33: 102–113.

Li WL, Berlyn GP, Ashton PMS (1996) Polyploids and their structural and physiological characteristics relative to water deficit in *Betula papyrifera* (Betulaceae). Am J Bot 83: 15–20.

Lynch M (2002) Genomics. Genes duplication and evolution. Science 297: 945–947.

Lysak MA, Dolezelova M, Horry JP, Swennen R, Dolezel J (1999) Flow cytometric analysis of nuclear DNA content in *Musa*. Theor Appl Genet 98: 1344–1350.

Mable BK (2003) Breaking down taxonomic barriers in polyploidy research. Trends Plant Sci 8: 582–590.

Marriott H, Lancaster PA (1983) Bananas and plantains. In: HT Chan (ed) Handbook of Tropical Foods. Marcel Dekker, New York, USA, pp 85–143.

Mason JB, Garcia M (1993) Micronutrient deficiency—the global situation. SCN News 9: 11–16.

May GD, Rownak A, Mason H, Wiecko A, Novak FJ, Arntzen CJ (1995) Generation of transgenic banana (*Musa acuminata*) plants via *Agrobacterium*-mediated transformation. Bio/Technology 13: 486–492.

Mbaka J, Ndumgu V, Mwangi M (2007) Outbreak of Xanthomonas wilt (*Xanthomonas campestris* pv *musacearum*) on banana in Kenya. ISHS/ProMusa symposium. Recent advances in banana crop protection for sustainable production and improved livelihoods. Greenway Woods Resort, White River, South Africa.

Mbaka JN, Nakata VG, Auma J, Odero B (2009). Status of banana Xanthomonas wilt in western Kenya and factors enhancing its spread. Afr Crop Sci Conf Proc 9: 673–676.

Ming R, Del Monte TA, Hernandez E, Moore PH, Irvine JE, Paterson AH (2002) Comparative analysis of QTLs affecting plant height and flowering among closely-related diploid and polyploid genomes. Genome 45: 794–803.

Nwakanma DC, Pillay M, Okoli BE, Tenkouano A (2003a) Sectional relationships I the genus *Musa* L. inferred from the PCR-RFLP of organelle DNA sequences. Theor Appl Genet 107: 850–865.

Nwakanma DC, Pillay M, Okoli BE, Tenkouano A (2003b) PCR-RFLP of the ribosomal DNA internal transcribed spacers (ITS) provides markers for the A and B genomes in *Musa* L. Theor Appl Genet 108: 154–159.

Ortiz R (1997) Occurrence and inheritance of 2n pollen in *Musa*. Ann Bot 79: 449–453.

Ortiz R, Vuylsteke D, Dumpe B, Ferris RSB (1995) Banana weevil resistance and corm hardness in *Musa* germplasm. Euphytica 86: 95–102.

Osborn TC, Pires JC, Birchler JA, Auger DL, Chen ZJ, Lee HS, Comai L, Madlung A, Doerge RW, Colot V, Martienssen RA (2003) Understanding mechanisms of novel gene expression in polyploids. Trends Genet 19: 141–147.

Oselebe HO, Tenkouano A, Pillay M, Obi IU, Uguru MI (2006) Ploidy and genome segregation in *Musa* breeding populations assessed by flow cytometry and randomly amplified polymorphic markers. J Hort Sci 131: 780–786.

Osuji JO, Vuylsteke D, Ortiz R (1997a). Ploidy variation in hybrids from interploid $3x$ x $2x$ crosses in *Musa*. Tropicultura 15: 37–39.

Osuji JO, Crouch J, Harrison G, Heslop-Harrison JS (1997b) Identification of the genomic constitution of *Musa* L. genotypes (bananas, plantains and hybrids) using molecular cytogenetics. Ann Bot 80: 787–793.

Osuji JO, Crouch J, Harrison G, Heslop-Harrison JS (1998) Molecular cytogenetics of *Musa* species, cultivars and hybrids: Location of 18S–5.8S –25S and 5S rDNA and telomere-like sequences. Ann Bot 82: 243–248.

Pavis C (1991) Etude des relatiuons plante-insecte chez le charancon du bananier *Cosmopolites sordidus* Germar (Coleoptera:Curculionidae). In: CS Gold, B Gemmill (eds) Proc of a Research Coordination Meeting on Biological and Integrated control of Highland Banana and Plantain Pests and Diseases. Cotonou, Benin, IITA, Ibadan, Nigeria, pp 171–181.

Pietersen G, Thomas JE (2001) Overview of *Musa* virus diseases. In: Plant virology in sub Saharan Africa: http://www.iita.org/info/virology/pdf_files/50-60.pdf

Pillay M, Adeleke MTV (2001) Silver Staining of *Musa* L. Chromosomes. Cytologia 66: 33–37.
Pillay M, Tripathi L (2007) Banana breeding. In: MS Kang, PM Priyadarshan (eds) Breeding Major Food Staples. Blackwell Science, Boston, MA, USA, pp 393–428.
Pillay M, Nwakanma DC, Tenkouano A (2000) Identification of RAPD markers linked to A and B genome sequences in *Musa*. Genome 43: 763–767.
Pillay M, Tenkouano A, Hartman J (2002) Future challenges in *Musa* breeding. In: MS Kang (ed) Crop Improvement: Challenges in the Twenty-first Century, Food Products Press, New York, USA, pp 223–252.
Pillay M, Tenkouano A, Ude G, Ortiz R (2004) Molecular characterization of genomes in *Musa* and its applications. In: S Mohan Jain, R Swennen (eds) Banana Improvement: Cellular, Molecular Biology, and Induced Mutations. Science Publishers, Inc., NH, USA, pp 271–286.
Pinochet J, Jaizme MC, Fernandez C, Jaumot M, De Waele D (1998) Screening banana for root-knot (*Meliodogyne* spp.) and lesion nematode (*Pratylenchus goodeyi*) resistance for the Canary Islands. Fund Appl Nematol 21: 17–23.
Ploetz RC (1990) Population biology of *Fusarium oxysporum* f.sp. *cubense*. In: RC Ploetz (ed) Fusarium Wilt of Banana. APS Press, St Paul, Minnesota, USA, pp 63–76.
Ploetz RC (1994) Panama disease: return of the first banana menace. Int J Pest Manag 40: 326–336.
Ploetz RC, Vazquez A, Nagel J, Benscher D, Sianglew P, Srikul S, Kooariyakul S, Wattanachaiyingcharoen W, Lertrat P, Wattanachaiyingcharoen D (1997) Current status of Panana disease in Thailand. Fruits 51: 387–395.
Price NS, McLaren CG (1996) Techniques for field screening of *Musa* germplasm. In: EA Frison, JP Horry, D De Waele (eds) Proceedings of the Workshop, New Frontiers in Resistance Breeding for Nematodes, Fusarium and Sigatoka, Kuala Lumpur, Malaysia, INIBAP, Montpellier, France, pp 87–107.
Purseglove JW (1972) Tropical Crops: Monocotyledons, John Wiley, New York, USA.
Reynolds PK (1927) The Banana, its History, Cultivation and Place among Staple Foods. Houghton Mifflin, Boston, MA, USA.
Rishi N (2009) Significant plant virus diseases in India and a glimpse of modern disease management technology. J Gen Plant Pathol 75: 1–18.
Robinson JC (1996) Bananas and plantains. Crop Production Science in Horticulture 5. CAB Intl. Wallingford, Oxon, UK.
Rossel G (1991) The diffusion of plantain (*Musa* sp. AAB) and banana (*Musa* sp. AAA) in Africa: a case for linguists, taxonomists and historians, focused on Nigerian crop names. In: RE Leakey, LJ Slikkerveer (eds) Origins and Development of Agriculture in East Africa: The Ethnosystems Approach to the Study of Early Food Production in Kenya, Iowa State University Research Foundation, Ames, Iowa, USA, pp 129–160.
Rossignol M, Peltier JB, Mock HP, Matros A, Maldonado AM, Jorrín JV (2006) Plant proteome analysis: A 2004–2006 update. Proteomics 6: 5529–5548.
Roux NS (2004) Mutation induction in *Musa*—A review In: S Mohan Jain, R Swennen. (eds) Banana Improvement: Cellular, Molecular Biology, and Induced Mutations. Science Publishers, Inc., NH, USA, pp 23–32.
Roux N, Toloza A, Radecki Z, Zapata-Arias FJ, Doležel J (2003) Rapid detection of aneuploidy in *Musa* using flow cytometry. Plant Cell Rep 21: 483–490.
Roux NS, Toloza A, Dolezel J, Panis B (2004) Usefulness of embrygenic cell suspension cultures for the induction and selection of mutants in *Musa* spp. In: S Mohan Jain, R Swennen (eds) Banana Improvement: Cellular, Molecular Biology, and Induced Mutations. Science Publishers, Inc., NH, USA, pp 33–43.
Rowe P, Rosales F (1990) Breeding bananas and plantains with resistance to black Sigatoka. In : RA Fullerton, RH Stover (eds) Proceeding of an International Workshop, Sigatoka Leaf Spot Disease of Bananas. San Jose, Costa Rica, pp 243–251.

Rowe P, Rosales F (1996) Bananas and plantains. In: J Janick, JN Moore (eds) Fruit Breeding, vol I: Tree and Tropical Fruits, John Wiley and Sons, Inc. New York, USA, pp 167–211.
Sagi L, Remy S, Panis B, Swennen R, Volckaert G (1994) Transient gene expression in electroporated banana (*Musa* spp, cv Bluggoe, ABB group) protoplasts isolated from regenerable embryogenetic cell suspensions. Plant Cell Rep 13: 262–266.
Sagi L, Panis B, Remy S, Schoofs H, De Smet K, Swennen R, Cammus B (1995) Genetic transformation of banana (*Musa* spp.) via particle bombardment. Bio/Technology 13: 481–485.
Samyn B, Sergeant K, Carpentier S, Debyser G, Panis B, Swennen R, Van Beeumen J (2007) Functional proteome analysis of the banana plant (*Musa* spp.) using de novo sequence analysis of derivatized peptides. J Proteome Res 6: 70–80.
Sarah JL, Pinochet J, Stanton J (1996) The burrowing nematode of bananas, *Radopholus similis* Cobb, 1913. *Musa* pest fact sheet No. 1. INIBAP, Montpellier, France
Sasser JN, Freckman DW (1987) A world perspective on nematology. In: DW Dickson, JAVeech (eds) Vistas in Nematology, Society of Nematologists, Hyattsville, Maryland, USA, pp 7–14.
Sathiamoorthy S, Balamohan TN (1993) Improvement of banana. Adv Hort (India) 1: 303–335.
Sharrock S (1989) Collecting *Musa* in Papua New Guinea. In: RL Jarret (ed) Identification of genetic diversity in the genus *Musa*, Los Baños, the Philippines, INIBAP, Montpellier, France, pp 140–157.
Shepherd K (1996) Cytogenetics of the Genus *Musa*. International Network for the Improvement of Banana and Plantain, Montpellier, France.
Shepherd K, Ferreira FR (1984) The PNG Biological Foundation Banana Collection at Laloki, Port Moresby, Papua New Guinea. IBPGR Regional Committee for South East Asia Newsl 8: 28–34.
Shepherd K, Da Silva KM (1996) Mitotic instability in banana varieties. Aberrations inconventional triploid plants. Fruits 51(2): 99–103.
Shepherd K, Dantas JLL, de Oliveira e Silva S (1994) Breeding Prata and Maca cultivars in Brazil, In: DR Jones (ed) Proceedings of the first Global Conference of the International Testing Program, The Improvement and Testing of *Musa:* a Global Partnership, Honduras, INIBAP, Montpellier, France, pp 57–168.
Simmonds NW (1948) Genetical and cytological studies of *Musa*, X: stomata size and plant vigour in relation to polyploidy. J Genet 49: 57–68.
Simmonds NW (1954) Isolation in *Musa* section Eumusa and Rhodochlamys. Evolution 8: 65–74.
Simmonds NW (1956) Botanical results of the banana collecting expedition 1954–1955. Kew Bull 3: 463–489.
Simmonds NW (1962) Evolution of the Bananas. Longmans, London, UK.
Simmonds NW (1995) Bananas *Musa* (*Musa*ceae). In: J Smartt, NW Simmonds (eds) Evolution of Crop Plants. 2nd edn. Longmans Scientific and Technical, Essex, UK, pp 370–375.
Simmonds NW (1966) Bananas. 2nd edn. Tropical Agri Series. Longman, London, UK.
Simmonds NW, Shepherd K (1955) The taxonomy and origins of the cultivated bananas. J Linn Soc Lond Bot 55: 302–312.
Someya S, Yoshiki Y, Okubo K (2002) Antioxidant compounds from bananas (*Musa* Cavendish). Food Chem 79: 351–354.
Stoffelen R, Verlinden R, Xuyen NT, Swennen R, DeWaele D (1999) Screening of Papua New Guinea bananas to root-lesion and root- knot nematodes. InfoMusa 8: 12–15.
Stover RH, Simmonds NW (1987) Bananas. 3rd Edn. Longman, London, UK.
Strosse H, Schoofs H, Panis B, André E, Reyniers K, Swennen R (2006) Development of embryogenic cell suspensions from shoot meristematic tissue in bananas and plantains (*Musa* spp.). Plant Sci 170: 104–112.
Swennen R, Rosales F (1994) Bananas. In: CJ Arntzen, EM Ritter (eds) Encyclopedia of Agricultural Science A-D vol, Academic Press, New York, USA, pp 215–232.

Swennen R, Wilson GF, De Langhe E (1984) Preliminary investigation of the effects of gibberellic acid (GA3) on sucker development in plantain (*Musa* cv. AAB) under field conditions. Trop Agri 61: 253–256.

Tate J, Soltis DE, Soltis PS (2005) Polyploidy in Plants. In: TR Gregory (ed) The Evolution of the Genome. Elsevier, San Diego, USA, pp 371– 426.

Taylor G (2002) Populus: *Arabidopsis* for forestry. Do we need a Model tree. Ann Bot 90: 681–689.

Tenkouano A (2005) Breeding banana and plantain: Integrating molecular techniques and conventional approaches to explore genetic polymorphisms and predict progeny performance. In: D Thangadurai, T Pullaiah, L Tripathi (eds) Genetic Resources and Biotechnology, vol 3. Regency Publications, New Delhi, India, pp 153–168.

Tenkouano A, Crouch JH, Crouch HK, Vuylsteke D (1998) Ploidy determination in *Musa* germplasm using pollen and chloroplast characteristics. HortScience 33: 889–890.

Tezenas du Montcel H (1988) *Musa acuminata* subspecies banksii: status and diversity. In: RL Jarret (ed) Identification of Genetic Diversity in the Genus *Musa*. Proc Int Workshop held at Los Banos, the Philippines. INIBAP, Montpellier, France, pp 211–218.

Tezenas du Montcel H, Carreel F, Bakry F (1995) Improve the diploids: the key for banana breeding. In: New Frontiers in Resistance Breeding for Nematode, Fusarium and Sigatoka. Proc of the Workshop on New frontiers in resistance breeding for nematode, fusarium and Sigatoka held at Kuala Lumpur, Malaysia.

Tripathi L, Tripathi JN, Hughes J (2005) *Agrobacterium*-mediated transformation of plantain cultivar Agbagba (*Musa* spp.). Afr J Biotechnol 4: 1378–1383.

Ude G, Pillay M, Nwakanma D, Tenkouano A (2002a) Analysis of genetic diversity and sectional relationships in *Musa* using AFLP markers. Theor Appl Genet 104: 1239–1245.

Ude G, Pillay M, Nwakanma D, Tenkouano A (2002b) Genetic diversity in *Musa acuminata Colla* and *Musa balbisiana* Colla and some of their natural hybrids using AFLP Markers. Theor Appl Genet 104: 1246.

US Department of Agriculture, Agricultural Research Service (USDA-ARS) (2004) USDA National Nutrient Database for Standard Reference, Release 17. Nutrient Data Laboratory:http://www.nal.usda.gov/fnic/foodcomp.

Valmayor R, Danh LD, Hakkinen M (2004) Rediscovery of *Musa splendida* A. Chevalier and description of two new species (*Musa viridis* and *Musa lutea*). Philipp Agri Sci 87: 110–118.

Vandenhout H, Ortiz R, Vuylsteke D, Swennen R, Bai KV (1995) Effects of ploidy on stomatal and other quantitative traits in plantain and banana hybrids. Euphytica 83: 117–122.

Van Duren M, Morpurgo R, Dolezel J, Afza R (1996) Induction and verification of tetraploids in diploid banana (*Musa acuminata*) by *in vitro* techniques. Euphytica 88: 25–34.

Vavilov NI (1935) Theoretical Basis for Plant Breeding In: D Love (transl.) Moscow. Origin and Geography of Cultivated Plants, vol 1. The Phytogeographical Basis for Plant Breeding. Cambridge Univ. Press, Cambridge, UK, pp 316–366.

Viaene N, Duran LF, Rivera JM, Duenas J, Rowe P, De Waele D (2003) Responses of banana and plantain cultivars, lines and hybrids to the burrowing nematode *Radopholus similis*. Nematology 5: 85–98.

Vuylsteke D, Ortiz R, Ferris RSB, Crouch JH (1997) Plantain improvement. Plant Breed Rev 14: 267–320.

Wall MM (2006) Ascorbic Acid, vitamin A, and mineral composition of banana (*Musa* sp.) and papaya (*Carica papaya*) cultivars grown in Hawaii. J Food Comp Anal 19: 434–445.

Wehunt EJ, Hutchinson DJ, Edwards DI (1978) Reaction of banana cultivars to the burrowing nematode (*Radopholus similis*). J Nematol 10: 368–370.

Wendel JF (2000)Genome evolution in polyploids. Plant Mol Biol 42: 225–249.

Wenkam NS (1990) Food of Hawaii and the Pacific Basin, Fruits and Fruit Products: Raw, Processed, and Prepared, vol. 4: Composition. Hawaii Agricultural Experiment Station Research and Extension Series 110. Hawaii.

WHO (1992) National Strategies for Overcoming Micronutrient Malnutrition. 45th World Health Assembly Provisional Agenda Item 21, doc A45/17. WHO, Geneva, Switzerland.
Wilson GB (1946) Cytological studies in the *Musae*. II. Meiosis in some diploid clones. Genetics 31: 475–482.
Wong C, Kiew R, Argent G, Set O, Lee SK, Gan YY (2002) Assessment of the Validity of the Sections in *Musa* (Musaceae) using AFLP. Ann Bot 90: 231–238.
World Bank (1994) Enriching Lives: Overcoming Vitamin and Mineral Malnutrition in Developing Countries. World Bank, Washington DC, USA.
Wullschleger SD, Sanderson MA, McLaughlin SB, Biradar DP, Rayburn AL (1996) Photosynthetic rates and ploidy levels among different populations of switchgrass. Crop Sci 36: 306–312.

2

Classical Genetics and Traditional Breeding in *Musa*

Michael Pillay

ABSTRACT

Traditional breeding in *Musa* is faced with several impediments primarily the sterile nature of the triploid cultivars. Classical and linkage mapping studies using morphological, cytogenetic and isozyme markers have not been exploited fully in *Musa* although segregation has been observed in these characters. The main reason for the lack of these maps is the difficulty of developing large segregating populations especially among the cultivated triploid varieties. New efforts in molecular mapping in *Musa* are increasing rapidly. Genome synteny could be used to bridge the genetic information between other monocotyledons and *Musa*. The molecular markers produced for closely related taxa of *Musa* may be used to accelerate the development of linkage maps of *Musa*. The breeding objectives of *Musa* include increased yield, resistance to biotic and abiotic stresses, changes to plant morphology, and shortening of the crop cycle. To date, banana breeding programs have concentrated primarily on developing varieties with disease and pest resistance and maintaining current yields. Achievements in classical breeding are discussed and the rationale for molecular breeding is advanced.

Key words: mapping, classical breeding achievements, molecular breeding

2.1 Introduction

Musa is a large species complex which includes diploid, triploid and some tetraploid synthetic cultivars. These largely sterile and vegetatively

Vaal University of Technology, Private Bag X021, Vanderbijlpark 1900, Gauteng, South Africa; e-mail: *mpillay@vut.ac.za*

List of abbreviations after the text.

propagated cultivars are based on the A genome from *Musa acuminata*, exclusively for sweet bananas such as Cavendish, or associated with the B genome (*Musa balbisiana*) in cooking bananas such as plantain varieties. Traditional breeding in *Musa* is faced with several impediments primarily the sterile nature of the triploid cultivars (Pillay et al. 2002). Currently the development of new cultivars in *Musa* is approached in several diverse ways (Pillay and Tenkouano 2011). In *M. acuminata* cultivars, structural heterozygosity is thought to be one of the main causes of sterility, which is essential for obtaining seedless fruits but hampers breeding (Hippolyte et al. 2010). There is great need to learn more about the *Musa* genome, its cytology, mapping efforts and genetics to facilitate future breeding strategies. This chapter examines some of the efforts made in this regard.

2.2 Classical Mapping Efforts in *Musa*

The aim of genetic mapping is to determine the linear sequence of genes in genetic material. The two main types of maps are known as genetic or linkage maps and physical maps. A genetic linkage map illustrates the order of genes on a chromosome and the relative distances between those genes and is analogous to signs or landmarks along a highway (Collard et al. 2005). The distance between the genes is expressed in centiMorgans (cM) and derived from the recombination rates between the loci using either Haldane or Kosambi mapping function. Physical maps provide the physical, DNA base pair distances from one gene or marker to another. Physical maps are assembled from the complete genome sequences, bacterial artificial chromosome (BAC) contigs, or based on *in situ* hybridization. There is no specific relationship between the recombination distance and the physical map distance, expressed in base pairs, because the rate of recombination varies along the length of the chromosome (Kumar 1999). Further, in heterochromatic regions, such as the centromeres, recombination frequency is quite reduced. Crossing over is suppressed in heterochromatin and centromeres, and crossing over is variable even in euchromatin where most crossing over occurs (Anderson et al. 2004). Therefore linkage maps cannot be simply overlaid on chromosomes to determine the physical position of genes. In such cases, a type of physical map called, cytogenetic or cytological map is very useful since it is based on the fine structure of the chromosomes and can provide complementary information. Cytogenetic maps allow association of linkage groups with chromosomes and determination of the orientation of the linkage groups with respect to chromosome morphology (Schneider 2005). The physical map that provides the most detail is the sequence map that shows genetic markers as well as sequence between markers as measured in base pairs (www.ncbinlm.nih.gov/about/primer/mapping.html). Mapping would not be possible without genetic markers.

Genetic markers represent genetic differences between individual organisms or species. There are three main types of genetic markers: (1) morphological, "classical" or "visible" markers, (2) biochemical markers, which include allelic variants of enzymes called isozymes, and (3) DNA or molecular markers (Kumar 1999; Collard et al. 2005).

Unlike the progress made in classical mapping in crops such as maize and rice, linkage mapping studies using morphological, cytogenetic and isozyme markers in *Musa* have not been exploited fully. This does not imply that segregation for morphological, cytogenetic and isozyme characters are not evident in *Musa*. Although there is no published information on the segregation of cytological traits, studies on the number of 5S rDNA sites in *Musa* suggest that segregation for cytological features may be present (Dolezelova et al. 1998; Bartos et al. 2005). Segregation of morphological traits and isozyme markers has been reported in *Musa* (Faure et al. 1993; Lebot et al. 1993; Shepherd 1999). Differences in morphological characteristics have formed the basis of taxonomic studies and classification systems in *Musa* (Simmonds and Shepherd 1955). For example, 15 morphological characters are used to separate *M. acuminata* from *M. balbisiana,* the two ancestral species of cultivated bananas that are morphologically distinct but are able to hybridize. Variation in morphological characteristics has been used widely in germplasm characterization, differentiation of subspecies, sections, genomic groups, clones and clone sets and subgroups in *Musa* (Swennen and Vuylsteke 1987; Swennen et al. 1995; Ortiz 1997; Osuji et al. 1997; Ortiz et al. 1998a; Karamura 1999). If one considers plantains, morphological analyses of this taxon clearly differentiated it into four subgroups: French, French Horn, False Horn and Horn (Tezenas du Montcel et al. 1983; Swennen and Vuylsteke 1987). Each of these subgroups was further subdivided into giant, medium and small on the basis of pseudostem height (De Langhe 1964). The extent of morphological variation in *Musa* has even led to the production of a list of morphological descriptors for the classification of bananas (Anon 1984). The point to note is that there are ample morphological markers in *Musa*. The inheritance of morphological markers can be monitored without specialized or molecular techniques (Kumar 1999). Therefore one has to identify possible reasons for the absence of morphological linkage maps in *Musa*. Perhaps, the primary reason is due to the difficulty of developing large segregating populations especially among the cultivated triploid varieties that are mainly male and female sterile. A second reason is that the major goal of banana researchers over the last century lay in the genetic improvement of the crop. The first major disease that affected export bananas was Fusarium wilt (Jones 2000). The disease devastated the export banana production in the Latin American-Caribbean regions and prompted the growth of the first banana breeding program in Trinidad in 1922. However, while the first export banana hybrids were being developed for resistance to Fusarium wilt, another more devastating disease called Sigatoka became a major concern. Consequently, banana researchers concentrated their efforts on gaining more insight into the diseases and pests that affected the crop while the few breeding programs that existed concentrated on developing new banana varieties with resistance to the major pests and diseases. It appears that little scope existed for basic research in the crop and the development of classical maps.

One of the problems with morphological markers is that they can be influenced by environmental and genetic factors such as epistatis. Perhaps another reason for the absence of morphological maps in *Musa* may be due to the nature of inheritance of the morphological characters. The alleles for morphological markers interact in a dominant-recessive manner making it impossible to distinguish the heterozygotes from the homozygous individuals (Kumar 1999).

There has been a marked increase in basic research in banana especially with wild diploids that are fertile and are capable of producing large segregating populations. A number of segregating populations for some of the important traits have been developed in *Musa* (http://www.musagenomics.org).

2.3 Cytological Maps

Cytological maps incorporate data from genetic maps with actual cytological features of chromosomes such as centromeres, knobs, chromosome bands and *in situ* and fluorescence *in situ* hybridization (FISH) signals (Harper and Cande 2000). Cytological maps are created by microscopic examination of visible structures on fixed and stained chromosomes. Cytogenetic maps could be highly informative to support the construction of physical maps and map-based cloning projects and to position genes in large heterochromatic regions where linkage distances are inaccurate due to low levels of meiotic recombination (Lambie and Roeder 1986; Zhong et al. 1999). Molecular cytogenetic maps are also useful for placing BACs and other DNA sequences along the chromosomes. Cytogenetic maps are vital in genome sequencing projects because they provide information on the distribution of heterochromatin, euchromatin, centromeres and genes across genomes (Cheng et al. 2001a; Kulikova et al. 2001). Relatively few studies have attempted cytological mapping of the *Musa* genome (Dolezelova et al. 1998; Osuji et al. 1998; D'Hont et al. 2000). The major problems associated with developing cytogenetic maps in *Musa* have been the lack of suitable analytical tools and the homogeneous karyotype of the *Musa* genome. FISH has been used to study karyotypes of many crop species since the technique is able to provide more detailed landmarks of chromosome structure, gene localization and the linking of physical and genetic maps (Dolezel et al. 2004). FISH was used to map repetitive DNA sequences, telomeres, and mobile genetic elements in *Musa* chromosomes (Osuji et al. 1998; Dolezelova et al. 1998; D'Hont et al. 2000). Two sites of hybridization were observed for the 18S-5.8S-26S rDNA in all banana genotypes examined to date. These sites corresponded with the nucleolar organizer regions of the chromosomes. These regions were also observed as satellites when banana chromosomes were silver stained (Pillay and Adeleke 2001). The number of 5S rDNA sites ranged from four to six in different banana genotypes and indicated the presence of structural chromosome heterozygosity in some of the AA genotypes (Dolezelova et al. 1998). The number of 5S-rDNA loci were also found to vary in many other plant species (Cerbah et al. 1999). It would be interesting to assess the inheritance of 5S-rDNA sites in *Musa* by examining the progeny from crosses between genotypes with differential numbers of these sites. FISH studies confirmed that *Musa* possesses *Arabidopsis*-like telomeres (Dolezel et al. 2004). Three fragments of a mobile genetic element known as "monkey" showed differential hybridization to different sites and chromosomes with FISH analysis in *Musa* (Balint-Kurti et al. 2000). Until recently, all FISH analyses in *Musa* have been carried on somatic metaphase chromosomes. Metaphase chromosomes are generally highly condensed and this limits the resolution of adjacent FISH targets to Megabases rather than kilobases of DNA (Kulikova et al. 2001). Cytogenetic studies rely on accurate and consistent chromosome identification, which is always a challenge in plant species with small chromosomes (Cheng et al. 2001b). Limitations of chromosome size can be overcome by alternative means such as the use of pachytene chromosomes. Pachytene chromosomes are 10–40 times longer than mitotic metaphase chromosomes and they also display a differentiated pattern of euchromatic and heterochromatic regions (Albini and Schwarzacher 1992). The heterochromatic banding, chromosome length and centromere positions of pachytene chromosomes are few of the landmarks that make it easier to identify individual chromosomes. New techniques are now available to stain the pachytene chromosomes of *Musa* (Adeleke et al. 2002). Further research is needed to identify the pachytene chromosomes and develop a pachytene karyotype for *Musa*. This would pave the way for detailed cytological characterization of the *Musa* genomes and integrate the DNA sequence based maps with cytological maps. The 18S-5.8S-26S rDNA sites are the only chromosome specific molecular marker in *Musa* (Dolezel et al. 2004). The first attempt to identify each of the banana chromosomes by FISH analysis using highly repetitive DNA was made by Valarik et al. (2002). Since then several BAC libraries have been constructed in banana, including a BAC library from *M. acuminata* cv. Calcutta 4 (MA4 BAC library, Vilarinhos et al. 2003), BIBAC library of *M. acuminata* cv. Tuu Gia (TGBIBAC library, Ortiz-Vazquez et al. 2005) and BAC library of *M. balbisiana* cv. Pisang Klutuk Wulung (MBP BAC library, Safar et al. 2004).

These clones could be used to develop new cytogenetic markers for *Musa* (Hribova et al. 2008). In a study using the MA4 BAC library for identification of the low- and single-copy BAC clones one chromosome specific BAC clone was identified in *M. acuminata* cv. Calcutta 4 (Hribova et al. 2008). FISH of BAC clones anchored to a genetic map were able to characterize translocations in *M. acuminata* (Vilarinhos et al. 2006). The identification of translocations will assist in mapping since they make it difficult to construct genetic maps, to study the inheritance of agronomical inheritance and to breed in general. A recent study has highlighted the potential of high resolution BAC-FISH for banana breeding (De Capdeville et al. 2008). The authors discuss the usefulness of this technique including the placement of BAC clones on *Musa* chromosomes, generation of cytogenetic maps, validating the position of markers and the detection of inversions and deletions.

2.4 Isozymes

Isozymes are differently charged protein molecules that can be separated by electrophoresis and detected by specific staining (Marker and Moller 1959). Protein markers reveal differences in the gene sequence and function as co-dominant markers (Kumar 1999). A few studies have used isozymes as genetic markers in bananas and plantains (Bonner et al. 1974; Jarret and Litz 1986a, b; Lebot et al. 1993). These studies demonstrated that variation in isozyme patterns do occur in *Musa*. It was also shown that seedling progenies of *M. acuminata* and *M. balbisiana* segregated for allozymes in different enzyme systems (Lebot et al. 1993). Segregation for isozyme patterns was also observed by Faure et al. (1993) who placed the location of malate dehyrodgenase (*Mdh*) in the first low density linkage map of *Musa*. These studies indicate that it may be possible to develop isozyme linkage maps in *Musa*. The only concern is that although polymorphisms detected by isozyme systems are high, very few enzyme systems have been optimized in banana and in plants in general. To date, only 40–50 reagent systems have been developed that permit staining of a particular protein. It will only be possible to map a very small segment of the genome with such a small number of markers. The paucity of isozyme loci and the fact that they are subject to post-translational modification also restricts their utility (Staub et al. 1982) in mapping. Isozymes also display tissue variability. Some isozymes are expressed well in roots while others are best sampled from leaves. This implies that several parts of a plant will have to be sampled from a segregating population to score all the available isozymes.

For mapping purposes the number of morphological, cytological and isozyme markers could theoretically be increased greatly by mutagenesis (Kumar 1999). For example, mutation mapping was achieved in barley by an amplified fragment length polymorphism (AFLP) technique (Castiglioni et al. 1998).

2.5 Limitations of Classical Mapping Endeavors and Utility of Molecular Mapping

The limitations of classical mapping efforts in *Musa* as well as for other crop plants have been adequately described (Escalant and Panis 2002). The advent of DNA markers has added a new dimension to genetic mapping. The ultimate genetic map is best represented by the complete nucleotide sequence of the genome of an organism. The complete genome sequence of several plants including *Arabidopsis*, rice, soybean, papaya, grape, peach, sorghum and poplar is now known and the genome of over 20 other plant species is currently being sequenced.

2.5.1 Molecular Mapping

Biotechnology is now making an important contribution to agriculture because it is being used to modify microorganism, plants, animals and agricultural processes (Kumar 1999). Biotechnology has found many applications in agriculture, including cell and tissue culture for rapid propagation of plant species; in diagnostics, for detecting plant pests and diseases based on the use of monoclonal antibodies and nucleic acid probes; and in genetic engineering of plant species, to introduce new traits and in aiding conventional plant breeding programs using molecular markers (Persley 1992).

With the advent of DNA markers genetic maps have been developed for a number of plant species. Unlike morphological or biochemical markers, DNA markers are practically unlimited and are not affected by environmental factors and/or the development stage of the plant (Winter and Kahl 1995). DNA markers can be classified into three categories based on their method of detection: (1) hybridization based, (2) polymerase chain reaction (PCR) based and (3) DNA sequence based (Gupta et al. 1994; Joshi et al. 1999). Hybridization based markers include restriction fragment length polymorphisms (RFLP) and variable number of tandem repeats (VNTR) loci where random genomic clones or probes are hybridized to filters containing DNA which has been digested with restriction enzymes. PCR is a technique, which is used to amplify the number of copies of a specific region of DNA in order to produce enough DNA to be adequately visible on a gel. A large number of variations in PCR technology are now available to generate polymorphisms in plants. One is only restricted by the apparatus available and funding in deciding what technology to use. It does appear that some PCR techniques are better able to uncover DNA polymorphisms in specific species than others. The different types of molecular markers and how they generate polymorphisms are described adequately in a number of reviews (Kumar 1999; Gostimsky et al. 2005; Doveri et al. 2008). Similarly the advantages and disadvantages of the most commonly used markers are described in many papers (Barone 2004; Collard et al. 2005; Doveri et al. 2008).

DNA markers have been recognized as having enormous potential in plant breeding. The most valuable application of DNA markers in plant breeding includes, (i) the construction of genetic linkage maps, and (2) marker assisted selection (MAS).

2.5.2 Construction of Genetic Linkage Maps

Linkage maps have been used to identify chromosomal regions that contain genes controlling simple traits (controlled by a single gene) and quantitative traits using quantitative trait loci (QTL) analysis. A genetic linkage map graphically represents the arrangement of the many loci, which include morphological, isozyme and DNA markers along the chromosome.

Molecular linkage maps have been developed for a large number of crop plants including *Musa*. The first linkage map of *Musa* was developed by Faure et al. (1993). This map is based on 58 RFLP, four isozyme and 28 random amplified polymorphic DNA (RAPD) markers segregating in an F_2 population of 92 individuals.

Recently, two parental maps composed of 167 simple sequence repeat (SSR) and 322 diversity array technology (DArT) markers covering 1,197 cM were established from 180 F_1 progeny from a cross between two genetically distant accessions of *M. acuminata*, "Borneo" and "Pisang Lilin" (P. Lilin) (Hippolyte et al. 2010). Based on the gametic recombination of each parent, a significant proportion of the markers (21.7%) deviated ($p < 0.05$) from the expected Mendelian ratios. These skewed markers were distributed in different linkage groups for each parent. This first saturated map is proposed as a "reference *Musa* map" for further analyses (Hippolyte et al. 2010).

Since *Musa* behaves as cross-pollinating species, heterozygous parents can be used to derive mapping populations such as F_1 and backcross lines. There is no information on the level of heterozygosity in *Musa* species or cultivars. However in potato, a vegetatively propagated crop, the heterozygosity of parental lines was evaluated to be approximately 57–59% (Gebhardt et al. 1989).

Mapping studies in *Musa* have not kept pace with the advances made in other crop species. One of the most useful applications of genetic linkage maps have been the comparison of genomes of distantly related and cross-incompatible species (Kumar 1999). It is anticipated that *Musa* researchers would take advantage of genome synteny, which is a major feature of several plants studied so far. Genome synteny implies that large regions of the genomes of plants have been advantageous and have not undergone any significant changes over millions of years. In the grass family (monocotyledons) large regions of colinearity have been observed. Since *Musa* is also monocotyledonous, genome synteny relationships could be used to bridge the genetic information between other monocotyledons and *Musa*. The molecular markers produced for closely related taxa of *Musa* may be used to accelerate the development of linkage maps of *Musa*. DNA markers from the conserved regions of related taxa can serve as heterologous probes for tagging genes of interest in *Musa*. Synteny relationships can help to predict the position of orthologous genes of agronomic interest in related species (Jena et al. 1994; Harrington et al. 1997). For example, the majority of cloned disease resistance genes (R-genes) in plant species encode a large family of nucleotide-binding site/leucine-rich repeat (NBS-LRR) proteins. The NBS-LRR genes share significant homology in nucleotide sequences, amino acid sequences and structural motifs. Therefore R-gene analogs (RGAs) can be identified from plants by using PCR amplification with degenerate primers targeting the short conserved regions in the NBS. By making use of this method, Pei et al. (2007) were able to isolate RGAs from five species of *Musa*. An interesting aspect of this study was that of the 12 distinct *Musa* RGAs, *Musa* RGA-I was found to be specific to an ABB ("Zhongshandajiao") and an AA ("Gongjiao") cultivar while *Musa* RGA-K was specific to "Zhongshandajiao".

Genetic linkage maps have been useful in tagging genes of economic importance with molecular markers. The likelihood of identifying a marker linked to a gene is inversely proportional to the distance between the marker and the gene (Kumar 1999). Linkage relationships between markers and genes are done using various statistical methods. In classical genetic improvement programs selection is based on the phenotype. Except in the case of major gene effects, the breeder is not aware of the genes that are actually being selected. The development of molecular markers was hailed as a breakthrough and promised to change the manner in which breeding and selection will be done.

However, most of the traits considered in genetic improvement of plants such as yield, quality, maturity and resistance to several biotic and abiotic stresses are controlled by many genes and are known as quantitative, "polygenic", "multifactorial" or "complex" traits (Collard et al. 2005). The effects of these genes or loci on the phenotype are influenced by environmental factors. Such polygenic clusters, know as quantitative trait loci (QTLs) make either a small positive or negative contribution to the final phenotypic value of a trait. The process of constructing linkage maps and conducting QTL analysis- to identify genomic regions associated with QTLs is known as QTL mapping (Paterson 1996). The preliminary aim of QTL mapping is to provide a "framework" linkage map that covers all chromosomes evenly to identify markers flanking those QTLs that control traits of interest (Collard et al. 2005). Framework maps can be saturated by mapping additional markers. High resolution QTL maps are necessary to develop reliable markers for marker-assisted selection.

Some molecular markers are not genes and do not have any biological effect but are landmarks in the genome. They are identifiable DNA sequences found at specific locations in the genome and transmitted by Mendelian laws of inheritance. A large number of molecular markers are now in existence including RFLPs, RAPDs, AFLPs, microsatellites, single nucleotide polymorphisms (SNPs).

Maps are essential for mapping traits of interest in a crop species. Markers can then be linked to traits of agronomic importance and used for genetic improvement by MAS.

Molecular genetic maps are essential to determine the link between QTLs and the underlying gene(s). Molecular mapping is essential for map-based cloning of genes. Map-based or positional cloning makes use of tightly linked markers on either side of the gene. Physical rather than linkage maps are required for map-based cloning.

2.6 Breeding Objectives

The major objectives in banana breeding are similar to those of most crops and include:

1. Increased yield and quality of the crop
2. Resistance/tolerance to pests and diseases
3. Changes to plant morphology such as dwarfing traits
4. Shortening of the crop cycle
5. Increased tolerance to environmental pressures such as drought, salinity, extreme temperature and metal ions.

To date, banana breeding programs have concentrated primarily on developing varieties with disease and pest resistance and maintaining current yields. This was due to the onset of serious fungal diseases that almost destroyed the entire commercial cultivation of bananas (Tenkouano et al. 2011). Although banana breeding programs were initiated primarily to improve commercial varieties, it was realized that varieties used by small-holder farmers were also affected by the same diseases and pests. Banana and plantain are the staple food for millions of people in the tropical regions of the world and a decline in banana production could threaten their livelihoods. The major banana breeding programs have developed a number of hybrids. However, a major stumbling block is the adoption of the new hybrids due to consumer acceptability. This scenario is now changing and new hybrids are now being increasingly adopted by farmers in many countries. With increasing evidence of global climate change and the current emphasis placed on the role of nutrition on human health, new banana breeding objectives will have to be incorporated to meet new demands.

2.6.1 Breeding for Stress Resistance in Musa

The most practical definition of a biological stress is an adverse force or a condition, which inhibits the normal functioning and well being of a biological system such as plants (Jones et al. 1998). Besides the biotic stresses that affect crop plants, abiotic stresses including drought, low temperature, heat, salinity, flooding, heavy metal toxicity and oxidative stresses also cause severe crop losses worldwide (Mahajan and Tuteja 2005). Drought and salinity are already widespread in many regions and are expected to cause serious salinization of more than 50% of all arable lands by the year 2050 (Ashraf 1994). Climate change due to greenhouse gas emissions is expected to increase temperature and alter rainfall patterns (Challinor et al. 2007).

Considering the world's population growth and various stresses that affect plants, technological innovations are required to increase the output of existing arable lands to feed the world's population. Small farmers are the most vulnerable to the impacts of abiotic stresses and generally do not have the means to do anything about them (Kijne 2006). Plant breeding for abiotic stresses appears to have the greatest likelihood of success than other interventions.

In contrast to plant resistance to biotic stresses, which are mainly controlled by single genes, the genetically complex responses to abiotic stress are multigenic and therefore more difficult to control and engineer (Vinocur and Altman 2005) through traditional breeding. The transgenic approach offers the best alternative to conventional breeding for crops including banana because it enables the introduction of one or more genes into the genome without affecting their genetic background (Jain and Jain 2000). There is a large amount of information on the ecological/physiological/cellular alterations, which take place in cell/tissues/whole plants in response to application of stress (Grover et al. 1999). The metabolism of carbohydrates, fats, proteins and nucleic acids is adversely affected when plant tissues are exposed to stresses. All major physiological processes such as photosynthesis, nitrogen metabolism and respiration as well as developmental stages of the plant are affected by stress. Abiotic stresses elicit multiple responses within the plant cell. The resistance to different abiotic stresses is contributed by a range of biochemical/physiological mechanisms (Grover et al. 1999). Understanding these different mechanisms, especially the primary events are vital for developing stress tolerant/resistant plants. Studies related to abiotic stress in banana are limited and appear to concentrate on the effects of low temperature of only the fruit because of its commercial value (Caamal-Velaquez et al. 2007). Differential display used to identify gene expression of banana fruit to low temperature showed that genes involved in response to pathogen attack, wounding and ripening were also affected by low temperature (Caamal-Velaquez et al. 2007) showing that tolerance to abiotic stress is very complex at the whole plant and cellular levels (Ashraf and Foolad 2007). Yano-Melo et al. (2003) showed that plantlets of the cultivar "Pacovan" showed increased salt tolerance in the presence of arbuscular mycorrhizal fungi under glasshouse conditions. Whether this study can be extended to field conditions remains questionable.

Efforts to improve plant stress tolerance by genetic transformation have made great strides; however, the genetically complex mechanisms of abiotc stress tolerance make the task extremely difficult (Vinocur and Altman 2005). The major limiting factor in the widespread application of genetic engineering is the availability of target genes. But compared with 10 years ago, breeders now have new perspectives for plant improvement for stresses. Soon markers for QTLs and single genes for plant transformation will be available.

Drought is regarded as the most important environmental stress in agriculture. In many crop plants, breeding for drought tolerance has led to some yield increases in drought environments. There is no indication in the literature of any breeding work for drought tolerance in *Musa*. One of the ancestral species of cultivated bananas, *M. balbisiana* (BB) is reported to have originated in drier monsoon regions (Robinson 1995). It appears that genotypes containing two B genomes such as the ABB cooking bananas are able to grow in relatively drier areas and are able to withstand long dry periods (M. Pillay, pers. observ.). On the other hand cultivated bananas require between 25 and 50 mm of water weekly (Stover and Simmonds 1987).

This section will aim to provide a brief overview of the progress in breeding for drought tolerance in other crops to ascertain if these methods would be applicable to banana.

In recent years crop physiology and genomics have provided new insights in drought tolerance providing breeders with new knowledge and tools for plant improvement (Tuberosa and Salvi 2006). Physiological studies have identified some characteristics that are associated with plant adaptability to drought prone environments. These include smaller plant size, reduced leaf area, early maturity and prolonged stomatal closure and reduced total evapo-transpiration. These characteristics may not be applicable easily to *Musa* because of the gross morphology of the banana plant and its long generation time. While breeding for a reduced stature is already a goal of banana breeding, reducing the number of leaves may have a negative influence on yield. In wheat and rice, it was shown that genotypes bred for high yield potential and high yield stability in the absence of stress sustained a significant yield under mild or moderate drought (Slafer et al. 2005). Experiments with barley (Rizza et al. 2004) and sugar beet (Ober et al. 2004, 2005) also showed similar results. Our unpublished results from a banana field trail in a relatively dry region in Uganda showed that some of the genotypes did not suffer any yield decline compared to those grown in a region where moisture was not lacking. This may suggest that selection under favorable environments has positive effects when plants are grown in stressed environments. There appears to be no simple solution in assessing the degree of drought tolerance of different genotypes. In many studies drought tolerance/sensitivity is based on a few physiological measures related to drought response. A number of indices have been proposed to describe the performance of a genotype under stress and non-stress conditions (Cattivelli et al. 2008). Any breeding work related to drought tolerance in *Musa* should take this into consideration. Different crop developmental stages show different sensitivity to drought stress.

Molecular markers can be used to identify drought tolerant/resistant traits. Drought tolerance is a typical quantitative trait. Over the last 10 years, many studies have used QTL analyses to identify chromosome regions that control variation in physiological, morphological and developmental changes observed during growth in water limiting conditions. This information may make selection for drought tolerance more efficient.

The disadvantages of molecular markers are that the linkage between the markers and the target gene locus can be broken by genetic recombination. The polygenic nature of drought tolerance and G x E influences make MAS for drought tolerance QTLs very difficult (Francia et al. 2005). When numerous genes are involved in a polygenic trait it means that each gene has a small effect on the phenotype. This implies that several regions of the chromosomes (QTLs) must be manipulated at the same time to obtain a significant impact, and the effect of individual regions is not readily identifiable (Cattivelli et al. 2008). The high cost of molecular marker technology and the low efficiency of applying MAS for QTLs must be considered to determine the effectiveness of QTL-MAS versus standard breeding strategies for drought resistance/tolerance.

An important step in the application of molecular markers in breeding for drought tolerance is the cloning of the DNA sequences (genes) underlying QTLs. So far no QTL for drought tolerance has been cloned in crop species (Cattivelli et al. 2008). However, many stress related genes have been isolated and characterized in a number of crop plants (Cattivelli et al. 2008; Ramanjulu and Bartels 2002). For example, Fu et al. (2007) identified functional candidate genes for drought tolerance in rice.

It is envisaged that information from model plants will be useful in searching for regulatory components of drought response in crop plants (Shen et al. 2003; Li et al. 2005). Transcription factors are also considered to be optimal targets for engineering complex traits such as stress tolerance. However, it appears that not all transcription factors involved in drought signal transduction are suitable targets for biotechnology (Cattivelli et al. 2008). It has been shown that overexpression of transcription factors could activate non-stress related genes and adversely affect the agronomic characters of a crop (Wang et al. 2003). When the *Arabidopsis* gene *GF14 lambda* was introduced into cotton plants a "stay green-like" phenotype was obtained. The transformed plants showed improved water stress tolerance and a higher photosynthetic rate due to an altered stomatal conductance under water deficit conditions (Yan et al. 2004). Therefore developing transgenic plants with drought avoidance mechanism may be a good strategy for crop improvement. A more robust root system may enable plants to absorb more water during water deficit stress. Banana genotypes show great variation in their rooting systems (Swennen et al. 1987). However, the vertical root zone is very shallow with about 40% of the root volume in the top 100 mm and 85% in the top 300 mm (Robinson 1996). In addition banana requires large quantities of water for maximum productivity.

Therefore developing a plant with suitable drought traits may not be easy in banana. Wild germplasm should be exploited to find new useful alleles for drought tolerance. It is believed that centuries of selective breeding of our crop plants have led to the loss of many useful alleles as well as useless ones. Successful breeding for high yield under drought conditions will be possible when a true integration of traditional breeding with physiology and genomics is achieved (Cattivelli et al. 2008).

2.7 Banana Breeding Achievements

Despite the availability of a spectrum of new technologies, the present breeding progress documented by the annual registration of new cultivars is the result of classical breeding (Wenzel 2006). Despite many problems that slow the genetic improvement of bananas and plantains, significant achievements and advances have been accomplished.

To overcome the problem of infertility in breeding bananas, a number of breeding strategies were developed to take advantage of the little fertility that existed in some of the cultivars (Tomekpe et al. 2004). Determining the fertility of the germplasm especially among the triploid cultivars for use in breeding was the first major task of any breeding program. This was largely achieved by crossing the 3*x* cultivars with fertile diploid plants. However, banana germplasm in many countries have not been characterized for fertility. Resistance to pests and diseases has been identified in many wild species. A large number of diploids with combinations of disease resistance and desirable agronomic qualities have been developed (Rowe 1984; Rowe and Rosales 1993; Vuylsteke and Ortiz 1995; Tenkouano et al. 2003). Tenkouano et al. (1999a, b) showed that most traits of economic importance are more predictably inherited from the diploid parents than from parents with higher ploidy level. The quantitative and qualitative breeding efforts have produced several banana and plantain hybrids, some of them with high levels of resistance to black Sigatoka and good bunch characteristics (Rowe and Rosales 1993; Vuylsteke et al. 1993a, b; Vuylsteke et al. 1995; Ortiz and Vuylsteke 1998b, c) and have been globally tested for possible adoption as new cultivars. The yields of most banana hybrids have doubled or in some cases as in the FHIA hybrids tripled that of their parent landraces. This has been attributed to characteristics such as shorter and robust plant stature, better suckering behavior and early maturity, which are all linked with yield. The high yield in plantain tetraploids is particularly attributed to improved ratooning as compared to their parents, which have generally low suckering behavior. Major gains in fruit quality have been achieved by restoration of the seedless character in resulting 3*x* offspring (Tenkouano et al. 1998). Seed set rates have tremendously improved in many *Musa* species and the germination percentage drastically enhanced using established tissue culture techniques. Consequently, a number of improved genotypes have been widely evaluated and knowledge on genotype-by-environment interaction and stability of the important traits gained. Insight into combining abilities, heterotic groups, and the genetics of qualitative and quantitative traits has been gained and is being applied to make breeding more efficient. A wide array of breeding schemes has been explored, combining conventional and innovative approaches, and producing potential cultivars from primary tetraploids, secondary triploids and other populations (Tenkouano et al. 2003). Embryo culture techniques developed for banana increased the rate of seed germination by a factor of 10 or more (Vuylsteke et al. 1990). Tissue culture techniques such as micropropagation have played a major role in banana improvement programs since it was possible to obtain multiple plants from a single seed. Somaclonal variation was able to introduce useful variability into the banana germplasm (Vuylsteke et al. 1991).

2.8 Limitations of Traditional Breeding and Rationale for Molecular Breeding

The limitations of traditional breeding in *Musa* have been described adequately in the past (Pillay et al. 2002; Pillay and Tripathi 2007). In summary the major limitations include:

1. Majority of the edible banana cultivars are triploids ($2n = 3x = 33$) and are male and/or female sterile or are characterized by low male and female fertility.
2. Most of the edible cultivars lack genes for disease and pest resistance.
3. Desired traits of interest such as resistance to diseases and pests must be introgressed from wild and cultivated diploid ($2n = 2x = 22$) species.
4. Since the wild species possess a number of undesirable characteristics, time consuming and long backcrossing programs are essential to select progeny that are as close as possible to the cultivated varieties.
5. The inherent sterility of many banana clones limits the number of varieties that can be used in a breeding program.
6. In varieties that do have female fertility seed set per bunch can be very low and germination of these seeds in soil is less than one 1% (Ortiz and Vuylsteke 1995).
7. The long generation time of bananas, up to 18 months in most cases, prolongs breeding efforts compared to annual crops.
8. The effect of multiploid and autopolyploid chromosome behavior results in unpredictable frequency of aneuploids and undesirable hyperpolyploids (>5x) in addition to $2x$, $3x$, and $4x$ euploids (Simmonds 1966). Taking the above into consideration there appears to be a strong case for molecular breeding in *Musa*.

10. Conclusion

The role of plant breeding in increasing food production and providing sustainable nutrition is well recognized. With increasing population and reducing land resources, it becomes even more important to breed plant varieties, which can sustain production under the varied agroclimatic conditions of different regions (Ahloowalia and Maluszynski 2001).

Plant breeding is an applied multidisciplinary science that uses knowledge from agronomy, botany, genetics, cytogenetics, molecular genetics, physiology, pathology, entomology, biochemistry and statistics (Schlegel 2003). The ultimate outcome of plant breeding is mainly improved cultivars. About 50% of crop productivity increases around the world is due to genetic improvement (Fehr 1984). Today, plant breeding is based on creating variation, evaluating and selecting desired genotypes and multiplication of these genotypes. Cultivated varieties of *Musa* are susceptible to a wide range of pests and diseases. On the contrary wild species are resistant to most pathogens of *Musa*. The introgression of useful genes from wild species to the cultivated varieties is hampered by a number of factors described earlier. Conventional breeding involves the screening of a large number of offspring over many years to find the few plants with desirable traits. Consequently, modern plant biotechnology is seen as the latest tool that can speed up the development of plant breeding technology. Plant biotechnology offers the potential of major improvements in almost every area of crop production. Although some progress in the field of molecular applications has been made in *Musa*, the full potential has not been exploited. In many other crop species (e.g., rice, wheat, sorghum, maize) multiple laboratories in many countries have teamed up to carry out intensive genomics research. Extensive resources have been devoted for genomics research in these crops. On the contrary, genomics research in *Musa* is being carried out by a few laboratories with meager resources.

Despite the strides that plant breeding had made in the production of food for humans over the last 100 years, there is an increasing concern around the world about the loss of plant breeding programs (Gepts and Hancock 2006). The loss of traditional plant breeding programs is, probably, due to the significant rise in modern plant techniques.

Perhaps, this just is one of the many waves that we have witnessed in science over the years. We believe that although biotechnology is in vogue currently, it will not completely replace traditional breeding.

Abbreviations

AFLP	:	amplified fragment length polymorphism
BAC	:	bacterial artificial chromosome
cM	:	centimorgans
FISH	:	fluorescence *in situ* hybridization
MAS	:	marker assisted selection
PCR	:	polymerase chain reaction
QTL	:	quantitative trait loci
RAPD	:	random amplified polymorphic DNA
RFLP	:	restriction fragment length polymorphisms
VNTR	:	variable number of tandem repeats

References

Adeleke MTV, Pillay M, Okoli BE (2002) An improved procedure for examining meiotic chromosomes in *Musa*. HortScience 37: 959–961.

Ahloowalia BS, Maluszynski M (2001) Induced mutations a new paradigm in plant breeding. Euphytica 118: 167–173.

Albini SM, Schwarzacher T (1992) *In situ* localization of two repetitive DNA sequences to surface-spread pachytene chromosomes of rye. Genome 35: 551–559.

Anderson LK, Salameh N, Bass HW, Harper LC, Cande WZ, Weber G, Stack SM (2004) Integrating genetic linkage maps with pachytene chromosome structure in maize. Genetics 166: 1923–1933.

Anonymous (1984) Revised banana descriptors IBGR. AGPG:IBPGR, Rome, Italy.

Ashraf M (1994) Breeding for salinity tolerance in plants. Crit Rev Plant Sci 13: 17–42.

Ashraf M, Foolad MR (2007) Roles of glycine betaine and proline in improving plant abiotic stress resistance. Environ Exp Bot 59: 206–216.
Balint-Kurti PJ, Clendennen SK, Dolezelova M, Valarik M, Dolezel J, Beetham PR, May GD (2000) Identification and chromosomal localization of the monkey retrotransposon in *Musa* sp. Mol Gen Genet 263: 908–915.
Barone A (2004) Molecular marker-assisted selection for potato breeding. Amer J Potato Res 81: 111–117.
Bartos J, Alkhimova O, DolezelovÃ¡ M, De Langhe E, Dolezel J (2005) Nuclear genome size and genomic distribution of ribosomal DNA in *Musa* and *Ensete* (Musaceae): taxonomic implications. Cytogenet and Genome Res 109: 50–57.
Bonner JW, Warner RM, Brewbaker JL (1974) A chemosystematic study of *Musa* cultivars. HortScience 9: 325–328.
Caamal-Velazquez JH, Chi-Manzanero BH, Canche-Yam JJ, Castano E, Rodriguez-Zapata LC (2007) Low temperature induces differential expression genes in banana fruits. Sci Hort 114: 83–89.
Castiglioni P, Pozzi C, Heun M, Terzi V, Miller KJ, Rohde W, Salamini F (1998) An AFLP-based procedure for the efficient mapping of mutations and DNA probes in barley. Genetics 149: 2039–2056.
Cattivelli L, Rizza F, Badeck F-W, Mazzucotelli E, Mastrangelo AM, Francia E, Mare C, Tondelli A, Stanca AM (2008) Drought tolerance improvement in crop plants: An integrated view from breeding to genomics. Field Crop Res 105: 1–14.
Cerbah M, Kevei Z, Siljak-Yakovlev S, Kondorosi E, Kondorosi A, Trinh TH (1999) FISH Chromosome Mapping Allowing Karyotype Analysis in *Medicago truncatula* Lines Jemalong J5 and R-108-1. Mol Plant-Micr Interact 12: 947–950.
Challinor A, Wheeler T, Garforth C, Craufurd P, Kassam A (2007) Assessing the vulnerability of food crop systems in Africa to climate change. Climatic Change 83: 381–399.
Cheng Z, Buell CR, Wing RA, Gu M, Jiang J (2001a) Toward a Cytological Characterization of the Rice Genome. Genome Res 11: 2133–2141.
Cheng Z, Presting GG, Buell CR, Wing RA, Jiang J (2001b) High-Resolution Pachytene Chromosome Mapping of Bacterial Artificial Chromosomes Anchored by Genetic Markers Reveals the Centromere Location and the Distribution of Genetic Recombination Along Chromosome 10 of Rice. Genetics 157: 1749–1757.
Collard BCY, Jahufer MZZ, Brouwer JB, Pang ECK (2005) An introduction to markers, quantitative trait loci (QTL) mapping and marker-assisted selection for crop improvement: The basic concepts. Euphytica 142: 169–196.
D'Hont A, Paget-Goy A, Escoute J, Carreel F (2000) The interspecific genome structure of cultivated banana, *Musa* spp. revealed by genomic DNA *in situ* hybridization. Theor Appl Genet 100: 177–183.
De Capdeville G, Souza Junior MT, Szinay D, Diniz LEC, Wijnker E, Swennen R, Kema GHJ, Jong H (2009) The potential of high-resolution BAC-FISH in banana breeding. Euphytica 166: 431–443.
De Langhe E (1964) The origin of variation in the plantain banana. State Agricultural University of Ghent, Ghent, Belgium 39: 45–80.
Dolezel J, Valarik M, Vrana J, Lysak MA, Hribova J, Bartos J, Gasmanova N, Dolezelova M, Safar J, Simkova H (2004) Molecular cytogenetics and cytometry of banana (*Musa* spp.). In: JS Mohan, R Swennen R (eds) Banana Improvement: Cellular, Molecular Biology, and Induced Mutations. Science Publishers, Inc., Enfield, NH, USA, pp 229–244.
Dolezelova M, Valarik M, Swennen R, Horry JP, Dolezel J (1998) Physical mapping of the 18S–25S and 5S ribosomal RNA genes in diploid bananas. Biol Plant 41: 497–505.
Escalant JV, Panis B (2002) Biotechnologies toward the genetic improvement in Musa. Memorias XV Reuni6n Internacional ACORBAT Cartagena de Indias, Colombia.
Faure S, Noyer JL, Horry JP, Bakry F, Lanaud C, Goazalez de Lean D (1993) A molecular marker-based linkage map of diploid bananas (*Musa acuminata*). Theor Appl Genet 87: 517–526.

Fehr WR (1984) Genetic Contributions to Yield Gains of Five Major Crop Plants. CSSA Spl Publ 7. ASA and CSSA, Madison, WI, USA.
Francia E, Tacconi G, Crosatti C, Barabaschi D, Bulgarelli D, Dall'Aglio E, Vale G (2005) Marker assisted selection in crop plants. Plant Cell, Tissue and Organ Culture 82: 317–342.
Fu BY, Xiong JH, Zhu LH, Zhao XQ, Xu HX, Gao YM, Li YS, Xu JL, Li ZK (2007) Identification of functional candidate genes for drought tolerance in rice. Mol Genet Genom 278: 599–609.
Gebhardt C, Ritter E, Debener T, Schachtschabel U, Walkemeier B, Uhrig H, Salamini F (1989) RFLP analysis and linkage mapping in *Solanum tuberosum*. Theor Appl Genet 78: 65–75.
Gepts P, Hancock J (2006) The future of plant breeding. Crop Sci 46: 1630–1634.
Gostimsky SA, Kokaeva ZG, Konovalov FA (2005) Studying plant genome variation using molecular markers. Russ J Genet 41: 378–388.
Grover A, Sahi C, Sanan N, Grover A (1999) Taming abiotic stresses in plants through genetic engineering: current strategies and perspective. Plant Sci 143: 101–111.
Gupta M, Chyi YS, Romero-Severson J, Owen JL (1994) Amplification of DNA markers from evolutionarily diverse genomes using single primers of simple-sequence repeats. Theor Appl Genet 89: 998–1006.
Harper LC, Zacheus Cande W (2000) Mapping a new frontier; development of integrated cytogenetic maps in plants. Funct Integr Genom 1: 89–98.
Harrington SE, Bligh HFJ, Park WD, Jones CA, McCouch SR (1997) Linkage mapping of starch branching enzyme III in rice (*Oryza sativa* L.) and prediction of location of orthologous genes in other grasses. Theor Appl Genet 94: 564–568.
Hippolyte I, Bakry F, Seguin M, Gardes L, Rivallan R, Risterucci A-M, Jenny C, Perrier X, Carreel F, Argout X, Piffanelli P, Khan I, Miller R, Pappas G, Mbeguie-A-Mbeguie D, Matsumoto T, De Bernardinis V, Huttner E, Kilian A, Baurens F-C, D'Hont A, Cote F, Courtois B, Glaszmann J-C (2010) A saturated SSR/DArT linkage map of *Musa acuminata* addressing genome rearrangements among bananas. BMC Plant Biol 10: 65–65.
Hribova E, Dolezelova M, Dolezel J (2008) Localization of BAC clones on mitotic chromosomes of *Musa acuminata* using fluorescence *in situ* hybridization. Biol Plant 52: 445–452.
Jain RK, Jain S (2000) Transgenic strategies for genetic improvement of Basmati rice. Indian J Exp Biol 38: 6–17.
Jarret RL, Litz RE (1986a) Enzyme polymorphism in *Musa acuminata* Colla. J Hered 77: 183–188.
Jarret RL, Litz RE (1986b) Isozymes as genetic markers in bananas and plantains. Euphytica 35: 539–549.
Jena KK, Khush GS, Kochert G (1994) Comparative RFLP mapping of a wild rice, *Oryza officinalis,* and cultivated rice, *O. sativa*. Genome 37:382–389.
Jones DR (2000) Introduction to banana, abaca and enset. In: DR Jones (ed) Diseases of Banana, Abaca and Enset. CAB International, Wallingford, UK, pp 1–36.
Jones HG, Flowers TJ, Jones MB (1998) Introduction: some terminology and common mechanisms. In: HG Jones, TJ Flowers, MB Jones (eds) Plants under Stress. Cambridge University Press, Cambridge, UK, pp 1–10.
Joshi S, Ranjekar P, Gupta VK (1999) Molecular markers in plant genome analysis. Curr Sci 77: 230–240
Karamura DA (1999) Numerical taxonomic studies of the East African Highland bananas (Musa AAA-East Africa) in Uganda, INIBAP, France.
Kijne JW (2006) Abiotic stress and water scarcity: identifying and resolving conflicts from plant level to global level. Field Crops Res 97: 3–18.
Kulikova O, Gualtieri G, Geurts R, Kim D-J, Cook D, Huguet T, De Jong JH, Fransz PF, Bisseling T (2001) Integration of the FISH pachytene and genetic maps of *Medicago truncatula*. Plant J 27: 49–58.
Kumar LS (1999) DNA markers in plant improvement: an overview. Biotechnol Adv 17: 143–182.

Lambie EJ, Roeder GS (1986) Repression of Meiotic Crossing over by a Centromere (CEN3) in *Saccharomyces cerevisiae*. Genetics 114: 769–789.
Lebot V, Aradhya KM, Manshardt R, Meilleur B (1993) Isozymes and classification of Pacific bananas and plantins. In: J Ganry (ed) Breeding Banana and Plantain for Resistance to Disease and Pests. CIRAD in collaboration with INIBAP, Montpellier, pp 43–53.
Li X-P, Tian A-G, Luo G-Z, Gong Z-Z, Zhang J-S, Chen S-Y (2005) Soybean DRE-binding transcription factors that are responsive to abiotic stresses. Theor Appl Genet 110: 1355–1362.
Mahajan S, Tuteja N (2005) Cold, salinity and drought stresses: an overview. Arch Biochem Biophys 444: 139–158.
Markert CL, Moller F (1959) Multiple forms of enzymes: tissue, ontogenetic, and species specific patterns. Proc Natl Acad Sci USA, 45: 753–763.
Ober ES, Clark CJA, Bloa ML, Royal A, Jaggard KW, Pidgeon JD (2004) Assessing the genetic resources to improve drought tolerance in sugar beet: agronomic traits of diverse genotypes under droughted and irrigated conditions. Field Crop Res 90: 213–234.
Ober ES, Bloa ML, Clark CJA, Royal A, Jaggard KW, Pidgeon JD (2005) Evaluation of physiological traits as indirect selection criteria for drought tolerance in sugar beet. Field Crop Res 91: 231–249.
Ortiz R (1997) Morphological variation in *Musa* germplasm. Genet Resour Crop Evol 44: 393–404.
Ortiz R, Vuylsteke D (1995) Factors Influencing Seed Set in Triploid *Musa* spp. L. and Production of Euploid Hybrids. Ann Bot 75: 151–155.
Ortiz R, Vuylsteke D (1998a) 'BITA-3': a starchy banana with partial resistance to black and tolerance to streak virus. Hort Sci 33: 358–359.
Ortiz R, Vuylsteke D (1998b) 'PITA-14': a black sigatoka-resistance tetraploid plantin hybrid with virus tolerance. HortScience 33: 360–3611.
Ortiz R, Madsen S, Vuylsteke D (1998) Classification of African plantain landraces and banana cultivars using a phenotypic distance index of quantitative descriptors. Theor Appl Genet 96: 904–911.
Ortiz-Vazquez E, Kaemmer D, Zhang HB, Muth J, Rodriguez-Mendiola M, Arias-Castro C, James A (2005) Construction and characterization of a plant transformation-competent BIBAC library of the black Sigatoka-resistant banana *Musa acuminata* cv. Tuu Gia (AA). Theor Appl Genet 110: 706–713.
Osuji JO, Okoli BE, Vuylsteke D, Ortiz R (1997) Multivariate pattern of quantitative trait variation in triploid banana and plantain cultivars. Sci Hort 71: 197–202.
Osuji JO, Crouch JH, Harrison G, Heslop-Harrison JS (1998) Molecular cytogenetics of *Musa* species, cultivars and hybrids: Location of 18S-5.8S-25S and 5S rDNA and telomere-like sequences. Ann Bot 82: 243–248.
Paterson AH (1996) Genome Mapping in Plants. Academic Press, California and R. G. Landes Company, Austin, USA.
Pei X, Li S, Jiang Y, Zhang Y, Wang Z, Jia S (2007) Isolation, characterization and phylogenetic analysis of the resistance gene analogues (RGAs) in banana (*Musa* spp.). Plant Sci 172: 1166–1174.
Persley GJ (1992) Beyond Mendel's Garden: Biotechnology in Agriculture. In: Biotechnology Enhancing Research on Tropical Crops in Africa. CTA/IITA co-publication, pp 11–19.
Pillay M, Adeleke MTV (2002) Silver staining of *Musa* chromosomes. Cytologia 66: 33–37.
Pillay M, Tripathi L (2007) Banana Breeding In: MS Kang, PM Priyadarshan (eds) Breeding Major Food Staples. Blackwell Publishing, Boston MA, USA, pp 393–428.
Pillay M, Tenkouano A (2011) Genomes, cytogenetics and flow cytometry of Musa. In: M Pillay, A Tenkouano (eds) Banana Breeding: Progress and Challenges. CRC Press, Boca Raton, FL, USA, pp 53–70.
Pillay M, Hartman J, Tenkouano A (2002) Future challenges in *Musa* breeding. . In: M Kang (ed) Crop Improvement for the 21st Century. Food Products Press, New York, USA, pp 223–252.

Ramanjulu S, Bartels D (2002) Drought- and desiccation-induced modulation of gene expression in plants. Plant Cell Environ 25: 141–151.
Rizza F, Badeck F-W, Cattivelli L, Lidestri O, Di Fonzo N, Stanca AM (2004) Use of water Stress Index to Identify Barley Genotypes Adapted to Rainfed and Irrigated Conditions. Crop Sci 44: 2127–2137.
Robinson JC (1996) Bananas and Plantains. CABI, Wallingford, UK.
Rowe P (1984) Breeding banana and plantains. Plant Breed Rev 2: 135–155.
Rowe P, Rosales F (1993) Diploid breeding at FHIA and the development of Goldfinger. InfoMusa 2: 9–11.
Safar J, Noa-Carrazana JC, Vrana J, Bartos J, Alkhimova O, Sabau X, Simkova H, Lheureux F, Caruana M-L, Dolezel J, Piffanelli P (2004) Creation of a BAC resource to study the structure and evolution of the banana (*Musa balbisiana*) genome. Genome 47: 1182–1191.
Schlegel RHJ (2003) Dictionary of Plant breeding. Food Products Press. The Haworth Reference Press, New York, USA
Schneider K (2005) Mapping Populations and Principles of Genetic Mapping. In: K Meksem, G Kahl (eds) The Handbook of Plant Genome Mapping Genetic and Physical Mapping. Wiley-VCH Verlag GmbH & Co., Weinheim, Germany.
Shen YG, Zhang WK, He SJ, Zhang JS, Liu Q, Chen SY (2003) An EREBP/AP2-type protein in *Triticum aestivum* was a DRE-binding transcription factor induced by cold, dehydration and ABA stress. Theor Appl Genet 106: 923–930.
Shepherd K (1999) Cytogenetics of the genus *Musa*. INIBAP, Montpellier, France.
Simmonds NW (1966) Bananas. Longmans, London, UK.
Simmonds NW, Shepherd K (1955) The taxonomy and origins of the cultivated bananas. J Linn Soc (London) Bot 55: 302–312.
Slafer GA, Araus JL, Royo C, Moral LFGA (2005) Promising eco-physiological traits for genetic improvement of cereal yields in Mediterranean environments. Ann Appl Biol 146: 61–70.
Staub JE, Kuhns LJ, May B, Grun P (1982) Stability of potato tuber isozymes under different storage regimes. HortScience 107: 405–408.
Stover RH, Simmonds NW (1987) Bananas. 3rd edn. Longmans, London, UK
Swennen R, Vuylsteke D (1987) Morphological taxonomy of plantains (*Musa* cultivars AAB) in West Africa. In: GJ Persley, EA De Langhe (eds) Banana and Plantain Breeding Strategies Proceedings of a workshop held at Cairns, Australia. ACIAR Proceedings, Cairns, Australia, pp 165–171.
Swennen R, Vuylsteke D, Ortiz R (1995) Phenotypic Diversity and Patterns of Variation in West and Central African Plantains (*Musa* spp., AAB group Musaceae). Econ Bot 49: 320–327.
Swennen R, De Langhe E, Janssen J, Decoene D (1987) Study of the root development of some *Musa* cultivars in hydroponics. Fruits 41: 515–524.
Tenkouano A, Ortiz R, Vuylsteke D (1998) Combining ability for yield and plant phenology in plantain-derived populations. Euphytica 104: 151–158.
Tenkouano A, Crouch JH, Crouch HK, Vuylsteke D, Ortiz R (1999a) Comparison of DNA marker and pedigree-based methods of genetic analysis in plantain and banana (*Musa* spp.) clones. II. Predicting hybrid performance. Theor Appl Genet 98: 69–75.
Tenkouano A, Crouch JH, Crouch HK, Vuylsteke D, Ortiz R (1999b) Comparison of DNA marker and pedigree-based methods of genetic analysis of plantain and banana (*Musa* spp.) clones. I. estimation of genetic relationships. Theor Appl Genet 98: 62–68.
Tenkouano A, Pillay M, Ortiz R (2011) Breeding Techniques. In: M Pillay, A Tenkouano (eds) Banana Breeding: Progress and Challenges. CRC Press, Boca Raton, FL, USA, pp 181–202.
Tezenas de Montcel H, De Langhe E, Swennen R (1983) Essai de classification des bananiers plantains (AAB). Fruits 38: 461–474.

Tomekpe K, Jenny C, Escalant JV (2004) A review of conventional improvement strategies for Musa. InfoMusa 13: 2–6.
Tuberosa R, Salvi S (2006) Genomics-based approaches to improve drought tolerance of crops. Trends Plant Sci 11: 405–412.
Valarik M, Simkova H, Hribova E, Safar J, Dolezelova M, Dolezel J (2002) Isolation, characterization and chromosome localization of repetitive DNA sequences in bananas (*Musa* spp.). Chromosome Research: An International Journal on the Molecular, Supramolecular and Evolutionary Aspects of Chromosome Biology 10: 89–100.
Vilarinhos AD, Piffanelli P, Lagoda P, Thibivilliers S, Sabau X, Carreel F, D'Hont A (2003) Construction and characterization of a bacterial artificial chromosome library of banana (*Musa acuminata* Colla). Theor Appl Genet 106: 1102–1106.
Vilarinhos AD, Carreel F, Rodier M, Hippolyte I, Benabdelmouna A, Triaire D, Bakry F, Courtois B, D'Hont A (2006) Characterization of translocations in Banana by FISH of BAC clones anchored to a genetic map. Plant Animal Genome XIV Conference, San Diego, CA, USA.
Vinocur B, Altman A (2005) Recent advances in engineering plant tolerance to abiotic stress: achievements and limitations. Curr Opin Biotechnol 16: 123–132.
Vuylsteke D, Ortiz R (1995) Plantain-derived diploid hybrids (TMP2x) with black Sigatoka resistance. HortScience 30: 147–149.
Vuylsteke D, Swennen R, De Langhe E (1990) Tissue culture technology for the improvement of African Plantains. In: RA Fullerton, RH Stover (eds) Sigatoka leaf spot disease of bananas. INIBAP, Montpellier, France, pp 316–337.
Vuylsteke D, Swennen R, De Langhe E (1991) Somaclonal variation in plantains (*Musa* spp., AAB group) derived from shoot-tip culture. Fruits 46: 429–439.
Vuylsteke D, Swennen R, Ortiz R (1993a) Registration of 14 improved tropical *Musa* plantain hybrids with black Sigatoka resistance. HortScience 28: 957–959.
Vuylsteke DR, Swennen RL, Ortiz R (1993b) Development and performance of balck sigatoka-resistant tetraploid hybrids of plantain (*Musa* spp., AAB group). Euphytica 65: 33–42.
Vuylsteke D, Ortiz R, Ferris S, Swennen R (1995) Pita-9—a black-Sigatoka-resistant hybrid from the false horn plantain gene pool. HortScience 30: 395–397.
Wang W, Vinocur B, Altman A (2003) Plant responses to drought, salinity and extreme temperatures: towards genetic engineering for stress tolerance. Planta 218: 1–14.
Wang Y, Chen J-Y, Jiang Y-M, Lu W-J (2007) Cloning and expression analysis of phenylalanine ammonia-lyase in relation to chilling tolerance in harvested banana fruit. Postharvest Biology and Technology 44: 34–41.
Wenzel G (2006) Molecular plant breeding: achievements in green biotechnology and future perspectives. Applied Microbiology and Biotechnology 70: 642–650.
Winter P, Kahl G (1995) Molecular marker technologies for plant improvement. World J Microbiol Biotechnol 11: 438–448.
Yan J, He C, Wang J, Mao Z, Holaday SA, Allen RD, Zhang H (2004) Overexpression of the *Arabidopsis* 14-3-3 protein GF14 lambda in cotton leads to a "stay-green" phenotype and improves stress tolerance under moderate drought conditions. Plant & Cell Physiology 45: 1007–1014.
Yano-Melo AM, Saggin OJ, Costa Maia L (2003) Tolerance of mycorrhized banana (*Musa* sp. cv. Pacovan) plantlets to saline stress. Agriculture, Ecosystems & Environment 95: 343–348.
Zhong XB, Bodeau J, Fransz PF, Williamson VM, van Kammen A, de Jong JH, Zabel P (1999) FISH to meiotic pachytene chromosomes of tomato locates the root-knot nematode resistance gene *Mi-1* and the acid phosphatase gene *Aps-1* near the junction of euchromatin and pericentromeric heterochromatin of chromosome arms 6S and 6L, respectively. Theor Appl Genet 98: 365–370.

3

Diversity Analysis of Banana to Decipher the Origin of Cultivars

Kornel Burg[1,*] and *Ratri Boonruangrod*[2]

ABSTRACT

Thousands of banana cultivars that were selected by local farmers over many years have become part of the human diet in different parts of the world. The majority of these edible genotypes are seedless and are propagated vegetatively via rhizomes. The first *Musa* description and classification systems were based on morphological observations. Flavonoid compounds of banana were used in the 50s in banana taxonomy. Thereafter isoenzymes were used to classify banana clones. The development of techniques for the direct investigation of the genome allowed the development of markers with genome-wide distribution. This chapter summarizes the recent literature in which the nuclear and organellar genome markers are used in describing genetic diversity in *Musa*.

Keywords: markers, nuclear genome, organellar genomes, diversity, *Musa*

3.1 Introduction

Thousands of different banana cultivars selected by local farmers during many years have become part of the human diet in different parts of the world. The majority of these edible genotypes are seedless and are propagated vegetatively via rhizomes. Their survival in nature and geographical

[1]Health & Environment Department Bioresources, AIT Austrian Institute of Technology GmbH, 2444 Seibersdorf, Austria; e-mail: *kornel.burg@ait.ac.at*
[2]Department of Horticulture, Faculty of Agriculture at Kamphaengsaen, Kasetsart University, Kamphaengsaen, Nakhon Pathom, 73140 Thailand; e-mail: *ratri.b@ku.ac.th*
*Corresponding author

List of abbreviations after the text.

dispersal is not possible without human intervention. The first description of banana plants was based on these edible bananas as early as 1747 by Georgius Everhardus Rumphius. Seed bearing wild banana plants were first recorded by Luigi Aloysius Colla in 1820 that recognized *M. balbisiana* and *M. acuminata* as wild species distinct from any other species described at that stage (Cheesman 1948). The relationship of these wild genotypes to the cultivars always puzzled breeders and scientists for centuries. It was Kurz (1865) who suggested that the majority of edible banana cultivars of the *Eumusa* section originated from two wild species, *M. acuminata* and *M. balbisiana*, either directly or in various hybrid combinations. These cultivars were generally of higher ploidy levels but usually triploid. This idea was forgotten for nearly a century and was subsequently revitalized and confirmed by Cheesman in 1948. Subsequently a scoring system was established by Simmonds and Shepherd (1955) based on morphological traits of *M. acuminata* and *M. balbisiana*. This system proved to be successful in characterizing various hybrid genotypes representing different ploidy levels (Stover and Simmonds, 1987) and described five main genetic groups (AA, AB, AAA, AAB, ABB) for cultivated bananas (Simmonds and Shepherd 1955; Simmonds 1962). New evidence suggests that additional genomes such as the S (*M. schizocarpa*) and T (*M. textilis*) also contributed to the genomes of present day cultivated bananas. The A, B and S genomes represent *Eumusa* (with new nomenclature *Musa*) species possessing 11 chromosomes per haploid set, while the T genome is representative of *Australimusa* with 10 chromosomes per haploid set. Since the majority of the cultivated bananas are derived from the A and B genomes this chapter will focus mainly on them.

3.2 History of Markers

The first *Musa* description and classification systems were based on morphological observations. However, morphological characterization similar to all phenotypic characterization suffers from the influence of environmental factors on the morphological/phenotypic descriptors (e.g., pseudostem color is highly affected by local field conditions). However careful and purposeful selection of morphological markers may result in very precise analysis of the raised question. For example, the morphological characterization system developed by Simmonds and Shepherd (1955) for characterizing the genome composition of *M. acuminata*/*M. balbisiana* hybrids was based on 15 descriptors clearly discriminating *M. acuminata* and *M. balbisiana* wild types and assuming that all these phenotypic markers show a proportional manifestation in the hybrids. In this system 13 diagnostic characters were confined to the phenotype of the inflorescence. Seven markers were related to the bract of the male bud (shoulder, curling, shape, apex, color, color fading, and scars), two were related to the male flower (free tepal and color of the flower), two were related to the female flower (stigma color, arrangement of ovules), two related to the peduncle and pedicels of the inflorescence, while another two characters were related to the pseudostem (color) and the petiolar canal of the leaves. Based on this scoring system the various hybrid genotypes representing different ploidy levels could be well characterized (Stover and Simmonds 1960) and five main genetic groups (AA, AB, AAA, AAB, ABB) have been described for cultivated bananas (Simmonds and Shepherd 1955; Simmonds 1962). Later detailed morphological analysis of wild types revealed little variation in *M. balbisiana*, while nine different subspecies (ssp. *banksii*, ssp. *burmannica*, ssp. *Burmannicoides*, ssp. *errans*, ssp. *malaccensis*, ssp. *microcarpa*, ssp. *siamea*, ssp. *truncata* and ssp. *zebrina*) with distinct geographic distribution area is at present recognized in *M. acuminata* (Fig. 3-1). The drawback of morphological characterization is that the quantitative/precise measurement of the descriptors is difficult. Additionally most of the morphological characters possibly have polygenic inheritance and consequently minor changes in the genotype will not necessarily manifest in the phenotype of the plant,

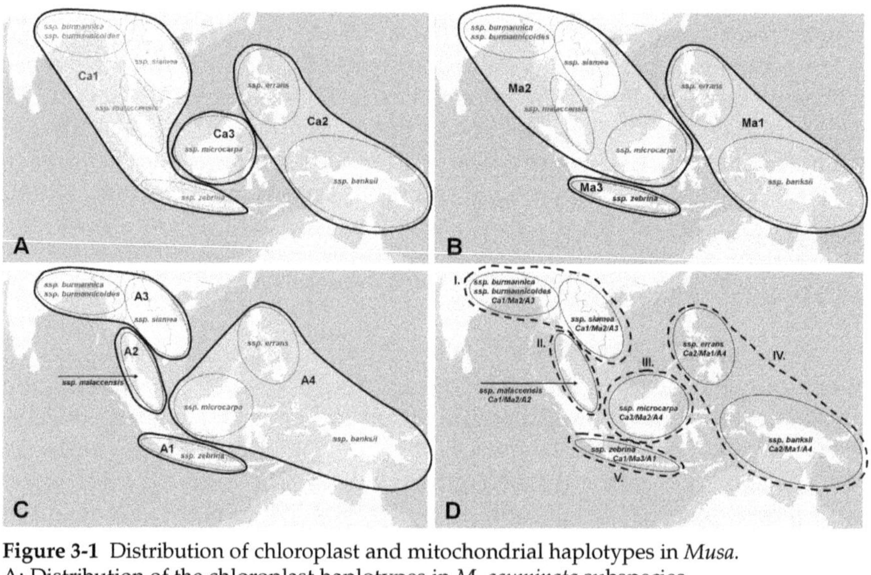

Figure 3-1 Distribution of chloroplast and mitochondrial haplotypes in *Musa*.
A: Distribution of the chloroplast haplotypes in *M. acuminata* subspecies.
B: Distribution of the mitochondrial haplotypes in *M. acuminata* subspecies.
C: Distribution of the 45S ribosomal genotypes of *M. acuminata*.
D: Combined genotypes of the morphological subspecies listed as: chloroplast/mitochondrial/45S ribosomal.
I-V.: suggested grouping of the morphological subspecies according to their combined genotypes.
The distribution area of *M. acuminata* ssp. *truncata* is within the ssp. *malaccensis* area (not shown).

making recognition of mutants and somaclonal variants difficult. Therefore, there was always a need for quantitatively measurable markers and for the development of standardized methodologies for description of the *Musa* germplasm.

Introduction of cytological methods, namely metaphase chromosome counting and analysis of meiotic chromosome pairing provided further details in understanding the genome structure of polyploid cultivars. Chromosome counting correctly confirmed the expected ploidy of the five main genetic groups mentioned above, while analysis of meiotic chromosome pairing and chromosome structural changes confirmed the existence of seven subspecies of *M. acuminata* (Shepherd 1990 and 1999).

Since morphological characterization is qualitative and also subject to changes based on environmental conditions, quantitatively measurable characters provided the next generation of markers in diversity analysis of bananas. With the advent of biochemistry the analysis of certain metabolites/ metabolic pathways and proteins and with the advent of molecular genetics direct analysis of the genome became possible. Analysis of metabolite and protein diversity reflects not only exclusively the genetic diversity but also the adaptation of the organism to the local environment, which may influence the observed variation via altered gene expression. However, diversity analysis of DNA directly gives the most precise distinction of the organisms, revealing differences on the DNA base pair level. Additionally while the number of available markers is limited in the case of metabolic and protein markers since these mirror only part of the expressed genetic information, DNA markers are limited only by the genome size.

3.2.1 Metabolome-Flavonoid Markers

Flavonoids are products of the secondary metabolism and show very high structural diversity by hydroxylation, methylation and glycosylation of the flavonol backbone dependent on the genetic background, e.g., availability of genes for modifying enzymes. This set of genes is mostly species/subspecies specific and by analyzing the flavonoid pattern complete evolutionary lineages can be assessed. The weakness of the system is that the genes involved in flavonoid biosynthesis represent only a small fraction of genes of the *Musa* genome. The analysis of flavonoid compounds of banana commenced in the 50s, when Simmonds (1954) drew attention to the possible use of flavonoid markers in banana taxonomy by showing the taxonomic value of leucoanthocyanins. Based on this marker system, the analysis of anthocyanins and flavonols by Horry and Jay (1988) clearly discriminated the wild species *M. acuminata* and *M. balbisiana* and were used to classify the subspecies of *M. acuminata* by suggesting three groups for the analyzed subspecies: 1./ ssp. *siamea* and ssp. *burmannica* 2./ssp. *malaccensis*, ssp. *microcarpa* and ssp. *burmannicoides* and 3./ ssp. *zebrina* (Horry and Jay 1990). The study also revealed evolutionary steps in anthocyanin biosynthesis based on the presence/absence of modifying enzymes among the *M. acuminata* subspecies suggesting that ssp. *zebrina* is possibly most similar to a putative ancestral form. Based on the complexity of the anthocyanin patterns *M. balbisiana* was identified as the least developed genome in this respect. A detailed analysis of the anthocyanin metabolites or pathway revealed variation in anthocyanin pigments among the edible diploids and suggested that ssp. *malaccensis*, ssp. *banksii* and ssp. *zebrina* were putative ancestors of the Cavendish cultivar (Horry and Jay 1990).

3.2.2 *Proteome—Enzyme Polymorphisms*

Allozyme/isoenzyme analysis is based on protein/enzyme polymorphism present in their amino acid composition, which eventually reflects the polymorphism of the genome. Allozymes are representing allelic variants of the same gene, while isoenzymes are enzymes with similar template specificity but representing different genes (consider gene families). Since discrimination of the two cases is difficult most authors mention isoenzymes/isozymes. The technology is based on the isolation and native electrophoresis of proteins based on their size, charge and structure followed by in-gel visualization of the specific enzyme activity by appropriate staining reaction based on the enzyme activity. The resulting banding patterns represent the genetic diversity of the analyzed enzyme/protein. One should note, however, that the resulting banding patterns for example may be different in the different tissues or may show seasonal fluctuations. The first isoenzyme study (Bonner et al. 1974) on various banana clones and species failed to show any differences in the peroxidase isoenzymes. Later Rivera (1983) could discriminate BBB and ABB genomes by peroxidase and polyphenoloxydase polymorphisms. Comparative analysis of *M. acuminata* and *M. balbisiana* by Jarret and Litz (1986) using 8 isozyme loci revealed species differentiation of *acuminata* and *balbisiana* and subspecies specific alleles in *M. acuminata* ssp. *microcarpa*, ssp. *burmannica*, ssp. *errans* and ssp. *zebrina* suggesting the usefulness of this marker system in elucidating the genetic structure of hybrid cultivars. Later Lebot et al. (1993) postulated that *M. acuminata* ssp. *banksii* is the A genome donor in Pacific plantains.

3.3 Genomes—DNA-based Analysis of Banana Genetic Diversity

Availability of techniques for the direct investigation of the genome allowed the development of markers with genome-wide distribution or on demand characterization of specific regions of interest within the genome in detail. Since the number of DNA sequence variations is immensely high the available sites for marker development is restricted only by the size of the analyzed genome itself. The discovery of the polymerase chain reaction (PCR) revolutionized marker development resulting in the development of numerous marker systems yielding marker densities that were not available in the past. These techniques include, random amplified polymorphic DNA (RAPD), amplified fragment length polymorphism (AFLP) and related techniques, including regions with high level of allelic variability such as simple sequence repeat (SSR). Other techniques such as PCR-restriction fragment length polymorphism (PCR-RFLP), and single nucleotide polymorphism (SNP) allow analysis of specific parts or selected genes of the genome. PCR based methods are relatively cheap and easy to adopt in any laboratory allowing high-throughput analysis of large amounts of samples with high number of marker sites. DNA sequence-based markers also made possible the independent analysis of the organellar and nuclear genomes of the plant cell.

There are three distinct genomes within a plant cell: the chloroplast, the mitochondrial and the nuclear genomes. The size of the nuclear genome in *M. acuminata* is 613 Mbps, while *M. balbisiana* harbors a somewhat smaller genome of 564 Mbps (Dolezel et al. 1994). The organellar genomes are significantly smaller, the chloroplast genome is in the range of 0.13–0.14 Mbp (Kornel Burg, unpubli.) while the exact size of the mitochondrial genome is not known at present and is expected to be 0.4–0.5 Mbps. In *Musa* the inheritance of the chloroplast genome is strongly biased towards a maternal lineage, while the mitochondrial genome is paternally inherited (Fauré et al. 1994). Uniparental inheritance excludes the possibility of recombination and the corresponding genome evolves only by mutations. Consequently the organellar genomes provide an excellent possibility to follow the maternal as well as the paternal lineages by use of chloroplast and mitochondrial markers, respectively.

Considering Kurz's hypothesis that the present day cultivars are the offspring of ancient wild types especial focus should be paid on the genetic diversity of the present day wild types. Extant wild types represent the progeny of ancestors, which yielded the modern sterile cultivars by hybridization; consequently these ancient wild genotypes are still present in a more or less unchanged form in the cultivars. Classifying the cultivar genotypes along with present day wild types opens the possibility to identify the subspecies/geographic regions contributing genome elements to the cultivars.

3.3.1 Genetic Analysis of the Uniparentally Inherited Organellar Genomes

Basically three methods are used to monitor organellar genome polymorphisms in banana. Firstly Southern hybridization was used applying heterologous DNA probes to identify organellar polymorphisms of banana. Later organelle specific universal PCR primer system was developed based on the high level sequence similarity of the chloroplast (Dumolin-Lapègue et al. 1997) and mitochondrial genomes (Duminil et al. 2002) in the plant kingdom. Based on this primer set either PCR-RFLP analysis or direct sequencing of the generated fragments was used to identify polymorphic regions/sequences.

The first chloroplast DNA-based markers and phylogenetic analyses on banana were published by Gawel and Jarret (1991a) using heterologous chloroplast probes (*Lactuca sativa* and *Oncidium excavatum*) and Southern analysis. It showed the cytoplasmic grouping of *Musa* for the first time clearly distinguishing the two wild species *M. acuminata* and *M. balbisiana*. Several triploid putative *M. acuminata* × *M. balbisiana* hybrids were not distinguishable, however despite the similar nuclear genotype some of those had *acuminata*, while others *balbisiana* type chloroplast genome. Additionally it was demonstrated that the chloroplast genome of one of the Cavendish type (Grande naine) was similar to that of *M. acuminata* spp. *malaccensis*. These results suggested that use of cytoplasmic markers "may be helpful in reconstructing the mechanisms of the ancestral hybridizations or mutations which gave rise to today's cultivars". Subsequently the same authors (Gawel and Jarret 1991b) compared different *Musa* species using the same methodology revealing that the chloroplast genome of *M. textilis* (*Australimusa*) is a close relative to *M. balbisiana* (*Eumusa*) suggesting a common ancestor. Later Carreel et al. (2002) published a comprehensive study based on Southern analysis on both the chloroplast and the mitochondrial genomes on a large number of banana accessions defining nine chloroplast and nine mitochondrial patterns in the *acuminata*, *balbisiana* and *schizocarpa* genomes. Umali (2002) using PCR amplification and subsequent sequencing of the intergene region between the *rpl16* and *rpl14* chloroplast genes identified SNPs discriminating various Cavendish subgroup cultivars. Subsequently Umali and Nakamura (2003) reported an SNP marker discriminating the *acuminata* and *balbisiana* chloroplast genomes in the *trn*L-F intergenic spacer sequence and described a deletion characteristic for the *M. acuminata* spp. *banksii*. Nwakanma et al. (2003) established sectional relationships of several *Musa* species based on both chloroplast and mitochondrial PCR-RFLP marker systems showing that *M. balbisiana* organellar genomes represent evolutionarily ancient forms compared to the *acuminata* subspecies. The existence of two evolutionary lines one including *Australimusa* and *Callimusa*, the other *Eumusa* and *Rhodoclamys* were suggested, *M. balbisiana* belonging to the *Australimusa*/*Callimusa* lineage indicating a putative common origin of their organellar genomes. A detailed population analysis of *M. balbisiana* partly based on chloroplast (PCR-RFLP) markers was published by Ge et al. (2005) revealing two geographic distribution areas with distinct chloroplast genomes in southern China. Swangpol et al. (2007) described the lineage analysis of cultivars by analyzing SNP variations in the chloroplast genome suggesting that at least two *M. balbisiana* donors were involved in the formation of hybrid cultivars and confirmed the previous observation of Gawel and Jarret (1991a) that Cavendish type Grande naine harbors a chloroplast genome possibly originating from *M. acuminata* ssp. *malaccensis*.

The above studies revealed that the organellar genomes *of M. acuminata* and *M. balbisiana* are quite distinct and that the diversity in *M. acuminata* was useful in partially classifying the morphological subspecies. No subspecies were identified in *M. balbisiana*. However, Ge et al. (2005) identified clades within Chinese *M. balbisiana* populations.

Uniparentally inherited genomes are subject to mutations in their evolution and the majority of the banana cultivars are sterile because of parthenocarpy. Therefore, it is possible to suppose that ancient organellar genomes are trapped in extant cultivars remaining more or less unchanged since initial cultivar formation. Considering that the present day wild types are the offspring of the ancient species their comparison to those present in the cultivars may identify genepools of common origin. This hypothesis was used by Boonruangrod et al. (2008) who identified three chloroplast genepools both for *M. acuminata* and *M. balbisiana* (Fig. 3-1b). Similarly three genepools were also observed for the mitochondrial genome (Fig. 3-1c). Chloroplast/mitochondria combinations, however, showed distinct geographic distribution (Fig. 3-1d). Analysis of some hybrids revealed that none of them had chloroplast/mitochondrial combinations similar to the wild types. The most common cytotype in the hybrid cultivars was a combination of the most abundant chloroplast type of the western region with the most abundant eastern type mitochondria, indicating hybridization of the "western" (*malaccensis, siamea, zebrina, burmannica, burmannicoides*) and "eastern" (*banksii, errans*) wild types (Boonruangrod et al. in press).

3.3.2 Analysis Based on the DNA Sequence Diversity of the Nuclear Genome

Numerous DNA-based methodologies have been used for revealing nuclear DNA sequence differences among banana individuals/cultivars or wild type species/subspecies. The simplest methodologies for generating multiband PCR patterns are based either on the use of arbitrary PCR primers (RAPD) or restriction enzyme cut site driven amplification of the genome (AFLP). These most popular dominant marker systems do not need prior knowledge of the genomic sequence but in these systems it is not possible to exclude the presence of organellar marker bands in the patterns since in most cases total DNA (containing nuclear as well as organellar DNA) was isolated and used for the analysis from the banana cells. However, it is assumed that the majority of the bands represent regions of the nuclear genome. On the other hand, using highly repeated nuclear sequence elements (Jarret et al. 1993) and sequence motifs of transposons (Nair et al. 2005) nuclear genome specific banding patterns were developed providing a dominant marker system. These latter marker systems, however, need prior sequence information of the analyzed regions of the genome. Similarly prior sequence information is needed for the marker systems such as SSR (simple sequence repeat), RFLP, SNP, which provide codominant marker systems. Recently high-throughput methods including diversity array technology (DArT, Risterucci et al. 2009) and targeting induced local lesions in the genomes (TILLING, Till et al. 2010) have been developed for the mass identification of genome regions with sequence diversity, which eventually may be used as genetic markers.

Many attempts were made to elucidate genetic variability based on the nuclear genome using the dominant marker systems such as RAPD and AFLP. These easily applicable techniques were used frequently in assessing local variability of banana populations (see for example Crouch et al. 1955; Ude et al. 2002, 2003; Wang et al. 2007; Brown et al. 2009) but also for species differentiation (Pillay et al. 2000). Analyzing *M. acuminata* wild types and cultivars using AFLP (Ude et al. 2002) distinguished three groups/genepools characterized by morphological subspecies ssp. *microcarpa*, ssp. *malaccensis* and ssp. *burmannica*. Previously Carreel et al. (1994) suggested four groups for the morphological subspecies (Pole1 *banksii, errans*, Pole2 *zebrina, microcarpa*, Pole3 *burmannica, burmannicoides, siamea*; Pole4 *malaccensis*) based on 30 nuclear markers representing all 11 chromosomes of *M. acuminata* genome using Southern-RFLP. Markers based on SSRs were also frequently used for genetic diversity studies in *Musa*; however, species specific alleles were not detected. Four groups/genepools represented by morphological subspecies (ssp. *malaccensis*, ssp. *zebrina*, ssp. *banksii and* ssp. *burmannica*) were distinguished by Grapin et al. (1998), while five genepools marked by morphological subspecies (ssp. *banksii*, ssp. *siamea/burmannica*, ssp. *malaccensis*, ssp. *zebrina* and ssp. *microcarpa*) were identified by Perrier et al. (2009) using nuclear SSR markers. Genome specific sequence variation was identified distinguishing *M. acuminata* and *M. balbisiana* using the 45S ribosomal region (Nwakanma et al. 2003). Species specific as well as *M. acuminata* subspecies specific markers (SNP-PCR) were developed by Boonruagrod et al. (2009). In the latter study the morphological subspecies were allocated in four genepools showing distinct geographical distribution (Fig. 3-1c) confirming and summarizing previous findings of Carreel (1994), Grapin et al. (1998), Ude et al. (2002) and Perrier et al. (2009). However, within-species analysis revealed little variation in *M. balbisiana*, despite Ge et al. (2005) description of clades within *M. balbisiana* based on SSR markers. In summary, the analysis of *M. acuminata* nuclear genome by several different methodologies revealed four genepools marked by geographically close morphological subspecies, e.g., 1/ *banksii, errans, microcarpa*, 2/ *burmannica, burmannicoides, siamea*, 3/*zebrina* and 4/ *malaccensis*.

3.3.3 Combined Use of Nuclear and Organellar Markers in Describing Genetic Diversity of M. acuminata *and* M. balbisiana

Previously the comparison of morphological data with that of molecular was mostly done by using either organellar or nuclear markers. Recently Boonruangrod et al. (2011) suggested the identification of genotypes according to their combined chloroplast, mitochondrial and nuclear genotypes. In this way all major morphological subspecies of *M. acuminata* were distinguished by molecular markers (Fig. 3-1d) and some diversity mostly based on the organellar genomes was identified in *M. balbisiana*. On the basis of this study the nine morphological subspecies may be reduced to five, keeping ssp. *malaccensis* and ssp. *zebrina*, while ssp. *banksii* and ssp. *errans;* ssp. *microcarpa and* ssp. *truncata;* ssp. *burmannica*, ssp. *burmannicoides* and ssp. *siamea* may be merged into three groups considering also their geographic distribution (Fig. 3-1d).

3.4 Analyzing Genome Structure of the Polyploid Cultivars

Since Kurz's suggestion that the majority of edible banana cultivars of the *Eumusa* section originated from the two wild species *M. acuminata* (AA) and *M. balbisiana* (BB) either directly or in various hybrid combinations, the participation of the S and T genomes in edible varieties was also revealed by Carreel et al. (1994) and visualized by D'Hont et al. (2000) using genomic *in situ* hybridization (GISH). The evolution of biochemical marker systems supported the main classification of bananas based on morphological markers and also identified the proportion of the A and B genomes in the cultivars (Simmonds 1962). Later, the analysis of chemical features of the phenotype provided more detailed insights into the genome structure of the cultivars and revealed putative donor subspecies in the formation of Cavendish cultivars (Horry and Jay 1990), while later Lebot et al. (1993) identified *M. acuminata* ssp. *banksii* as donor of the A genome in Pacific plantains using isozyme markers. The appearance of DNA based molecular markers allowed the generation of precise genomic "fingerprints" of cultivars as well as wild types. The first detailed analysis identifying putative donor wild type genotypes was published by Raboin et al. (2005) using 30 single copy nuclear markers and revealed the putative ancestors of the triploid export bananas "Cavendish" and "Gros Michel".

The majority of the edible cultivars are the intra- or interspecific hybrids of *M. acuminata* and *M. balbisiana* as was shown by morphological as well as molecular tools. Since *M. acuminata* shows a high level of diversity represented also by subspecies showing specific geographical distribution monitoring the participation of these subspecies in present day cultivars would provide valuable information for banana breeding. Recently Boonruangrod et al. (2009) developed subspecies specific PCR methods based on the 45s rDNA locus for hybrid analysis, which allowed the identification of the *M. acuminata* subspecies participating in cultivar formation. Combining organellar and nuclear inheritance data, however, allowed the identification of putative hybridization events that occurred during the evolution of today's cultivars (Boonruangrod et al. 2011). Frequently morphological markers indicate the unexpected unbalanced appearance of phenotypic traits in *acuminata* x *balbisiana* hybrids. Possible explanation of this effect may be intergenome informational change at an early stage in cultivar development. Based on organellar and nuclear diversity data De Langhe et al. (2010) suggested the significance of possible backcrossing events in the early evolution of banana hybrids as a putative source in unbalanced genome development.

3.5 Molecular Taxonomy of Bananas

The distribution areas of species of the Musaceae family cover the tropical regions of Asia, Australia and Africa. In this family there are three genera recognized; *Musa, Musella* and *Ensete* based on morphological markers as well as chromosome number. The haploid chromosome number of *Musa* is 10–11, while only 9 chromosome pairs are found in *Musella* and *Ensete*. The largest genus *Musa* consists of over 60 wild species (OECD 2009) and is divided into four sections; *Australimusa, Callimusa, Musa* (formerly *Eumusa*) and *Rhodochlamys*. The haploid chromosome number is 11 in *Musa* and *Rhodoclamys*, and 10 in *Australimusa* and *Callimusa*. Recent studies (Li et al. 2010; Liu et al. 2010) based on comparative sequence analysis of chloroplast and nuclear genomic regions revealed two major clades - one including the genera *Musella* and *Ensete*, while the other harbored all the *Musa* species. Instead of the four sections previously considered clade/genus *Musa* was divided into two clades—one including the sections *Musa* and *Rhodochlamys* and the other *Australimusa* and *Callimusa*. This distribution confirms the classification based on the chromosome number.

3.6 Conclusions

Morphological discrimination of *Musa* species and *M. acuminata* subspecies was completely confirmed by molecular analysis especially with DNA based marker systems confirming the excellent work done by morphologists. Simplification of genera within Musaceae family and the section structure within genus *Musa* was suggested to be based on DNA based molecular results. Additionally rethinking subspecies categorization within *M. acuminata* may be suggested based on molecular data.

Molecular markers, particularly the DNA based markers allowed the identification of the contribution of *M. acuminata* to hybrid genotypes on subspecies/accession level providing valuable data for banana breeding.

Abbreviations

AFLP	:	amplified fragment length polymorphism
GISH	:	genomic *in situ* hybridization
PCR	:	polymerase chain reaction
RAPD	:	random amplified polymorphic DNA
RFLP	:	restriction fragment length polymorphims
SNP	:	single nucleotide polymorphism detection
SSR	:	simple sequence repeat

References

Bonner JW, Warner RM, Brewbaker JL (1974) A chemosystematic study of *Musa* cultivars. HortScience 9: 325–328.

Boonruangrod R, Desai D, Fluch S, Berenyi M, Burg K (2008) Identification of cytoplasmic ancestor gene-pools of *Musa acuminata* Colla and *Musa balbisiana* Colla and their hybrids by chloroplast and mitochondrial haplotyping. Theor Appl Genet 118: 43–55.

Boonruangrod R, Desai D, Berenyi M, Fluch S, Burg K (2011) Application of cytoplasmic and nuclear DNA-based marker systems for elucidation of the phylogenetic relationships of *Musa acuminata* and *Musa balbisiana* and their hybrids Proc. Int'l ISHS-ProMusa Symp. on Global Perspectives on Asian Challenges Eds.: I. Van den Bergh et al. Acta Hort. 897 pp 133–137 ISHS 2011.

Boonruangrod R, Fluch S, Burg K (2009) Elucidation of origin of the present day hybrid banana cultivars using the 5'ETS rDNA sequence information. Mol Breed 24: 77–91.

Brown N, Venkatasamy S, Khittoo G, Bahorun T, Jawaheer S (2009) Evaluation of genetic diversity between 27 banana cultivars (*Musa* spp.) in Mauritius using RAPD markers. Afr J Biotechnol 8: 1834–1840.

Carreel F, Faura S, de Lean D, Lagoda PJL, Perrier X, Bakry F, du Montcel H, Lanaud C, Horry JP (1994) Evaluation de la diversite genetique chez les bananiers diploids (*Musa* sp.). Genet Sel Evol 26: 1–12.

Carreel F, Gonzalez de Leon D, Lagoda P, Lanaud C, Jenny C, Horry JP, Tezenas du Montcel H (2002) Ascertaining maternal and paternal lineage within *Musa* by chloroplast and mitochondrial DNA RFLP analyses. Genome 45: 679–692.

Cheesman EE (1948) Classification of the bananas IIIc *Musa paradisiaca* L. sp., *Musa sapientum* L. syst Kew Bull 3: 145–153.

Crouch JH, Crouch HK, Constandt H, Van Gysel A, Breyne P, Van Montagu M, Jarret RL, Ortiz R (1999) Comparison of PCR-based molecular marker analyses of *Musa* breeding populations. Mol Breed 5: 233–244.

D'Hont A, Paget-Goy A, Escoute J, Carreel F (2000) The interspecific genome structure of cultivated banana, *Musa* spp. revealed by genomic DNA *in situ* hybridization. Theor Appl Genet 100: 177–183.

De Langhe E, Hribova E, Carpentier S, Dolezel J, Swennen R (2010) Did backcrossing contribute to the origin of hybrid edible bananas? Ann Bot 106: 849–857.

Dolezel J, Dolezelova M, Novak FJ (1994) Flow cytometric estimation of nuclear DNA amount in diploid bananas (*Musa acuminata* and *M. balbisiana*). Biol Planta 36: 351–357.

Duminil J, Pemonge Ma H, Petit RJ (2002) A set of 35 consensus primer pairs amplifying genes and introns of plant mitochondrial DNA. Mol Ecol Notes 2: 428–430.

Dumolin-Lapegue S, Demesure B, Fineschi S, Corre VL, Petit RJ (1997) Phylogeographic Structure of White Oaks Throughout the European Continent. Genetics 146: 1475–1487.

Faure S, Noyer JL, Carreel F, Horry JP, Bakry F, Lanaud C (1994) Maternal inheritance of chloroplast genome and paternal inheritance of mitochondrial genome in bananas (*Musa acuminata*). Curr Genet 25: 265–269.

Gawel N, Jarret RL (1991a) Cytoplasmic genetic diversity in bananas and plantains. Euphytica 52: 19–23.

Gawel NJ, Jarret RL (1991b) Chloroplast DNA restriction fragment length polymorphisms (RFLPs) in *Musa* species. Theor Appl Genet 81: 783–786.

Ge XJ, Liu MH, Wang WK, Schaal BA, Chiang TY (2005) Population structure of wild bananas, *Musa balbisiana*, in China determined by SSR fingerprinting and cpDNA PCR-RFLP. Mol Ecol 14: 933–944.

Grapin A, Noyer JL, Carreel F, Dambier D, Baurens FC, Lanaud C, Lagoda PJ (1998) Diploid *Musa acuminata* genetic diversity assayed with sequence-tagged microsatellite sites. Electrophoresis 19: 1374–1380.

Horry J, Jay M (1988) Distribution of anthocyanins in wild and cultivated banana varieties. Phytochemistry 27: 2667–2672.

Horry J, Jay M (1990) An evolutionary background of bananas as deduced from flavonoid diversification. In: RL Jarret (ed) Identification of Genetic Diversity in the Genus *Musa* Proceedings of an International Workshop, Los Banos (1989), the Philippines. INIBAP, Montpellier, France, pp 41–55.

Jarret RL, Litz RE (1986) Enzyme polymorphism in *Musa acuminata* Colla. J Hered 77: 183–188.

Jarret RL, Vuylsteke DR, Gawel NJ, Pimentel RB, Dunbar LJ (1993) Detecting genetic diversity in diploid bananas using PCR and primers from a highly repetitive DNA sequence. Euphytica 68: 69–76.

Kurz S (1865) Note on plantains of the Indian Archipelago J Agric Hort Soc India 14: 295–301.

Lebot V, Aradhya KM, Manshardt R, Meilleur B (1993) Genetic relationships among cultivated bananas and plantains from Asia and the Pacific. Euphytica 67: 163–175.

Li L-F, Hakkinen M, Yuan Y-M, Hao G, Ge X-J (2010) Molecular phylogeny and systematics of the banana family (Musaceae) inferred from multiple nuclear and chloroplast DNA fragments, with a special reference to the genus Musa. Mol Phylogenet Evol 57: 1–10.

Liu A-Z, Kress WJ, Li D-Z (2010) Phylogenetic analyses of the banana family (Musaceae) based on nuclear ribosomal (ITS) and chloroplast (trnL-F) evidence. Taxon 59: 20–28.

Nair AS, Teo CH, Schwarzacher T, Harrison PH (2005) Genome classification of banana cultivars from South India using IRAP markers. Euphytica 144: 285–290.

Nwakanma DC, Pillay M, Okoli BE, Tenkouano A (2003) Sectional relationships in the genus *Musa* L. inferred from the PCR-RFLP of organelle DNA sequences. Theor Appl Genet 107: 850–856.

OECD (2009) Series on Harmonisation of Regulatory Oversight in Biotechnology No.48 Consensus Document on the Biology of Bananas and Plantains (*Musa* spp.).

Perrier X, Bakry F, Carreel F, Jenny C, Horry JP, Lebot V, Hippolyte I (2009) Combining biological approaches to shed light on the evolution of *Musa* complex. Ethnobot Res Appl 7: 199–216.

Pillay M, Nwakanma DC, Tenkouano A (2000) Identification of RAPD markers linked to A and B genome sequences in *Musa* L. Genome 43: 763–767.

Raboin L-M, Carreel FO, Noyer J-L, Baurens F-C, Horry J-P, Bakry F, Montcel HTD, Ganry J, Lanaud C, Lagoda PJL (2005) Diploid Ancestors of Triploid Export Banana Cultivars: Molecular Identification of 2n Restitution Gamete Donors and n Gamete Donors. Mol Breed 16: 333–341.

Risterucci A-M, Hippolyte I, Perrier X, Xia L, Caig V, Evers M, Huttner E, Kilian A, Glaszmann J-C (2009) Development and Assessment of Diversity Arrays Technology for High-Throughput DNA Analyses in *Musa*. Theor Appl Genet 119: 1093–1103.

Rivera FN (1983) Protein and isoenzyme banding patterns among Philippine cooking bananas and their wild parents (*Musa* species). Paradisiaca 6: 7–12.

Rumphius GE (1749) Herbarium Amboinense: 125–139.

Shepherd K (1990) Observations on *Musa* Taxonomy: a note on *Musa* germplasm in the Philippines. In: RL Jarret (ed) Identification of Genetic Diversity in the genus *Musa* Proceedings of an international workshop held at Los Banos (1989), the Philippines. INIBAP, Montpellier, France, pp 158–165.

Shepherd K (1999) Cytogenetics of the genus *Musa*. INIBAP, Montpellier, France.

Simmonds NW (1954) Anthocyanins in Bananas. Nature 173: 402–403.

Simmonds NW (1962) The Evolution of the Bananas. Longmans London, UK.

Simmonds NW, Shepherd K (1955) The taxonomy and origins of the cultivated bananas. J Linn Soc (London) Bot 55: 302–312.

Stover RH, Simmonds NW (1987) Bananas, Tropical Agriculture Series, 3rd edn. Longman, London, UK.

Swangpol S, Volkaert H, Sotto RC, Seelanan T (2007) Utility of selected non-coding chloroplast DNA sequences for lineage assessment of *Musa* interspecific hybrids. Journal of Biochemistry and Molecular Biology 40: 577–587.

Till BJ, Jankowicz-Cieslak J, Sagi L, Huynh OA, Utsushi H, Swennen R, Terauchi R, Mba C (2010) Discovery of nucleotide polymorphisms in the *Musa* gene pool by Ecotilling. Theor Appl Genet 121: 1381–1389.

Ude G, Pillay M, Nwakanma D, Tenkouano A (2002) Genetic Diversity in *Musa acuminata* Colla and *Musa balbisiana* Colla and some of their natural hybrids using AFLP Markers. Theor Appl Genet 104: 1246–1252.

Ude G, Pillay M, Ogundiwin E, Tenkouano A (2003) Genetic diversity in an African plantain core collection using AFLP and RAPD markers. Theor Appl Genet 107: 248–255.

Umali RP (2002) Development of PCR-based fingerprinting tool in banana (*Musa* sp., AAA) and conversion of negative to positive DNA marker. HortScience 37: 1108–1111.

Umali R, Nakamura I (2003) Identification of dCAPS markers that discriminate A and B cytoplasms in banana (*Musa* Spp.). Plant Biotechnol 20: 159–164.

Wang X-L, Chiang T-Y, Roux N, Hao G, Ge X-J (2007) Genetic diversity of wild banana (*Musa balbisiana* Colla) in China as revealed by AFLP markers. Genet Resour Crop Evol 54: 1125–1132.

4

Molecular Marker Techniques in *Musa* Genomic Research

Michael Pillay,[1,*] *Kaliyaperumal Ashokkumar,*[2] *Andrew James,*[3] *Silvas Jabekumar Prince Kirubakaran,*[4] *Robert Miller,*[5] *Rodomiro Ortiz*[6] and *Elayabalan Sivalingam*[7]

ABSTRACT

Banana is one of the most important subtropical crops. Its genetic system is relatively unknown and complicated by inter- and intra-specific hybridization, heterozygosity, and polyploidy. Thus molecular markers have been widely used among *Musa* germplasm to detect genetic variation and relationships, duplicate accessions identification, monitoring genetic stability, analyzing irradiated genotypes and identification of trait markers for use in breeding programs. Various markers based morphology, protein, and DNA are in existence. Among them the most widely used markers suitable for marker assisted selection (MAS) are RFLP, random amplified polymorphic DNA (RAPD), amplified fragment length polymorphisms (AFLP), and

[1]Vaal University of Technology, Private Bag X021, Vanderbijlpark 1900, Gauteng, South Africa; e-mail: *mpillay@vut.ac.za*

[2]Department of Plant Sciences, University of Saskatchewan, 51 Campus Drive, Saskatoon, SK, Canada, S7N 5A8; e-mail: *ashok.kumar@usask.ca*

[3]Centro de Investigacion Cientifica de Yucatan, Calle 43 No. 130, Col. Churburna de Hidalgo, Merida, C.P. 97200, Yucatan, Mexico; e-mail: *andyj007@cicy.mx*

[4]Department of Plant Molecular Biology and Biotechnology, Center for Plant Molecular Biology, Tamil Nadu Agricultural University, Coimbatore-641 003-India; e-mail: *symsil07@gmail.com*

[5]Laboratório de Microbiologia: Interação Planta-Praga, Instituto de Ciências Biológicas, Departamento de Biologia Celular, Universidade de Brasília, Campus Universitário Darcy Ribeiro, Asa Norte, 70910-900, Brasília, D.F., Brasil; and Universidade Católica de Brasília, SGAN Qd. 916 Módulo B, Av. W5 Norte CEP: 70.790-160, Brasília/DF, Brazil; e-mail: *rmiller@pos.ucb.br or bruna@tba.com.br*

[6]Department of Plant Breeding and Biotechnology, Swedish University of Agricultural Sciences, Box 101, SE-230 53, Alnarp, Sweden; e-mail: *rodomiroortiz@gmail.com*

[7]Crop Tech Ltd., P.O. Box 1367, Arusha, Tanzania; e-mail: *balabiotech@gmail.com*

*Corresponding author

List of abbreviations after the text.

simple sequence repeat (SSR). This chapter compares the efficiency of available markers, its application and development for *Musa* genome analysis. High-throughput technologies based on single nucleotide polymorphisms (SNPs) or small scale indels are efficient alternatives with desired dosage sensitivity and the ability to distinguish haplotypes in polyploid crops. However, the choice of markers differ and depends on the labor required for their detection, possibility of revealing single or multiple loci, dominant or co-dominant nature, and costs. Recent developments reported in various crops like high resolution DNA melting (HRM) analysis has great potential for cultivar identification, candidate gene identification, mapping (genomic and association) studies. Thus the identified genomic regions in banana with these approaches has the potential to dramatically enhance the pace and efficiency of genetic improvement in *Musa*.

Keywords: molecular marker systems, *Musa*, genetic improvement

4.1 Introduction

Advances in molecular biology gave rise to a number of new technologies that held great promise for crop breeding. One of such innovations was the development of molecular markers for tagging genes of interest. Genetic markers are landmarks on a chromosome that assist in genome analysis (Lefebvre et al. 1995). Markers are generally described as "genetic tags" that identify particular locations within a plant's DNA sequence and are transmitted to the offspring following the laws of inheritance from one generation to the next (Drew 1997). The use of DNA-based markers has allowed researchers to mark genes or chromosome regions that are related to genetic traits such as host plant resistance to pathogens and pests (Drew 1997). Once genetic markers are defined, they can be used directly in breeding programs since they allow researchers to predict phenotypes based on the presence or absence of the marker. Alternatively, once genes for particular trait have been identified by genetic marker techniques, they can be cloned. A particular advantage of such techniques is that complex multigenic traits can be analyzed. Mapping techniques can also be used to isolate genes based entirely on their genetic behavior. Once these have been identified, sequenced and cloned, gene transfer techniques can be used to transfer them to other species (Drew 1997).

Molecular markers are based on naturally occurring polymorphisms in DNA sequences (i.e., base pair deletions, substitutions or additions, among others). Different kinds of molecular markers exist, some are morphological, or protein-based while the most common ones are DNA-dependent. Genetic markers that reveal polymorphisms at the protein level are known as biochemical markers, while DNA markers reveal polymorphisms at the DNA level. Biochemical markers are proteins, produced from gene expression, which can be separated by electrophoresis to identify the alleles. The most commonly used protein markers are isozymes, which are variant forms of the same enzyme. Protein markers reveal differences in the gene sequence and function as codominant markers. However, their use is restricted due to their limited number in any crop species and also because they are subject to post-translational modifications (Staub et al. 1982). There are two categories of DNA markers according to how their polymorphism is expressed: hybridization-based and PCR-based polymorphisms. The hybridization-based polymorphisms include restriction fragment length polymorphisms (RFLP) (Sambrook et al. 1989) and variable number tandem repeats (VNTR) loci (Rogstad 1993; Weising et al. 1998), where probes such as random genomic clones, and probes for microsatellite (or simple sequence repeats, SSR) and mini-satellite sequences are hybridized to filters containing DNA that has been digested with restriction enzymes. The polymorphisms in the case of VNTR loci are due to a difference in the number of repeats, while RFLPs are generated due to events such as point mutations, inversions, deletions, or translocations.

The use of molecular markers has greatly improved breeding approaches since these markers directly reveal genetic variability through DNA analysis (Staub et al. 1996) and therefore their detection is not influenced by environmental effects. The development of numerous molecular markers for plant genome analysis makes it possible to select the genotype instead of the phenotype, leading to the concept of marker-assisted selection or MAS (Paterson et al. 1991). The most widely used markers suitable for MAS are RFLP, random amplified polymorphic DNA (RAPD), amplified fragment length polymorphism (AFLP), and simple sequence repeat (SSR) (Barone 2004). The choice of the most suitable markers for MAS however differs and depends on the labor required for their detection, possibility of revealing single or multiple loci, dominant or codominant nature, and costs.

Molecular markers have been widely used in *Musa* for detecting genetic variation and genetic relationships in banana germplasm, identification of duplicate accessions in field and tissue culture germplasm banks, monitoring genetic stability of tissue culture material, analyzing irradiated banana genotypes and identification of trait markers for use in breeding programs. This chapter examines some of the more common markers used in *Musa* and mentions some new marker systems that may be potentially useful in *Musa*.

4.2 Isozyme Markers

Many studies in banana and plantain were undertaken to characterize isozymes as genetic markers for estimating genetic diversity (Bhat et al. 1992a) within the genus, to measure among others somaclonal variation, to identify protoplast fusion products, and for clonal identification (Bhat et al. 1992b).

However, the sensitivity of this technique is a function of the number of polymorphic loci that can be resolved. The use of isozymes as genetic markers for variety identification in banana dates back to 1980s and received considerable attention (Tanksley and Orton 1983). Few studies conducted to examine the peroxidase isozymes in various banana clones and species noted a lack of polymorphism (Bonner et al. 1974). Rivera (1983) examined the peroxidase and polyphenoloxidase systems for polymorphism and was able to distinguish between the "Saba" (ABB/BBB) and "Bluggoe" (ABB) banana genome groups. Jarret and Litz (1986a, b) studied the efficiency and application of various isozymes such as *shikimate dehydrogenase* (*SKDH*), *malate dehydrogenase* (*MDH*), *peroxidase* (*PRX*), *phosphoglucomutase* (*PGM*) and *glutamate oxaloacetate transminase* (*GOT*). Some isozymes were useful in discriminating clones in some genomic groups. Simmonds (1966) used *esterase* (*EST*), *tetrazolium oxidase* (*TO*), *6-phosphogluconate dehydrogenase* (*6PGDH*), *malic enzyme* (*ME*), *phosphoglucose isomerase* (*PGI*), *alcohol dehydrogenase* (*ADH*), *triosephosphate isomerase* (*TPI*), *galactose dehydrogenase* (*GDH*), *succinate dehydrogenase* (*SUDH*), *sorbitol dehydrogenase* (*SDH*) and *glutamate dehydrogenase* (*GUDH*) to classify bananas. Espino and Pimentel (1990) indicated that *MDH* was useful in distinguishing the AAB and ABB cultivars from those with BB/BBB genomes. The isozymes of *MDH*, *PRX* and *GOT* were used to analyze the variation of 100 Indonesian *Musa* cultivars belonging to the AA, AAA, AAB, ABB, and BB genomic groups (Megia et al. 2001). A high degree of polymorphism was recognized for *MDH* and *PRX*. *GOT* was the least polymorphic isoenzyme. The numbers of cultivars that exhibited specific isozyme profiles were 16 for *MDH*, 20 for *PRX*, and 8 for *GOT*. These isozyme profiles were able to distinguish 31 cultivars. Isozymes were also useful for assessing the induction of genetic variability by γ-radiation in banana (Megia et al. 2001). Isozymes were used to identify resistance to *Banana Bract Mosiac Virus* (BBrMV) (Dhanya et al. 2006). The *BBrMV*-infected samples showed the presence of a definite single peroxidase band that was absent in non-infected samples.

Most isozyme markers used in banana research helped elucidating clonal identification in phylogenetic studies. Nevertheless, the use of isozymes as genetic markers will provide estimates of genetic diversity within the cultivated and wild relatives of the genus *Musa*, facilitate clonal identification and should provide valuable evidence on the origins of triploid cultivars. Differentiation of somatic mutations from their parental clones may occasionally be detected by isozyme analysis (Brettell et al. 1986).

4.3 Hybridization Based Systems

4.3.1 Restriction Fragment Length Polymorphisms (RFLP)

RFLPs are variations in DNA fragment length banding patterns of electrophoresed restriction digests of DNA samples. These variations are frequent due to the presence of a restriction enzyme cleavage site at one site in the genome of one individual and the absence of the site in another individual. RFLP is a codominant marker meaning that it is able to distinguish homozygotes and heterozygotes. RFLP is robust, easily transferred between laboratories, and requires no prior sequence information about an organism for its use. It is, therefore, useful for constructing genetic linkage maps, especially in *Musa*.

As it is locus-specific, estimations of conservation of synteny are possible. Using single locus probes it allows discrimination at the population level. The disadvantages are that it is expensive to develop, requires large amounts of DNA, is not possible to automate as compared to AFLP and diversity array technology (DArT) or multiplexing of SSRs, needs a suitable probe library, may require radioactive labeling and is laborious and time consuming. RFLPs have been used in all *Musa* mapping projects to date and most recently have been linked to polymorphisms in resistance gene analogs (Hippolyte et al. 2010).

RFLPs are considered highly repeatable and specific since a probe of known sequence is used. RFLPs of diverse germplasm have been used to study the taxonomy and phylogeny of *Musa* species (Gawel and Jarret 1991a, b; Gawel et al. 1992; Jarret et al. 1992) and variation in the chloroplast genome (Gawel and Jarret 1991a, b; Nwakanma et al. 2003). However, there is only one report of their usefulness in distinguishing closely related material (Bhat et al. 1995). The relatively high cost and technically demanding nature of this technique makes it inappropriate for routine breeding applications.

4.3.2 Variable Number of Tándem Repeats (VNTR)

VNTRs are tandem repeats of short DNA sequences and include minisatellites (10–45 bp) and microsatellites (2–6 bp). While initially VNTRs were detected involving radioactive probes presently they can be detected by employing PCR technology. VNTRs are abundant and ubiquitous in the eukaryotic genomes (Tourmente et al. 1994). These sequences hybridize to multiple loci and are hypervariable in nature, thus are highly informative for genetic analysis. Tandem repeats are usually classified into microsatellites (the length of the repeat unit being 5 bp or less) and minisatellites (more than 5 bp and up to 100 bp in length) (Buard and Vergnaud 1994). Crouch et al. (1998; 1999a, b) used VNTR analysis of microsatellite loci to compare the genetic similarity of full-sib 2*x* and 4*x* plantain-banana hybrids and their parental genotypes.

4.4 PCR-based Marker Systems

All PCR-based molecular markers appear to detect a high level of polymorphism within a range of *Musa* breeding populations. PCR-based assays are amenable to the large-scale throughput demands of screening breeding populations. Thus, researchers have concentrated on the applications of PCR for *Musa* genome analysis. Various molecular markers, especially different PCR-based molecular markers including AFLP, RAPD, microsatellites, have been frequently used for assessing genetic diversity and phylogenetic relationship in wild banana and cultivated accessions (Grapin et al. 1998; Loh et al. 2000; Wong et al. 2001; Carreel et al. 2002; Ude et al. 2002a, b, 2003; Creste et al. 2003a,b; Nwakanma et al. 2003).

A large variety of PCR amplification methods has emerged since its discovery. These include single primer amplification reaction (SPAR) (Gupta et al. 1994; Weising et al. 1995). The SPAR approach to minisatellite analysis has been described as direct amplification of minisatellite-region DNA (DAMD), which directs the amplification to regions rich in minisatellite repeats by using the core sequence of minisatellites as single primer (Heath et al. 1993; Zhou et al. 1997). Single-strand conformation polymorphism analysis (SSCP) is a rapid method for detection of minor sequence changes in polymerase chain reaction-amplified DNA (Orita et al. 1989). RAMPO combines arbitrarily or semispecifically primed PCR with microsatellite hybridization to produce several independent and polymorphic genetic fingerprints per electrophoretic gel (Richardson et al. 1995). Recently the sequence-related amplified polymorphism (SRAP) technique, aimed for the amplification of open reading frames (ORFs), was used to analyze the genetic variation and relationships among 40 *Musa* accessions (Muhammad et al. 2010).

4.4.1 Random Amplified Polymorphic DNA (RAPD)

RAPD has the advantage that it is easy to carry out with relatively little costs, but reproducibility is sometimes limited and reliability depends on the skill of the operator. Different results have been reported when RAPD analysis is carried out in different laboratories using identical PCR parameters and conditions. The RAPD technique requires no prior knowledge of the genome of an organism since it depends on the use of short arbitrary primers. The advantages of the RAPD technique include its speed, low DNA template requirement and technical simplicity. It is a convenient tool for detecting genetic variation within the germplasm. RAPDs have been widely used to examine genetic relationships in banana germplasm (Uma et al. 2006; Jain et al. 2007; Nsabimana and van Staden 2007) for identification of duplications among accessions in field and tissue culture germplasm banks and monitoring of genetic stability of tissue culture material (i.e., somaclonal variation) (Bairu et al. 2006, 2008; Ray et al. 2006; Lakshmanan et al. 2007), analyzing and differentiating irradiated banana genotypes (Toruan-Mathius and Haris 1999; Finalet et al. 2000; Imelda et al. 2001; Hautea et al. 2004) and identification of trait markers for use in cross- and mutation-breeding programs (Damasco et al. 1996). DNA profiling would also be important for policing plant patents and for legal protection of newly bred cultivars in asexual crops like *Musa* (Kaemmer et al. 1997). Highly informative RAPD markers can be converted to sequence characterized amplified regions (SCAR) primers for specific applications. RAPD assays have proven to be powerful and efficient means of marker-assisted introgression and backcross breeding. Specific RAPD markers for the A and B genomes of *Musa* have been identified and were used routinely (Pillay et al. 2000; Oselebe et al. 2006; Pillay et al. 2006).

However, RAPD analysis has several disadvantages including the dominant nature of the marker system and reproducibility problems, which may limit their application in marker-assisted selection. The RAPD technique has been successfully used to distinguish diverse *Musa* germplasm (Howell et al. 1994; Bhat and Jarret 1995; Pillay et al. 2001; Uma et al. 2006; Jain et al. 2007; Nsabimana and van Staden 2007). In addition, a molecular linkage map has also been developed using a variety of marker systems including RAPD (Faure et al. 1993). RAPD analysis has also been used to differentiate *Musa* genome groups (Howell et al. 1994; Pillay et al. 2000), closely related *Musa* germplasm (Bhat and Jarret 1995) and full-sib hybrids in plantain breeding populations (Crouch et al. 1998, 2000).

True-to-the-type clonal fidelity is one of the most important prerequisites in the micropropagation of any banana species. A major problem often encountered with *in vitro* cultures is the presence of somaclonal variation amongst subclones of one parental line. RAPD and inter-simple sequence repeat (ISSR) were adopted for evaluation of clonal fidelity in banana plantlets (Bennici et al. 2004; Carvalho et al. 2004; Martins et al. 2004; Ray et al. 2006). The somaclones confirmed through PCR-based tests were relatively stable without generally reverting to normal phenotypes (Ramage et al. 2004). A stringent RAPD technique called high annealing temperature (HAT)-RAPD was used with Thai banana cultivars (Sitthipron et al. 2005). Despite the criticism of the technique, RAPD is still being used in genomic analysis of banana as evidenced from recent publications (Das et al. 2009; Miri et al. 2009; Khatri et al. 2009).

4.4.2 Simple Sequence Repeats (SSR)

Limitations in RAPD markers led to the development and utilization of microsatellite primers for *Musa* (Jarret et al. 1994; Kaemmer et al. 1997;

Grapin et al. 1998; Creste et al. 2003a, b). Microsatellite or SSR loci are abundant, randomly dispersed, locus specific, codominant and multi-allelic markers. They are tandemly repeated DNA motifs (generally less than or equal to 4–6 bp) with an overall length in the order of tens of base pairs. Microsatellites or SSRs have been reported to be highly abundant and randomly dispersed throughout the genomes of many plant species (Crouch et al.1998). Variation in the motif is thought to arise through slippage errors during DNA replication. SSRs were shown to be reproducible markers in other systems due to their abundance, polymorphism and reliability (Campbell et al. 2003).

Microsatellites have been widely used to detect the genetic diversity of plant species. These markers have been widely used in plants for fingerprinting, mapping, and genetic analysis. Simple sequence repeat length polymorphism (SSRLP) analysis has been shown to detect a high level of polymorphism between individuals of *Musa* breeding populations (Crouch et al. 1998, 2000). However, isolation of microsatellites is time consuming and expensive. Nonetheless, with the availability of automated DNA sequencing facilities, improved techniques for the construction of genomic libraries enriched for SSR and improved techniques for the screening of appropriate clones, the isolation of SSR is becoming routine. This has recently allowed the rapid isolation of several hundred microsatellites from the *Musa* B genome (Buhariwalla et al. 2005). To date, several hundred SSR markers have been identified across *M. acuminata* and *M. balbisiana* material (Jarret et al. 1994; Kaemmer et al. 1997; Crouch et al. 1998; Creste et al. 2003a, b; Wang et al. 2009), with numbers likely to increase with advances in genomic technologies.

Numerous uses exist for such markers, although perhaps their most important applications reside in genetic map saturation and MAS, following co-localization with specific traits. For example, 352 SSR loci were recently discovered in *M. acuminata* Calcutta 4 bacterial artificial chromosome (BAC) end-sequences (Cheung and Town 2007). In comparison with other crop species, however, novel SSR loci available today for genetic analyses remain relatively limited in *Musa*, given that alleles can be absent or monomorphic when applied across cultivars. Over 40 microsatellite loci have recently been identified from five fully sequenced *M. acuminata* Calcutta 4 BAC consensi datasets (Genbank accession numbers AC186748, AC186749, AC186954, AC186747 and AC186750), with validation for polymorphism conducted on genotypes contrasting in host plant resistance to Sigatoka diseases (Miller et al. 2010).

4.4.3 Inter-Retrotransposon Amplified Polymorphism (IRAP)

IRAP is a complementary dominant marker system used to detect polymorphism in retro-transposon insertions. Retrotransposons are abundant and dispersed components of most plant genomes comprising over 50% of the nuclear DNA content in many species (Kumar 1999).

They act as mutagenic agents thereby providing putative sources of biodiversity (Heslop-Harrison 2000). The major classes of retroelements include long interspersed elements (LINES), short interspersed elements (SINES), copia and gypsy-like elements and retroviruses. Teo et al. (2005) used IRAP markers for identification and characterization of banana cultivars and classification of *Musa* genome constitutions. Ty1-copia-like retrotransposons have been extensively used as molecular marker in other plant species (Kalender et al. 1999; Pearce et al. 2000; Yu and Wise 2000). Balint-Kurti et al. (2000) suggested that Ty3-gypsy-like retrotransposons were introduced into *Musa* genus prior to the divergence of *Musa acuminata*, *M. balbisiana* and *M. vellutina* and they were able to distinguish the A and B genome using Ty3-gypsy-like retrotransposons. Aert et al. (2004) found highly homologous sequences to the "monkey" sequence from diploid *M. acuminata* Pisang Mas and *M. acuminata* Calcutta 4 and pointed out the relationship between A genomes in these three cultivars. Nair et al. (2005) used IRAP markers to identify the genome composition of bananas in India and identified a PCR-based molecular marker specific for the B genome. The B-specific band was absent in the AA and AAA cultivars (except in one AAA and AAB cultivar) but present in all other AB, AAB and ABB cultivars. Four cultivars with ABB genomes were identified as AAB, and one AAA cultivar showed a pattern similar to cultivars with the ABB genome. The authors suggested that the plants were misidentified.

4.4.4 Amplified Fragment Length Polymorphisms (AFLP)

AFLP is a DNA fingerprinting technique based on selective PCR amplification of DNA restriction fragments under stringent conditions (Vos et al. 1995). It can be used for DNA of any origin and complexity and is reported to be both reproducible and reliable (Vos et al. 1995). AFLP combines the reliability of RFLP with the power of PCR. AFLP is based on the selective amplification of restriction fragments obtained from the digestion of total genomic DNA. Given their dominant and biallelic nature, AFLP markers have been increasingly applied to various plants, mainly due to its capability of detecting a very high number of polymorphisms in a single assay, considerable repeatability and reasonable coverage of the genome (Vos et al. 1995; Cervera et al. 1998; Vuylsteke et al. 1999; Shim and Jørgensen 2000). AFLP has been used to detect the genetic diversity of both cultivated accessions and wild progenitors of *Musa* (Wong et al. 2001; Ude et al. 2002a, b, 2003; Opara et al. 2010). Unfortunately, the information content of these banding patterns is restricted, as they must initially be treated as dominant markers. However, when AFLP analysis is applied to large populations, they can be detected as codominant markers in a segregating population (Masiga and Turner 2004). Software has been developed to distinguish the nature of individuals (homozygotes or heterozygotes) on the basis of band intensity. AFLP assays are also technically demanding and expensive in that they require a number of DNA manipulations and a complex visualization procedure. In addition, they require relatively large amounts of reasonably high quality DNA as the use of poor quality DNA may lead to incomplete digestion, which can result in spurious polymorphisms. Using AFLP markers, Lheureux et al. (2003) found that 10 markers were co-segregating with the presence or absence of banana streak badnavirus infection in *Musa* hybrids. AFLPs and SSRs are now being used to identify markers for fruit parthenocarpy, dwarfism and apical dominance in banana and plantain. Microsatellite markers and AFLP analysis appear to be the most appropriate technologies for marker-assisted breeding in *Musa* (Crouch et al. 1999; Hautea et al. 2004).

An AFLP-based technique for surveying cytosine methylation at CCGG sites has been developed (Reyna-Lopez et al. 1997; Fraga and Esteller 2002). DNA methylation plays an essential role in regulating plant development through its influence on gene transcription, and is involved in a number of specific biological processes such as gene silencing or mobile element control (Finnegan et al. 1996). These methylation-based molecular markers are different from others regarding their genetics, transmission of characters, and the evolution/mutation process (Xiong et al. 1999; Cervera et al. 2002). These markers can target different genomic regions and help in understanding epigenetic processes, which could be involved in phenotypic diversity of plantain landraces. However, as no strong correlation exists between major phenotypic and genotypic traits (Crouch et al. 2000) and methylation diversity in plantain, the question of predictability of characters through crosses involving plantains as parents is still unresolved. Even under the hypothesis of methylation–phenotype correlation, evolution, transmission and control process of methylation is unclear and different from random mutation process (Cervera et al. 2002) has to be revealed. AFLP bands can also be sequenced and converted to codominant sequence characterized amplified region (SCAR) markers for PCR analysis for use with agarose gel electrophoresis (Bradeen and Simon 1998; Xu et al. 2001). These markers can also be used to screen pooled BAC libraries. AFLP has been shown to be a useful tool to generate linkage maps in *Musa* and has been used in the generation of the M53 (selfed) and the AFCAM (AFCAM refers to a F_2 population from a "Calcutta 4" x "Madang" cross) mapping populations.

4.4.5 Diversity Arrays Technology (DArT)

DArT is a DNA hybridization-based genotyping technology, which enables low-cost whole-genome profiling of crops without prior sequence information. DArT reduces the complexity of a representative sample (such as pooled DNA representing the diversity of *Musa*) using the principle that the genomic "representation" contains two types of fragments: constant fragments, found in any "representation" prepared from a DNA sample from an individual belonging to a given cultivar or species, and variable (polymorphic) fragments called molecular markers, only found in some but not all of the "representations" (www.diversityarray.com). DArT markers are biallelic and may be dominant (present or absent) or codominant (2 doses vs. 1 dose or absent). This approach enables quick profiling of segregating populations for quantitative trait loci (QTL) analysis with hundreds of polymorphic markers. Sequenced DArT markers can also be cheaply and rapidly anchored into BAC contigs, and therefore can facilitate the construction of high quality physical maps of the banana genome, which is a critical step in a sequencing project. Within the framework of the Generation Challenge Program, ca. 1,500 DArT markers have been developed using a wide array ("metagenome") of *Musa* accessions (Kilian 2007) and are now being used for the *Musa* framework map being developed at CIRAD additionally, 380 of these markers have recently been used in the construction of the BORLI map at CIRAD (Hippolyte et al. 2008, 2010). The application of DArT for bulked segregant analysis has been demonstrated in a *Hordeum* array (Wenzl et al. 2007). The disadvantage of this technology is that it depends on the availability of the array, a microarray printer and scanner, and computer infrastructure to analyze, store and manage the data produced. However, the markers are sequence ready and therefore if sequenced, can be developed for a PCR analysis using standard electrophoresis equipment.

4.4.6 Single Nucleotide Polymorphisms (SNPs)

On average, SNPs will occur in an organism's DNA more than 1% of the time (www.biotech.iastate.edu). Most SNPs are found outside the regions of genes of interest because about 3–5% of an organism's DNA codes for proteins (Sherry et al. 2001). SNPs found in a gene of interest are of particular interest to researchers because they are directly associated with a desired trait. Because of the recent advances in technology, SNPs are playing a greater role in selection and analysis of genetic traits. SNPs have now been added to the repertoire of useful markers, especially when they appear in coding regions and hence in the messages (so called expressed SNPs or eSNPs, or cSNPs for SNPs as part of a cDNA), since only then can they be expected to have an impact on protein function, and consequently on the phenotype of the organism (Brookes 1999). To date, not one report has appeared on detection of SNPs in the banana genome.

4.4.7 Ecotilling

Till et al. (2010) has applied ecotilling for discovering polymorphisms in the *Musa* genome. They provide the following account of ecotilling, which is a high-throughput method for the discovery and characterization of SNPs and small insertions/deletions (indels) (Till et al. 2010). It is an adaptation of the enzymatic mismatch cleavage and fluorescence detection methods originally developed for the targeting induced local lesions in genomes (TILLING) reverse-genetic strategy (Colbert et al. 2001; Comai et al. 2004). The technique was first described for *Arabidopsis* ecotypes (it was therefore named ecotilling). It is an accurate, low-cost and high-throughput method for the discovery and evaluation of nucleotide diversity and has been used in many organisms. About 700–1,600 bp gene target regions are amplified using gene-specific primers that are fluorescently labeled for ecotilling using enzymatic mismatch cleavage. After PCR, samples are denatured and annealed, and heteroduplexed molecules are created through the hybridization of polymorphic amplicons. Mismatched regions in otherwise double-stranded duplex are then cleaved using a crude extract of celery juice containing the single-strand specific nuclease CEL I. Cleaved products are resolved by denaturing polyacrylamide gel electrophoresis (PAGE) and observed by fluorescence detection (Till et al. 2006).

Denaturing PAGE provides base pair resolution allowing grouping of accessions based on shared banding patterns indicative of haplotype grouping (Comai et al. 2004). Sequence validation can be performed on only one or a small number of samples to provide base polymorphism data for the whole group, providing a savings in cost and informatics load over sequencing approaches. Alternatively, banding patterns alone can be used to evaluate genetic diversity and similarity between accessions on a gene-specific scale. When samples are screened alone, ecotilling provides a catalog of heterozygous nucleotide diversity between samples. Reference DNA can be added to each sample prior to screening to uncover homozygous polymorphisms. Additionally, the high sensitivity of the assay allows for pooling of multiple samples for the specific discovery of rare polymorphisms.

4.5 Efficiency of Molecular Marker Systems

In a molecular breeding study on *Musa*, Crouch et al. (1999) reported poor correlation between estimates of genetic similarities derived from different types of markers. The main reason is that data inconsistency stemmed from differences among molecular techniques that selectively screened complementary, but not overlapping, regions of the genome. Therefore, integration of genetic estimates from different molecular techniques was proposed to provide a clearer picture of *Musa* genetic relationship and generate highly accurate estimates of genetic similarity in germplasm analysis (Crouch et al. 1999; Wong et al. 2001). A better understanding of the population structure in *Musa* can, therefore, be obtained with simultaneous use of a different marker system, which will be very informative. Thus, it is necessary to utilize a range of marker systems in order to generate highly accurate estimates of genetic similarity in genetic diversity studies.

The relative advantages and disadvantages of the various molecular markers are summarized in Table 4-1.

4.6 Conclusion

New DNA technologies are constantly being developed especially for human genomic research and some of them are being used in plant research (Pillay et al. 2011). High-throughput technologies based on SNPs or small scale indel are efficient alternatives for traditional markers (RFLP, RAPD or AFLP) because of their greater abundance, high polymorphism, ease of measurement and ability to reveal hidden polymorphisms where other methods fail (Dillon et al. 2007). SNPs also allow easy and unambiguous identification of alleles or haplotypes.

A good marker system for polyploid crops should be dosage sensitive and have the ability to distinguish heterozygous genotypes with multiple haplotypes (de Koeyer et al. 2010). High resolution DNA melting (HRM) analysis has been shown to have several advantages over other genotyping methods (Montgomery et al. 2007; Reed et al. 2007; Erali et al. 2008). The advantages include a short analysis time and the absence of post-PCR sample processing or separation (de Koeyer et al. 2010). The three ways in which HRM can be used for genotyping and or variant screening are discussed in de Koeyer et al. (2010). HRM has been used in many crops and holds great potential for cultivar identification especially in polyploids, mapping, polymorphism discovery, mapping candidate genes, and in combination with QTL or association studies for identifying genomic regions involved in important traits. Such techniques have not been utilized in *Musa* as yet and eagerly awaited by scientists.

Table 4-1 Molecular markers, level of polymorphism and factors influencing their use (from Pillay and Tripathi 2007).

Markers	Quantity of DNA required	Level of polymorphism	Locus specificity	Reproducibility	Labor intensity	Operational costs	Development costs	Amenability to automation
Isozymes	-	Low	Yes	High	Low	Low	Low	No
RFLP	High	Medium	Yes	High	High	High	Medium-high	No
Minisatellite	High	High	No/yes	High	High	High	Medium-high	No
RAPD	Low	Medium	No	Low	Low	Low	Low	Yes
SSR	Low	High	Yes	High	Low	Low-medium	High	Yes
SNPs	Low	Low	Yes	Medium	Low-medium	Low-medium	Low	No
AFLP	Medium	Medium	No	High	Medium	Medium	Medium	Yes

Source: Pillay and Tripathi (2007)

Abbreviations

6-PGDH	:	6-phosphogluconate dehydrogenase
ADH	:	alcohol dehydrogenase
AFLP	:	amplified fragment length polymorphisms
BAC	:	bacterial artificial chromosome
BBrMV	:	Banana Bract Mosiac Virus
CIRAD	:	Centre de Coopération Internationale en Recherche Agronomique pour le Dévelopement
EST	:	esterase
GDH	:	galactose dehydrogenase
GOT	:	glutamate oxaloacetate transminase
GUDH	:	glutamate dehydrogenase
LINES	:	long interspersed elements
MAS	:	marker-assisted selection
MDH	:	malate dehydrogenase
ME	:	malic enzyme
PAGE	:	polyacrylamide gel electrophoresis
PCR	:	polymerase chain reaction
PGI	:	phosphoglucose isomerase
PGM	:	phosphoglucomutase
PRX	:	peroxidase
QTL	:	quantitative trait loci
RAPD	:	random amplified polymorphic DNA
RFLP	:	restriction fragment length polymorphisms
SCAR	:	sequence characterized amplified repeat
SDH	:	sorbitol dehydrogenase
SINES	:	short interspersed elements
SKDH	:	shikimate dehydrogenase
SSR	:	simple sequence repeats
SUDH	:	succinate dehydrogenase
TILLING	:	Targeting Induced Local Lesions IN Genomes
TO	:	tetrazolium oxidase
TPI	:	triosephosphate isomerase
VNTR	:	variable number tandem repeats

References

Aert R, Volckaert G, Sagi L (2004) Gene content and density in banana (*Musa acuminata*) as revealed by genomic sequencing of BAC clones. Theor Appl Genet 109: 129–139.

Bairu MW, Fennell CW, van Staden J (2006) The effect of plant growth regulators on somaclonal variation in Cavendish banana (*Musa* AAA cv. `Zelig'). Sci Hort 108: 347–351.

Bairu MW, Stirk WA, Dolezal K, Staden J (2008) The role of topolins in micropropagation and somaclonal variation of banana cultivars 'Williams' and Grand Naine (*Musa* spp. AAA). Plant Cell Tiss Org Cult 95: 373–379.

Balint-Kurti PJ, Clendennen SK, Dolezelova M, Valarik M, Dolezel J, Beetham PR, May GD (2000) Identification and chromosomal localization of the monkey retrotransposon in *Musa* sp. Mol Gen Genet 263: 908–915.
Barone A (2004) Molecular marker-assisted selection for potato breeding. Amer J Potato Res 81: 111–117.
Bennici A, Anzidei M, Vendramin GG (2004) Genetic stability and uniformity of *Foeniculum vulgare* Mill. regenerated plants through organogenesis and somatic embryogenesis. Plant Sci 166: 221–227.
Bhat KV, Jarret RL (1995) Random amplified polymorphic DNA and genetic diversity in Indian *Musa* germplasm. Genet Resour Crop Evol 42: 107–118.
Bhat KV, Bhat SR, Chandel KPS (1992a) Survey of isozyme polymorphism for clonal identification in *Musa*. II. Peroxidase, superoxide dismutase, shikimate dehydrogenase and malate dehydrogenase. J Hort Sci 67: 737–743.
Bhat KV, Bhat SR, Chandel KPS (1992b) Survey of isozyme polymorphism for clonal identification in *Musa*. I. Esterase, acid phosphatase and catalase. J Hort Sci 61: 507–507.
Bonner JW, Warner RM, Brewbaker JL (1974) A chemosystematic study of *Musa* cultivars. HortScience 9: 325–328.
Bradeen JM, Simon PW (1998) Conversion of an AFLP fragment linked to the carrot Y 2 locus to a simple, codominant, PCR-based marker form. Theor Appl Genet 97: 960–967.
Brettell RIS, Dennis ES, Scowcroft WR, Peacock WJ (1986) Molecular analysis of a somaclonal mutant of maize alcohol dehydrogenase. Mol Gen Genet 202: 235–239.
Brookes AJ (1999) The essence of SNPs. Gene 234: 177–186.
Buard J, Vergnaud G (1994) Complex recombination events at the hypermutable minisatellite CEB1 (D2S90). The EMBO Journal 13: 3203–3210.
Buhariwalla HK, Jarret RL, Jayashree B, Crouch JH, Ortiz R (2005) Isolation and characterization of microsatellite markers from *Musa balbisiana*. Mol Ecol Notes 5: 327–330.
Campbell D, Duchesne P, Bernatchez L (2003) AFLP utility for population assignment studies: analytical investigation and empirical comparison with microsatellites. Mol Ecol 12: 1979–1991.
Carreel F, Gonzalez de Leon D, Lagoda P, Lanaud C, Jenny C, Horry JP, Tezenas du Montcel H (2002) Ascertaining maternal and paternal lineage within *Musa* by chloroplast and mitochondrial DNA RFLP analyses. Genome 45: 679–692.
Carvalho LC, Goulao L, Oliveira C, Goncalves JC, Amancio S (2004) RAPD Assessment for Identification of Clonal Identity and Genetic Stability of *in vitro* Propagated Chestnut Hybrids. Plant Cell, Tiss Org Cult 77: 23–27.
Cervera MT, Cabezas JA, Sancha JC, de Toda Martınez F, Martınez-Zapater JM (1998) Application of AFLPs to the characterization of grapevine *Vitis vinifera* L. genetic resources. A case study with accessions from Rioja (Spain). Theor Appl Genet 97: 51–59.
Cervera MT, Ruiz-Garcia L, Martanez-Zapater JM (2002) Analysis of DNA methylation in Arabidopsis thaliana based on methylation-sensitive AFLP markers. Mol Genet Genom 268: 543–552.
Cheung F, Town CD (2007) A BAC end view of the *Musa acuminata* genome. BMC Plant Biol 7: 29–29.
Colbert T, Till BJ, Tompa R, Reynolds S, Steine MN, Yeung AT, McCallum CM, Comai L, Henikoff S (2001) High-Throughput Screening for Induced Point Mutations. Plant Physiol 126: 480–484.
Comai L, Young K, Till BJ, Reynolds SH, Greene EA, Codomo CA, Enns LC, Johnson JE, Burtner C, Odden AR, Henikoff S (2004) Efficient discovery of DNA polymorphisms in natural populations by Ecotilling. Plant J 37: 778–786.
Creste S, Neto TA, Silva SO, Figueira A (2003a) Genetic characterization of banana cultivars (*Musa* spp.) from Brazil using microsatellite markers. Euphytica 132: 259–268
Creste C, Neto TA, Sebastino OS, Figuira A (2003b) Genetic characterization of banana cultivars (*Musa* spp.) from Brazil using microsatellite markers. Euphytica 132: 259–268.

Crouch HK, Crouch JH, Jarret RL, Cregan PB, Ortiz R (1998) Segregation at microsatellite loci in haploid and diploid gametes of *Musa*. Crop Science 38: 211–217.

Crouch JH, Crouch HK, Tenkouano A, Ortiz R (1999a) VNTR-based analysis of 2x and 4x full-sib Musa hybrids. Elect J Biotechnol 2: 99–108.

Crouch JH, Crouch HK, Constand H, Van Gysel A, Breyne P, Van Montagu M, Jarret RL, Ortiz R (1999b) Comparison of PCR-based molecular marker analysis of Musa Breeding populations. Mol Breed 5: 233–244.

Crouch JH, Ortiz R, Crouch HK (2000) Utilization of molecular genetic techniques in support of plantain and banana improvement. Acta Hort 540: 185–191.

Damasco OP, Graham GC, Henry RJ, Adkins SW, Smiths MK, Godwin ID (1996) Random amplified polymorphic DNA (RAPD) detection of dwarf off-types in micropropagated Cavendish (*Musa* spp. AAA) bananas. Plant Cell Reports 16: 118–123.

Das BK, Jena RC, Samal KC (2009) Optimization of DNA isolation and PCR protocol for RAPD analysis of banana/plantain (*Musa* spp.). Int J Agric Sci 1: 21–25.

De Koeyer D, Douglass K, Murphy A, Whitney S, Nolan L, Song Y, De Jong W (2010) Application of high-resolution DNA melting for genotyping and variant scanning of diploid and autotetraploid potato. Mol Breed 25: 67–90.

Dhanya MK, Rajagopalan B, Umamaheswaran K, Ayisha R (2006) Isozyme variation in banana (*Musa* sp.) in response tobract mosaic virus infection. Indian J Crop Sci 1: 140–141.

Dillon SL, Shapter FM, Henry RJ, Cordeiro G, Izquierdo L, Lee LS (2007) Domestication to crop improvement: genetic resources for *Sorghum* and *Saccharum* (Andropogoneae). Ann Bot 100: 975–989.

Drew RA (1997) The Application of Biotechnology to the Conservation and Improvement of Tropical and Subtropical Fruit Species. Food and Agriculture Organization of the United Nations, Rome, Italy.

Erali M, Voelkerding KV, Wittwer CT (2008) High resolution melting applications for clinical laboratory medicine. Exp Mol Pathol 85: 50–58.

Espino RRC, Pimentel RB (1990) Electrophoretic analysis of selected isozymes in BB cultivars of Philippine bananas. In: RL Jarret (ed) Identification of Genetic Diversity in the Genus *Musa* INIBAP, Los Banos, the Philippines.

Faure S, Noyer JL, Horry JP, Bakry F, Lanaud C, Goazalez de Lean D (1993) A molecular marker-based linkage map of diploid bananas (*Musa acuminata*). Theor Appl Genet 87: 517–526.

Finalet J, Peteira B, Leon O, Ventura JCM, Lopez J (2000) Application of RAPD markers to the differentiation of *Musa* spp. mutants. Revista de Proteccion Vegetal 15: 60–63.

Finnegan EJ, Peacock WJ, Dennis ES (1996) Reduced DNA methylation in *Arabidopsis thaliana* results in abnormal plant development. Proc Natl Acad Sci USA 93: 8449–8554.

Fraga MF, Esteller M (2002) DNA methylation: a profile of methods and applications. BioTechniques 33: 632–634.

Gawel N, Jarret RL (1991a) Cytoplasmic genetic diversity in bananas and plantains. Euphytica 52: 19–23.

Gawel NJ, Jarret RL (1991b) Chloroplast DNA restriction fragment length polymorphisms (RFLPs) in *Musa* species. Theor Appl Genet 81: 783–786.

Gawel NJ, Jarret RL, Whittemore AP (1992) Restriction fragment length polymorphism (RFLP)-based phylogenetic analysis of *Musa*. Theor Appl Genet 84: 286–290.

Grapin A, Noyer JL, Carreel F, Dambier D, Baurens FC, Lanaud C, Lagoda PJ (1998) Diploid *Musa acuminata* genetic diversity assayed with sequence-tagged microsatellite sites. Electrophoresis 19: 1374–1380.

Gupta M, Chyi YS, Romero-Severson J, Owen JL (1994) Amplification of DNA markers from evolutionarily diverse genomes using single primers of simple-sequence repeats. Theor Appl Genet 89: 998–1006.

Hautea DM, Molina GC, Balatero CH, Coronado NB, Perez EB, Alvarez MTH, Canama AO, Akuba RH, Quilloy RB, Frankie RB, Caspillo CS (2004) Analysis of induced mutants of Philippine with molecular markers In: S Jain, R Swennen (eds) Banana Improvement:

Cellular, Molecular Biology and Induced Mutations. Kluwer Academic Publishers, Dordrecht, The Netherlands, pp 45–57.
Heath DD, Lwama GK, Devlin RH (1993) PCR primed with VNTR core sequences yields species specific patterns and hypervariable probes. Nucl Acid Res 21: 5782–5785.
Heslop-Harrison JS (2000) Comparative Genome Organization in Plants: From Sequence and Markers to Chromatin and Chromosomes. Plant Cell 12: 617–636.
Hippolyte I, Seguin M, Bakry F, Gardes L, Baurens FC, Miller RN, Khan IA, Jenny C, Carreel F, Huttner E, Perrier X, Kilian A, Risterucci A-M, Glaszmann J-C (2008) *Musa* Genetic Mapping. XVI International Plant and Animal Genome Conference (PAG), San Diego, CA, USA.
Hippolyte I, Bakry F, Seguin M, Gardes L, Rivallan R, Risterucci A-M, Jenny C, Perrier X, Carreel F, Argout X, Piffanelli P, Khan I, Miller R, Pappas G, Mbeguie-A-Mbeguie D, Matsumoto T, De Bernardinis V, Huttner E, Kilian A, Baurens F-C, D'Hont A, Cote F, Courtois B, Glaszmann J-C (2010) A saturated SSR/DArT linkage map of *Musa acuminata* addressing genome rearrangements among bananas. BMC Plant Biol 10: 65–65.
Howell EC, Newbury HJ, Swennen RL, Withers LA, Ford-Lloyd BV (1994) The use of RAPD for identifying and classifying *Musa* germplasm. Genome 37: 328–332.
Imelda M, Estiati A, Hartati NS (2001) Induction of mutation through gamma radiation in three cultivars of banana. Ann Bogorien 7: 75–82.
Jain PK, Saini MK, Pathak H, Gupta VK (2007) Analysis of genetic variation in different banana (*Musa* species) variety using random amplified polymorphic DNAs (RAPDs). Afri J Biotechnol 6 1987–1989.
Jarret RL, Litz RE (1986a) Enzyme polymorphism in *Musa acuminata* Colla. J Hered 77: 183–188.
Jarret RL, Litz RE (1986b) Isozymes as genetic markers in bananas and plantains. Euphytica 35: 539–549.
Jarret R, Gawel N, Whittemore A, Sharrock S (1992) RFLP-based phylogeny of *Musa* species in Papua New Guinea. Theor Appl Genet 84: 579–584.
Jarret RL, Bhat KV, Cregan PB, Ortiz R, Vuylsteke D (1994) Isolation of microsatellite DNA markers in *Musa*. InfoMusa 3: 3–4.
Kaemmer D, Fischer D, Jarret RL, Baurens FC, Grapin A, Dambier D, Noyer JL, Lanaud C, Kahl G, Lagoda PJL (1997) Molecular breeding in the genus *Musa*: a strong case for STMS marker technology. Euphytica 96: 49–63.
Kalendar R, Grob T, Regina M, Suoniemi A, Schulman A (1999) IRAP and REMAP: two new retrotransposon-based DNA fingerprinting techniques. Theor Appl Genet 98: 704–711.
Khatri A, Dahot MU, Khan I, Raza S, Bibi S, Yasmin S, Nizamani DGS (2009) Use of RAPD for the assessment of genetic diversity among exotic and commercial banana clones. Pak J Bot 41: 2995–2999.
Kilian A (2007) Towards Effective Deployment of Diversity Arrays Technology (DArT) in Banana Genomics and Sequencing. Plant and Animal Genomes XV Conference, January 13–17, San Diego, USA.
Kumar LS (1999) DNA markers in plant improvement: an overview. Biotechnol Adv 17: 143–182.
Lakshmanan V,Venkataramareddy SR, Neelwarne B (2007) Molecular analysis of genetic stability in long-term micropropagated shoots of banana using RAPD and ISSR markers. Elec J Biotechnol 10: 106–113.
Lefebvre V, Palloix A, Caranta C, Pochard E (1995) Construction of an intraspecific integrated linkage map of pepper using molecular markers and doubled-haploid progenies. Genome 38: 112–121.
Lheureux F, Carreel F, Jenny C, Lockhart BEL, Iskra C (2003) Identification of genetic markers linked to banana streak disease expression in inter-specific *Musa* hybrids. Theor Appl Genet 106: 594–598.
Loh JP, Kiew R, Set O, Gan LH, Gan Y-Y (2000) Amplified Fragment Length Polymorphism Fingerprinting of 16 banana cultivars (*Musa* cvs.). Mol Phylogenet Evol 17: 360–366.

Martins M, Sarmento D, Oliveira MM (2004) Genetic stability of micropropagated almond plantlets, as assessed by RAPD and ISSR markers. Plant Cell Rep 23: 492–496.

Masiga DK, Turner CMR (2004) Amplified (restriction) fragment length polymorphism (AFLP) analysis. In: SE Melville (ed) Methods in Molecular Biology Humana Press, USA, pp 173–186.

Megia R, Hadisunars, Sulistyaningsih YC, Djuitai NR (2001) Isozyme polymorphisms for cultivar identification in Indonesian banana. Hayati 8: 81–85.

Miller RNG, Passos MAN, Menezes NNP, Souza Jr. MT, Mota M, Azevedo VCR, Amorim EP, Pappas Jr. GJ , Ciampi AY (2010) Characterization of novel microsatellite markers in *Musa acuminata* subsp. *burmannicoides*, var. Calcutta 4. BMC Res. Notes 3: 149.

Miri SM, Mousavi A, Naghavi MR, Mirzaii M, Talaei AR, Khianabi BN (2009) Analysis of induced mutants of salinity resistant banana (*Musa acuminata* cv. Dwarf Cavendish) using morphological and molecular markers. Iran J Biotechnol 7: 86–92.

Montgomery J, Wittwer CT, Palais R, Zhou L (2007) Simultaneous mutation scanning and genotyping by high-resolution DNA melting analysis. Nature Protocols 2: 59–66.

Muhammad Y, James AC, Rivera-Madrid R, Ortiz R, Escobedo-Graciamedrano RM (2010) *Musa* Genetic Diversity Revealed by SRAP and AFLP. Mol Biotechnol. (August). doi:10.1007/s12033-010-9328-8.

Nair AS, Teo CH, Schwarzacher T, Harrison PH (2005) Genome classification of banana cultivars from South India using IRAP markers. Euphytica 144: 285–290.

Nsabimana A, Van Staden J (2007) Assessment of genetic diversity of highland bananas from the National Banana Germplasm Collection at Rubona, Rwanda using RAPD markers. Sci Hort 113: 293–299.

Nwakanma DC, Pillay M, Okoli BE, Tenkouano A (2003) PCR-RFLP of the ribosomal DNA internal transcribed spacers (ITS) provides markers for the A and B genomes in *Musa* L. Theor Appl Genet 108: 154–159.

Opara UL, Jacobson D, Al-Saady NA (2010) Analysis of genetic diversity in banana cultivars (*Musa* cvs.) from the South of Oman using AFLP markers and classification by phylogenetic, hierarchical clustering and principal component analyses. J Zhejiang Univ Sci 11: 332–341.

Orita M, Suzuki Y, Sekiya T, Hayashi K (1989) Rapid and sensitive detection of point mutations and DNA polymorphisms using the polymerase chain reaction. Genomics 5: 874–879.

Oselebe HO, Tenkouano A, Pillay M, Obi UI, Uguru MI (2006) Ploidy and genome segregation in *Musa* breeding populations assessed by flow cytometry and randomly amplified polymorphic DNA markers. HortScience 131: 780–786.

Paterson AH, Damon S, Hewitt JD, Zamir D, Rabinowitch HD, Lincoln SE, Lander ES, Tanksley SD (1991) Mendelian Factors Underlying Quantitative Traits in Tomato: Comparison across Species, Generations, and Environments. Genetics 127: 181–197.

Pearce SR, Knox M, Ellis TNM, Flavell AJ, Kumar A (2000) Pea *Tyl-copia* group retrotransposon: Transitional activity and use as makers to study genetic diversity in *Pisum*. Mol Genet Genomics 263: 898–907.

Pillay M, Tripathi L (2007) Banana breeding. In: MS Kang, PM Priyadarshan (eds) Breeding Major Food Staples. Blackwell Science, Boston, USA, Asia, pp 393–428.

Pillay M, Nwakanma DC, Tenkouano A (2000) Identification of RAPD markers linked to A and B genome sequences in *Musa* L. Genome 43: 763–767.

Pillay M, Ogundiwin E, Nwakanma DC, Ude G, Tenkouano A (2001) Analysis of genetic diversity and relationships in East African banana germplasm. Theor Appl Genet 102: 965–970.

Pillay M, Ogundiwin E, Tenkouano A, Dolezel J (2006) Ploidy and genome composition of *Musa* germplasm at the International Institute of Tropical Agriculture (IITA). Afr J Biotechnol 5: 1224–1232.

Pillay M, Tenkouano A, Ortiz R (2011) Molecular breeding of other vegetatively propagated crops: lessons for banana. In: M Pillay, A Tenkouano (eds) Banana Breeding: Progress and Challenges. CRC Press Boca Raton, FL, USA, pp 321–350.

Ramage CM, Borda AM, Hamill SD, Smith MK (2004) A simplified PCR test for early detection of dwarf off-types in micropropagated Cavendish bananas (*Musa* spp. AAA). Sci Hort 103: 145–151.
Ray T, Dutta I, Saha P, Das S, Roy SC (2006) Genetic stability of three economically important micropropagated banana (*Musa* spp.) cultivars of lower Indo-Gangetic plains, as assessed by RAPD and ISSR markers. Plant Cell Tiss Org Cult 85: 11–21.
Reed GH, Kent JO, Wittwer CT (2007) High-resolution DNA melting analysis for simple and efficient molecular diagnostics. Pharmacogenomics 8: 597–608.
Reyna-Lopez GE, Simpson J, Ruiz-Herrera J (1997) Differences in DNA methylation patterns are detectable during the dimorphic transition of fungi by amplification of restriction polymorphisms. Mol Gen Genet 253: 703–710.
Richardson T, Cato S, Ramser J, Kahl G, Weising K (1995) Hybridization of microsatellites to RAPD: a new source of polymorphic markers. Nucl Acid Res 23: 3798–3799.
Rivera FN (1983) Protein and isoenzyme banding patterns among Philippine cooking bananas and their wild parents (*Musa* species). Paradisiaca 6: 7–12.
Rogstad SH (1993) Surveying plant genomes for variable number of tandem repeat loci. Meth Enzymol 224: 278–294.
Sambrook J, Fritsch EF, Maniatis T (1989) Molecular cloning—A laboratory course manual. Cold Spring Harbor Laboratory Press, Cold Spring Harbor, New York. USA.
Sherry ST, Ward MH, Kholodov M, Baker J, Phan L, Smigielski EM, Sirotkin K (2001) dbSNP: the NCBI database of genetic variation. Nucl Acid Res 29: 308–311.
Shim SI, Jorgensen RB (2000) Genetic structure in cultivated and wild carrots (*Daucus carota* L.) revealed by AFLP analysis. Theor Appl Genet 101: 227–233.
Simmonds NW (1966) Bananas. 2nd edn. Longmans, London, UK.
Sitthipron S, Anutalabhochai S, Dumampai N, Thakumphu B, Dasanonda M (2005) Investigation of genetic relationships and hybrid detection in Longan by high annealing temperature RAPD. Acta Hort 665: 161–169.
Staub JE, Kuhns LJ, May B, Grun P (1982) Stability of potato tuber isozymes under different storage regimes. J Amer Society Hort Sci 107: 405–408.
Staub JE, Sequen FC, Gupta M (1996) Genetic markers, map construction, and their application in plant breeding. HortScience 31: 742–762.
Tansley SD, Orton TJ (1983) Isozymes in Plant Genetics and Breeding, part A and B. Elsevier, Amsterdam, The Netherlands.
Teo CH, Tan SH, Ho CL, Faridah QZ, Othman YR, Heslop-Harrison JS, Kalendar R, Schulman AH (2005) Genome constitution and classification using retrotransposon-based markers in the orphan crop banana. J Plant Biol 48: 96–105.
Till BJ, Zerr T, Comai L, Henikoff S (2006) A protocol for TILLING and Ecotilling in plants and animals. Nat Protocols 1: 2465–2477.
Till BJ, Jankowicz-Cieslak J, Sagi L, Huynh OA, Utsushi H, Swennen R, Terauchi R, Mba C (2010) Discovery of nucleotide polymorphisms in the *Musa* gene pool by Ecotilling. Theor Appl Genet 121: 1381–1389.
Toruan-Mathius N, Haris N (1999) Induction of genetic variation of banana cv. Nangka by gamma Co-60 irradiation and fusaric acids. Menara Perkebunan 67: 13–22.
Tourmente S, Deragon JM, Lafleuriel J, Tutois S, Pelissier T, Cuvillier C, Espagnol MC, Picard G (1994) Characterization of minisatellites in *Arabidopsis thaliana* with sequence similarity to the human minisatellite core sequence. Nucl Acid Res 22: 3317–3321.
Ude G, Pillay M, Nwakanma D, Tenkouano A (2002a) Analysis of genetic diversity and sectional relationships in *Musa* using AFLP markers. Theor Appl Genet 104: 1239–1245.
Ude G, Pillay M, Nwakanma D, Tenkouano A (2002b) Genetic Diversity in *Musa acuminata* Colla and *Musa balbisiana* Colla and some of their natural hybrids using AFLP Markers. Theor Appl Genet 104: 1246–1252.
Ude G, Pillay M, Ogundiwin E, Tenkouano A (2003) Genetic diversity in an African plantain core collection using AFLP and RAPD markers. Theor Appl Genet 107: 248–255.

Uma S, Siva SA, Saraswathi MS, Manickavasagam M, Durai P, Selvarajan R, Sathiamoorthy S (2006) Variation and intraspecific relationships in Indian wild *Musa balbisiana* (BB) population as evidenced by Random Amplified Polymorphic DNA. Genet Resour Crop Evol 53: 349–355.

Vos P, Hogers R, Bleeker M, Reijans M, van de Lee T, Hornes M, Frijters A, Pot J, Peleman J, Kuiper M (1995) AFLP: a new technique for DNA fingerprinting. Nucl Acid Res 23: 4407–4414.

Vuylsteke M, Mank R, Antonise R, Bastiaans E, Senior ML, Stuber CW, Melchinger AE, Lubberstedt T, Xia XC, Stam P, Zabeau M, Kuiper M (1999) Two high-density AFLP linkage maps of *Zea mays* L.: analysis of distribution of AFLP markers. Theor Appl Genet 99: 921–935.

Wang JY, Zheng LS, Huang BZ, Liu WL, Wu YT (2009) Development, characterization, and variability analysis of microsatellites from a commercial cultivar of *Musa acuminata*. Genet Resour Crop Evol 57: 553–563.

Weising K, Weigand F, Driesel AJ, Kahl G, Zischler H, Epplen JT (1989) Polymorphic simple GATA/GACA repeats in plant genomes. Nucl Acid Res 17: 10128.

Weising K, Atkinson RG, Gardner RC (1995) Genomic fingerprinting by microsatellite-primed PCR: a critical evaluation. Genome Res 4: 249–255.

Wenzl P, Raman H, Wang J, Zhou M, Huttner E, Kilian A (2007) A DArT platform for quantitative bulked segregant analysis. BMC Genomics 8: 196–196.

Wong C, Kiew R, Loh JP, Gan LH, Set O, Lee SK, Lum S, Gan YY (2001) Genetic diversity of the wild banana *Musa acuminata* Colla in Malaysia as evidenced by AFLP. Ann Bot 88: 1017–1025.

Xiong LZ, Xu CG, Saghai Maroof MA, Zhang Q (1999) Patterns of cytosine methylation in an elite rice hybrid and its parental lines, detected by a methylation-sensitive amplification polymorphism technique. Mol Gen Genet 261: 439–446.

Xu M, Huaracha E, Korban SS (2001) Development of sequence-characterized amplified regions (SCARs) from amplified fragment length polymorphism (AFLP) markers tightly linked to the *Vf* gene in apple. Genome 44: 63–70.

Youssef M, James AC, Rivera-Madrid R, Ortiz R, Escobedo-GraciaMedrano RM (2010) *Musa* genetic diversity revealed by SRAP and AFLP. Mol Biotechnol 47: 189–199.

Yu GX, Wise RP (2000) An anchored AFLP- and retrotransposon-based map of diploid *Avena*. Genome 43: 736–749.

Zhou Z, Bebeli PJ, Somers DJ, Gustafson JP (1997) Direct amplification of minisatellite-region DNA with VNTR core sequences in the genus *Oryza*. Theor Appl Genet 95: 942–949.

5

Molecular Linkage Maps: Strategies, Resources and Achievements

Cory Johnson[1,a] and *Christopher Cullis*[1,b,*]

ABSTRACT

The origin of genetic mapping lies in the discoveries by Mendel that the inheritance of phenotypic traits could be predicted. For the next 70 years mapping efforts were carried out using phenotypic characteristics. The combination of many of the characters within a single pair of parents was frequently impossible since one trait masked or interfered with the phenotypic scoring of others. The use of DNA-based markers removed these restrictions since most of the variants were silent as far as the phenotype of the individual, therefore an unlimited number of markers could be assayed in a single cross. DNA based markers have since then revolutionized the whole process of generating genetic maps. This chapter highlights the strategies, resources and achievements in molecular linkage mapping.

Keywords: markers, linkage maps, mapping

5.1 Introduction

The origin of genetic mapping lies in the discoveries by Mendel that the inheritance of phenotypic traits could be predicted. The original characteristics in pea that Mendel followed all segregated independently

[1]Department of Biology, Case Western Reserve University, 10900 Euclid Avenue, Cleveland, Ohio 44106-7080.
[a]e-mail: *clj18@case.edu*
[b]e-mail: *cac5@case.edu*
*Corresponding author

List of abbreviations after the text.

so that the rules of inheritance that Mendel developed were followed. The first actual genetic map was published in 1913 (Sturtevant 1913) with six sex-linked genes in *Drosophila*. The important aspects of genetic mapping that were identified in this paper were the concepts of the difference between genetic and physical distances between genes and the concept of interference—the presence of one recombination event affecting the possibility of another crossover nearby. For the next 70 years mapping efforts were carried out using phenotypic characteristics. In plants this has frequently involved color variation and easily scored morphological markers. However, the combination of many of the characters within a single pair of parents was frequently impossible since one trait masked or interfered with the phenotypic scoring of others. Therefore genetic maps were compiled from data of many crosses involving different parents. Additionally, the alternative forms of many of the traits were the result of mutations. The use of DNA-based markers removed these restrictions since most of the variants were silent as far as the phenotype of the individual so an unlimited number of markers could be assayed in a single cross.

DNA-based markers have since revolutionized the whole process of generating genetic maps , for the first time, a large number of loci could be followed in a single segregating population. The range of genetic markers that are available includes restriction fragment length polymorphism (RFLP) (Botstein et al. 1980), random amplified polymorphic DNA (RAPD) (Williams et al. 1990), amplified fragment length polymorphisms (AFLP) (Vos et al. 1995), single-nucleotide polymorphisms (SNPs), and simple sequence repeats (SSRs) or microsatellites (Senior and Heun 1993). In every case, after a polymorphism is identified its usefulness will depend on how frequently variation is observed. Therefore if a particular region has three polymorphic states, but one of them is present 99% of the time, then it will not be a useful polymorphism as two individuals are unlikely to differ in this region. The usefulness of polymorphic locus is best described by the polymorphism information content (PIC). This is the probability of any two individuals differing for this locus. Thus, if the PIC is 0.5 then there is a 50% chance that any two individuals will differ at this locus. Identifying polymorphisms is only the first step in constructing a molecular map; identifying those that will segregate in the germplasm being used to construct the map is equally important (Table 5-1). An additional consideration is to ascertain that this germplasm does not only contain very different individuals as the applicability of the markers may be limited when applications of the map are being considered.

The effectiveness of all mapping efforts depends on the accuracy of the typing of the individuals. The precise molecular and physical characterization is essential to get the correct relationships between markers

Table 5-1 Genetic segregation ratios at marker loci in different mapping populations.

Marker	Type	Genetic segregation ratio				
		F2	RIL	DH	NIL	BC1
RFLP	Co-dominant	1:2:1	1:1	1:1	1:1	1:1
RAPD	Dominant	3:1	1:1	1:1	1:1	1:0
AFLP	Dominant	3:1	1:1	1:1	1:1	1:0
SSR	Co-dominant	1:2:1	1:1	1:1	1:1	1:1
SNP	Co-dominant	1:2:1	1:1	1:1	1:1	1:1

and genes. If an individual is incorrectly entered into the mapping software it will skew the data and will be very difficult to subsequently find. Therefore accuracy in scoring and data entry is a priority.

5.2 Evolution of Marker Types: RFLPs to SNPs

The development of molecular tools resulted in the ability to use them to distinguish between individuals. The tools included the characterization of isozymes, the identification of distinctive secondary products and finally the use of DNA fragments. It is this latter class that will be the focus of this chapter.

5.2.1 Restriction Fragment Length Polymorphism

The first of the DNA markers was based on restriction fragment length polymorphism (RFLP).

As the name implies these are based on variations that are identified through the use of restriction enzymes. As shown in Fig. 5-1, the polymorphisms can be the result of a single base change in the restriction site or by either an insertion or a deletion between adjacent restriction sites. These have traditionally been assayed by hybridization with single (low) copy probes. If the polymorphism was generated by the loss of a restriction site then two outcomes are possible. If the probe is confined to a region of the genomic DNA completely within one of the restriction fragments then a larger band would be observed (probe y, Fig. 5-1). However, if the probe spanned the altered restriction site (probe x, Fig. 5-1) then a new band with a size of the sum of the other two bands would be observed. When an insertion or deletion is the cause of the RFLP then no matter what probe is used only a single band is identified with different sizes in various individuals (assuming they are homozygous). Another indication that the RFLP is a result of an insertion or deletion is that an RFLP is identified with a single probe with multiple restriction enzymes.

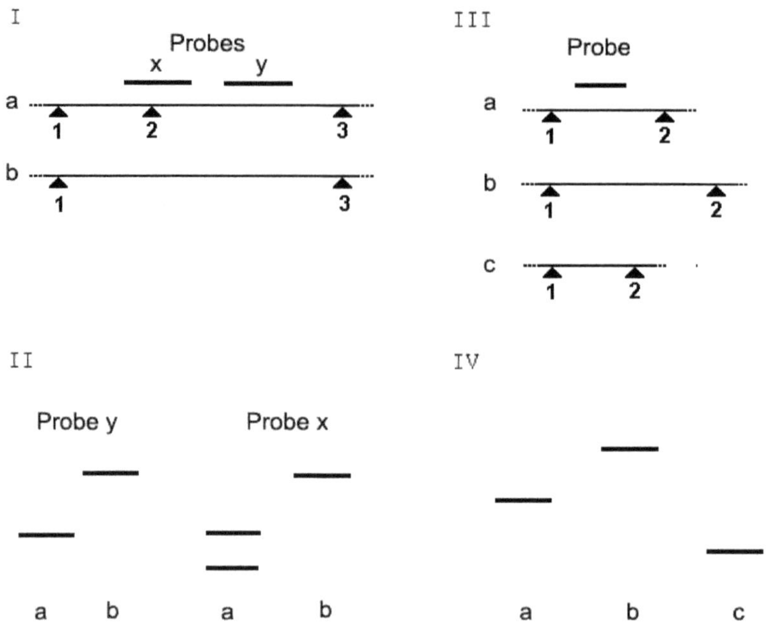

Figure 5-1 The various outcomes of hybridization with different probes to identify the basis of various RFLPs.
I. An RFLP caused by a loss of a restrictions site. II. The banding patterns after Southern blotting and hybridization with probes X and Y.
II. RFLPs caused by an insertion (b) or a deletion (c) in the original allele (a). IV. The banding patterns after southern blotting and hybridization with the probe for these three alleles.

Since traditional RFLPs are identified through Southern blots and hybridization they are time-consuming and require large amounts of starting DNA. Therefore they are now rarely used in the standard form. These have essentially been superseded by polymerase chain reaction (PCR) amplification methods such as random amplified polymorphic DNA (RAPD), amplified fragment length polymorphism (AFLP) and simple sequence repeat (SSR) markers, and these are frequently RFLPs assayed by an alternative method.

5.2.2 Random Amplified Polymorphic DNA

The availability of the polymerase chain reaction (PCR) opened the door to many applications. However, the drawback to PCR was that sequence information was required for the primers used in the reaction. The statistical calculation for the presence of a 10 bp (base pair) sequence is that it should appear once every 10^6 nucleotides. PCR amplification requires a primer at each end of the amplified fragment. Therefore a single 10 base primer present in reverse orientation with 50 to 2,000 bp between the two sites would occur with a probability of $5x10^{-8}$. Since most plant genomes are greater than this amplifying genomic DNA with a single decamer should give a small number of bands and was confirmed to be very successful in identifying large numbers of polymorphisms (Williams et al. 1990). The methodology can therefore be used when no other genomic information is known since hundreds of random decamers are available. The technique has some drawbacks. Firstly, the technique has been reported to suffer from a lack of reproducibility between laboratories and sources of thermostable enzyme, although, within a laboratory, reproducible results can be achieved (Jones et al. 1997). Newer versions of the Taq polymerases have been developed which do not function as well at low temperatures, and so are not useful for producing RAPD patterns since the length of the primers requires an annealing temperature of about 40°C. Finally, the assignment of RAPD bands to a molecular map is genome specific, thus for every individual, a unique RAPD map needs to be produced since amplified fragments cannot be correlated across genomes. Finally, RAPDs are dominant markers—a band is produced whether the fragment is homozygous or heterozygous so larger segregating populations are needed to produce accurate molecular maps. The basis for RAPDs can be a single nucleotide change in the primer sequence (since 1 base change is equivalent to a 10% mismatch in the primer sequence which will cause a 10°C change in the stability of the annealed primer) or insertion or deletion between the primer sites. Thus the same genomic variations that produce RFLPs are also responsible for RAPD variations except that the single base changes do not need to be in a restriction enzyme recognition site.

5.2.3 Amplified Fragment Length Polymorphism

These markers are essentially RFLPs detected by PCR amplification. The need for known specific primers with which to perform the PCR reaction is avoided by adding known adaptors to the ends of the restriction fragments. The method requires that the genomic DNA is first digested with two different restriction enzymes. Adaptors are added to the ends of restriction fragments, and these adaptors are then used as primers in a PCR reaction. A very large number of amplified fragments are observed and the polymorphic fragments are identified against this background of all of the possible sized restriction fragments that can be amplified. The complex mixture of amplified bands is separated on gels or using automated DNA sequencers. The polymorphic bands can be cloned and sequenced to generate sequence-tagged site (STS) markers, which now use the actual genomic sequences for subsequent characterization. Potential epigenetic effects resulting from hyper- or hypomethylated regions of the genome can be investigated by using methylation-sensitive and -insensitive restriction enzyme isoschizomers. As with RAPDs, AFLPs are also dominant markers.

5.2.4 Microsatellites and SSRs

Microsatellites or SSRs are genetic markers that are derived from short (usually <6 bp) tandemly repeated sequences such as $(GA)_n$, $(AAT)_n$, $(GT)_n$. The terms microsatellite and SSR are often used interchangeably, although microsatellites are generally longer than the 2- to 3-bp unit of the SSRs. This type of sequence is widely dispersed through most animal and plant genomes and polymorphisms are due to the variability in the number of repeats at a given site (Fig. 5-2). The length polymorphisms are produced by the changes in the number of the repeats as shown in Fig. 5-2. The three repeat classes shown in Fig. 5-2 are: a, with 5 repeats, b, with 2 repeats and c, with 10 repeats. If the SSR is a dinucleotide repeat, then the actual size differences between a and b is 6 bp, a and c is 10 bp and b and c is 16 bp. Therefore the separation of the amplified products needs to be done with high resolving power. Therefore the early ABI automatic sequencers are finding a renewed lease of life in use for characterizing SSR length polymorphisms.

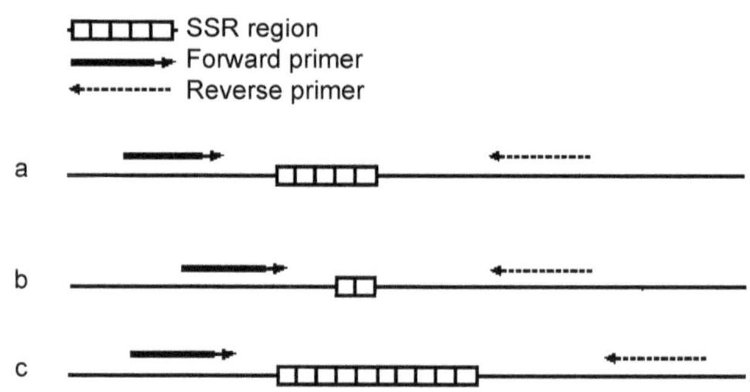

Figure 5-2 Length polymorphisms resulting from differences in the number of SSR repeats at a locus.

As with RFLPs the primers for each SSR need to be isolated for each species. The current technology using the second generation sequencing has made their isolation more rapid from genomic libraries or enriched genomic libraries (Panaud et al. 1995; Santana et al. 2009) or generated from an analysis of cDNA sequences. The enrichment techniques involve the use of biotinylated SSR fragments that are used to physically remove complementary sequenced from a PCR amplified set of genomic fragments. The early iterations of the technique used the AFLP primers to provide the genomic representation. Initially all of the recovered fragments needed to be cloned and sequenced. However, using the 454 next generation sequencing methodology the complete set of fragments can be sequenced and assembled. The contigs are then analyzed using the program SSRIT (http://www.gramene.org/db/markers/ssrtool) to identify the contigs containing SSRs. Where there is sufficient sequence surrounding the SSR region, primers are designed and the germplasm of interest is screened to find those that are polymorphic. Thus although the identification of the SSRs regions is now much more rapid than before, each of them needs to be screened to find those that are polymorphic which can be time consuming.

A second source of SSRs is from expressed sequence tag (EST) sequencing data. Frequently triplet repeats can occur in mRNAs since an additional amino acid will be inserted into the protein with no other changes in the structure and sequence of the protein. They can also be present in the 5′ and 3′ untranslated regions of the mRNA. The identification of these SSRs obviously needs a large EST sequence database and so is restricted to those species where a substantial sequence database is available. It is also possible that the level of polymorphisms present in EST-derived SSRs is lower than those identified from the genomic enrichment methods.

5.2.5 Single-Nucleotide Polymorphisms

Single-nucleotide polymorphisms (SNPs) are DNA sequence variations that occur when a single nucleotide (A, T, C, or G) in the genome sequence differs between two individual DNA samples. For example, a SNP might change the DNA sequence AGGATTCA to AGGATTTA. SNPs can occur in both coding (gene) and non-coding regions of the genome. As detailed above, SNPs are responsible for that class of RFLPs that result from the loss or gain of a restriction site. Many SNPs have no effect on cell function because they may not change protein structure (in fact, any SNP that occurs at the third position in the amino acid codon will have no effect if it does not change the amino acid sequence of the resulting protein). Their high frequency (perhaps as high as 2–3% in plant DNAs) means that they can be particularly useful in linkage mapping (Kristensen et al. 2001; Lai 2001). Again, the identification of SNPs depends on having sequence information available. Additionally, the sequence information needs to be obtained either from different varieties or from outbred individuals so that heterozygosity will be present. Informatics tools can be used to compare the sequences and identify variations, but the raw data in the form of trace files may be important in deciding which polymorphisms may be real. When ESTs, for example, are built into unigene sets, any differences that may be present are eliminated in the formation of the consensus sequence and would need to be retrieved. Since there is no *a priori* way of differentiating between a true SNP and sequencing errors, each potential SNP must be validated. Even at a frequency of 1% these polymorphisms would generate an exceptionally large number of haplotypes if every polymorphism could be inherited independently. However, relatively few haplotypes are observed, indicating that perhaps the rate of SNP production is similar to the rate at which recombination occurs across the regions of the genome making up the haplotype blocks. Therefore, SNPs are most likely to be useful for defining haplotypes, rather than for their information individually, and so the use of SNPs is likely to involve linkage disequilibrium studies using the haplotype rather than the use of specific SNPs as individual molecular markers.

5.3 Mapping Populations

These are usually derived from controlled crosses in which the selection of the parents, crossing schemes and marker systems are dependent on the objectives of a mapping experiment. The parents chosen need to have sufficient differences in both the molecular markers to be employed as well as in phenotypic traits. The commercial importance of any molecular map is usually dependent on the association between the marker and the phenotypic trait so as many phenotypic and quantitative traits as possible should be segregating in the mapping population. It is of particular importance to recognize that when wide crosses are involved there is likely to be recombination suppression due to chromosome rearrangements and the genetic and physical distances between markers might be substantially different.

5.3.1 Types of Mapping Populations

5.3.3.1 F_2 Population

This population is generated by crossing two parents and then selfing or sib-mating the resulting F_1 plants. Therefore this population is the result of a single meiotic event between the genomes of the two contributing parents. This population is the easiest to construct and is ideal for a preliminary study. However, since it is a series of individuals which are the result of random recombination events between the original parental genomes it cannot be reconstructed. Since the F_2s are the result of only a single cycle of meiosis fine mapping cannot be done unless large numbers of individuals are scored. In addition, since each individual is the result of a different genome, quantitative trait loci (QTL) cannot be mapped and growth by environment (GxE) interactions cannot be addressed. An F_3 population can be obtained by selfing the F_2 individuals. This will enable the genetic status of any F_2 expressing a dominant trait to be determined. Continued selfing will eventually result in recombinant inbred lines. Table 5-2 gives the rate at which the loci become homozygous with increasing numbers of selfed or backcrossed generations.

Table 5-2 The level of inbreeding with increasing selfing or backcrossing generations. These values do not take into account linkage drag so the actual level of homozygosity is lower at each generation.

Selfed generation	% within-line homozygosity at each locus	Backcross	% of recurrent parental genome present
F_3	75.0	BC_2	75.0
F_4	87.5	BC_3	87.5
F_5	92.25	BC_4	92.25
F_6	96.875	BC_5	96.875
F_7	98.4375	BC_6	98.4375
F_8	99.21875	BC_7	99.21875

5.3.2.2 Backcross

A backcross population is obtained by crossing the F_1 plants to one of the parents. Usually the parent that is recessive for traits of interest is used as the backcrossing parent. Here recombination only occurs in one of the parents—the F_1. Multiple rounds of backcrossing results in the regeneration of the recurrent parent's genome and if associated with the selection of a particular trait can be used to develop near isogenic lines (NILs). These lines have different alleles conditioning a particular trait all in the same genetic background and therefore the interaction between the trait and background can be ascertained. In addition, if NILs for different genes are developed, then these can be combined to understand the interaction between different genes across various environments without the influence of different genetic backgrounds.

5.3.2.3 Doubled Haploids (DH)

Double haploids are usually derived from anther cultures. However, in some specialized systems such as *Hordeum*, doubled haploids can result from chromosome elimination following interspecific crosses. DH plants are derived from the anther culture of F_1 individuals will all be different as they are independent results of the F_1 meiosis. They will also all be homozygous at every locus and so the DH population is an immortalized mapping population. Each of the DH individuals will breed true and so replicated trials in multiple environments are possible allowing the detection of QTL and genotype by environment (GxE) effects. One of the major drawbacks is that not all plants are amenable to anther culture. Even those in which anther culture is possible the problem of culture-induced variation needs to be monitored.

5.3.2.4 Recombinant Inbred Lines (RILs)

These are developed by successive selfing or sib mating of the F_1 generation. Each succeeding generation will be homozygous at an increasing number of loci until each of the lines is completely homozygous. These lines are equivalent to the doubled haploid lines but take much longer to derive but can originate from any plant species. Generally these lines are derived by a single seed descent method. A single seed from the selected plants are grown in each generation. The time to complete the isolation of these RILs depends on the life cycle of the plant—if it takes multiple years to first flowering then the time to develop RILs could be measured in centuries. When these lines have been developed they are permanent and can be distributed to other researchers so that all can add markers to the same map. They can also be multiplied up and, as with other immortalized mapping populations, can be used for replicated trials in multiple environments allowing the detection of QTL and GxE effects. The recombination intervals that can be resolved depend on the number of RILs that are screened.

5.3.2.5 Near Isogenic Lines (NILs)

Near isogenic lines are developed by the repeated selfing or backcrossing of the F_1. At each generation there is usually selection for two forms of a specific trait so that RILs are the same at nearly all loci except for the desired one. An example is the development of NILs for the round/wrinkled phenotype of pea. In this case round is dominant and wrinkled is recessive. A cross between an inbred round parent and an inbred wrinkled parent gives an F_1 that is all round. When this F_1 is backcrossed to the wrinkled parent and the BC1 (backcross generation 1) segregated 1:1 round to wrinkled. A round seed is planted and again backcrossed to the wrinkled parent. Again the progeny segregate 1:1 for round and wrinkled seeds. After several rounds of backcrossing the plants are essentially homozygous and identical to the wrinkled parent except for the locus that determines round and wrinkled. The final generation is a self of a round seeded individual (which is still heterozygous at this locus) and a round and wrinkled family selected. The round seeded individual is selfed once more to ensure that it is homozygous at the locus (does not segregate wrinkled seeds) and the two lines can be multiplied up and should be identical at every locus except that determining the round/wrinkled phenotype. This is a substantial investment in time to produce the NILs and are only useful for the trait that is under selection.

However, they can be important in functional genomics studies where pleiotropic effects are associated with the trait.

5.3.2.6 Test Crosses

A test cross is equivalent to a backcross to the recessive parent so that the data is really focused on the recombination in one of the parents. A pseudo test cross is one in which two parents are crossed and one of the parents is heterozygous for a particular marker and the other parent is a null. Therefore, this marker will segregate 1:1 in the F_1 progeny following a testcross configuration. These crosses can be used to map traits more rapidly.

5.4 Bulk Segregant Analysis

Bulk segregant analysis is an approach that can be used with any of the above populations except the NILs. It is based on the isogenic line principle. For example, individuals in an F_2 are typed for a trait such as disease resistance/susceptibility and grouped according to their phenotype. DNAs are isolated from 10 individuals from each group and the DNAs pooled. The two pools are then screened with a set of markers that are polymorphic between the parents. If the marker is not linked to the trait each pool will show the same pattern of amplification. For a marker closely linked to the trait the pattern for the two pools will resemble that of the original parents. Ten plants in a pool have been determined as being sufficient as a first screening.

5.5 Mapping Software Used

Many types of software are available for making genetic maps out of raw data of a cross. The earliest programs that were developed are still in use, with some modifications to handle larger data sets with greater speed. Other programs have been developed that handle more complex data than just simple linkage mapping, in particular the identification of quantitative trait loci.

5.5.1 Linkage-1

This program was developed in 1983 to facilitate the analysis of linked loci. It is a basic program for generating linkage scores among markers, which can be used to assemble a map. This program can analyze data from F_2s, backcrosses, and different combinations of single-factor segregation ratios between loci—loci that show different modes of inheritance within the same segregating population. Both dominant and codominant genes and markers can be analyzed with this program. This program utilizes the Chi square analysis to test for independent assortment among pairs of segregating loci. The ratio of genotypes for each marker is compared to the expected segregation ratio at that locus for the predetermined segregation pattern. When the *P* values show a deviation from the expected frequencies for independent assortment then recombination frequencies are calculated for each pair of loci. This early version gives the results as the 2-locus table, Chi-square and *P* values, and the recombination fraction and its standard error (Fig. 5-3). These data then need to be complied into a map either by another program or manually. The original Linkage-1 program is available at the Rockefeller University website, ftp://linkage.rockefeller.edu/software/

A Total number of individuals compared = 40 out of 40

The two loci being compared are PGM3 and DOM1

DOM1	PGM3 BB	PGM3 BC	PGM3 CC
A-	2	17	10
aa	11	0	0

Chi-square	Df	P	r +/- SE
3.15119E+01	2	<0.01	5.0000E-02 +/ 3.5811 E-02

B Total number of individuals compared = 40 out of 40

The two loci being compared are PGM3 and PG11

PG11	PGM3 BB	PGM3 BC	PGM3 CC
AA	12	8	0
AB	1	9	10

Chi-square	Df	P	r +/- SE
1.93665E+01	2	<0.01	4.0000E-02 +/ 4.3818 E-02

C Total number of individuals compared = 40 out of 40

The two loci being compared are PGM2 and PG11

PG11	PGM2 AB	PGM2 BB
AA	10	10
AB	7	13

Chi-square	Df	P	r +/- SE
9.20716E-01	2	0.5 - 0.25	None

This valuer is not significant

Figure 5-3 Sample output for two-factor linkage analysis. Three different family types are shown (A, B, C). The expected ratios for each gene are the same as in Fig. 5-1 (Taken from Suiter 1983).

linkage. This software has been improved to accommodate larger data sets by running multiple algorithms in parallel and can now also be found under the name FASTLINK at the NCBI's genetic analysis software website (http://www.ncbi.nlm.nih.gov/CBBresearch/Schaffer/genetic_analysis.html). The same LINKAGE programs are used but the current iteration gives the results much more rapidly, especially for large data sets.

5.5.2 *Mapmaker*

The MapMaker suite of software includes the original MapMaker, MapMaker/EXP and MapMaker/QTL. MapMaker can analyze 3-generation pedigrees, while MapMaker/EXP is used for constructing maps from experimental data derived from crosses using Mendelian markers. MapMaker/EXP can analyze data including F_2s, backcrosses, RILs, and F_3-self populations. EXP analyzes data in a manner similar to LINKAGE-1; for every pair of markers, the maximum-likelihood distance and a likelihood of linkage (LOD) score are calculated. When the LOD is greater than some threshold (the default is normally 3.0, which represents a likelihood of occurrence by chance of 10^{-3}), and the distance is less than some threshold (30 cM), the markers are considered to be linked. Loci are then assigned to linkage groups, and a map is constructed based on these data (Fig. 5-4). Mapmaker/QTL is used for mapping polygenes involved in quantitative traits using a genetic map, and is designed to work with maps generated by Mapmaker/EXP. Links to the software for MapMaker/EXP and QTL can be found at http://linkage.rockefeller.edu/soft/list.html. A detailed online tutorial is also available at this site.

```
15> map
==========================================================================
Map:
   Markers              Distance
     4 ' T24              14.8 cM
    11   C15               6.4 cM
     8   T125             18.9 cM
    12   T71              24.0 cM
     9   T83              18.1 cM
     6   T209             28.6 cM
    10   T17           ----------
                         110.8 cM     7 markers    log-likelihood= -688.99
==========================================================================
```

Figure 5-4 Output from MapMaker/EXP (Lincoln et al. 1992).

5.5.3 *JoinMap*

This mapping software was developed to combine map data from independent experiments into one integrated map. Different types of data, such as F_2s, backcrosses, RILs, and pairwise recombination frequencies can be integrated, as long as these data sets have some markers in common. This program can also be used to create maps using the data from single experiments. The coding of the information files is similar to that of MapMaker, so that MapMaker data can also be easily incorporated. JoinMap integrates mapping data in the following fashion.

The data is read, and for marker pairs common to all data sets, the estimated recombination frequencies are weighted and replaced by a single value. LOD scores are then calculated from the recombination frequency. Linkage groups are established for the common data based on LOD scores, and the map is then sequentially built up to include the rest of the data by comparing LOD scores to those that have already been placed on the map. JoinMap is available for evaluation and for purchase from Kyazma, at http://www.kyazma.nl/. Information about courses on mapping and QTL analysis is also available at this site.

JoinMap is the easiest to install and has the most user-friendly interface as compared to Linkage 1 and MapMaker suite software.

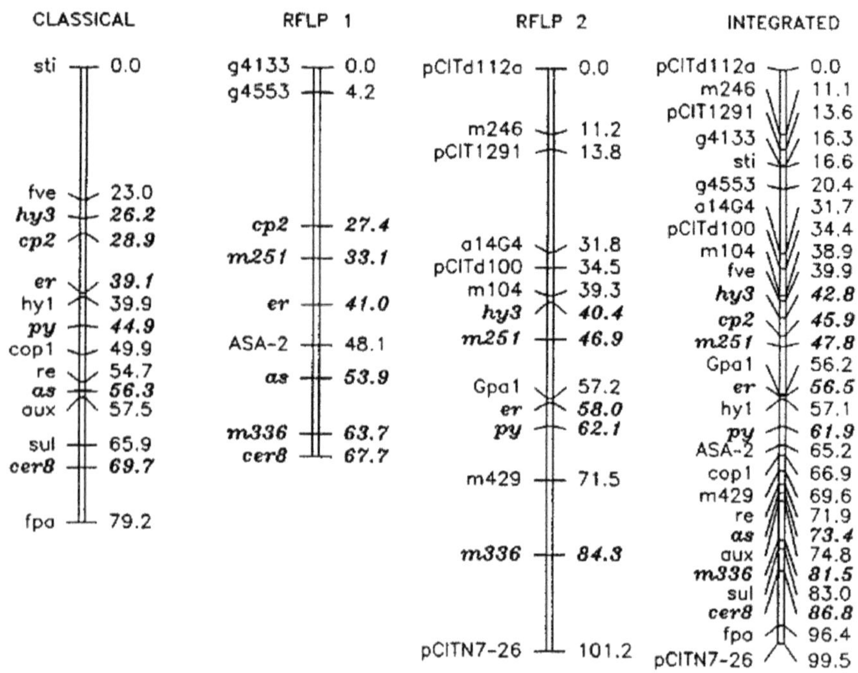

Figure 5-5 Three separate linkage maps and the integrated map of chromosome 2 of *Arabidopsis*. Markers designated in bold face occur on more than one of the individual maps. For the sake of clarity, only a fraction of the total number of markers that were actually mapped is shown here (from Stam 1993).

5.5.4 Other Software

Other mapping programs mentioned in the recent literature are Carthagene, which, like JoinMap, can both integrate data and create new maps, Surveyor, which uses Chi-square analysis like Linkage, CPROP, GMendel, CriMap, GeneHunter, Vitesse, Allegro, Merlin, Solar, SimWalk2, and Superlink. Some of these programs and the data they are best used for have been reviewed (Dudbridge 2003). The Rockefeller University website also lists many additional programs. As of August 2010, there were 580 programs listed for use in the basic manipulations of data from genetic crosses including animal and plant breeding and QTL analysis, population genetics and human pedigree analysis. The programs are listed alphabetically with a short description of the features and application of each, as well as links to the software and references.

5.6 Development of Molecular Genetic Maps

The foregoing sections have described various aspects of mapping independently from each other. Genetic and physical mapping is usually an integrated activity that is contributed by all sectors of the interested scientific and agricultural community. Nearly all plant improvement efforts depend on making crosses and selection of specific outcomes from those crosses. Possible exceptions include identifying useful somaclonal variation and mutation breeding, although in the latter case breeding and selection efforts can be more successful using genetic maps.

The first characters usually followed are phenotypic ones that are easy to score. However, since the expression of many phenotypic traits can interfere with the scoring of others only a small number of these can be followed in a specific cross. Therefore, early genetic maps were compilations of linkage data derived from many different crosses. The advent of molecular markers essentially permitted an unlimited number of loci that could be placed on the linkage map in a single cross.

A preliminary genetic map would only contain a few markers, ideally distributed over all the chromosomes. This would allow traits and additional markers to be sited on a specific chromosome. However, for higher density mapping many more markers would need to be added, and the law of diminishing returns would begin to apply. If random markers were used then the number that would need to be screened to fill a particular gap in a linkage group would steadily rise as the interval became smaller. Therefore some form of targeting of regions would become more useful.

The first generation molecular genetic linkage maps are usually based on markers that are separated by about 20 cM (centiMorgans) (two sites are 1 centiMorgan apart when the recombination frequency between them is ~1%). With coverage at this level, any new marker or trait would not be more than 10 cM from a known position. Thus the probability that that marker and the gene or trait of interest would be co-inherited at least 90% of the time. As the need for more accuracy increases more markers are placed on the genetic map. To be certain of the order of a series of closely linked markers it is necessary to increase the size of the mapping population. If the markers are 20 cM apart then, in a segregating population of 100 individuals, 20 recombination events will occur on an average. But the same population is insufficiently large to map markers at the 1 cM level.

All of the discussion so far has been related to genetic maps. Physical maps are the linear order of the DNA sequences in the chromosomes and distances between positions are known in base pairs. The reconciliation of genetic and physical maps is sometimes difficult as regions of the chromosomes may have reduced recombination activity so 1 cM of genetic distance does not have an exact equivalent in terms of base pairs. It can differ within a chromosome, between chromosomes and even be different depending on the sex of the individual in which it is measured. The integration of multiple maps can also pose problems. When the mapping populations are different then chromosomal rearrangements will cause specific markers to have different positions. For example, if two groups each make a molecular map using different starting populations and a translocation is present in one of the populations and not in the other then some markers will not even map to the same chromosomes. One method of resolving this is to develop a molecular-cytogenetic map.

The development of *in situ* hybridization has allowed the physical placement of sequences on the chromosomes. The development of the physical maps have allowed large fragments of low copy DNA sequences to be labeled and hybridized to the chromosomes and then the sites of hybridization visualized. If the chromosomes can be counter-stained and recognized, then each of the fragments can be physically placed on specific chromosome sites.

5.7 Conclusions

Genetic mapping is a necessity for plant improvement. The development of a whole suite of marker types that can be used on plants for which little other molecular information is available has allowed modern breeding techniques to be employed. However, many of these investigations are based on long term crossing programs that are not generally supported at present. The plant material takes a substantial period of time to develop and improvements will be hampered if this basic infrastructure is not supported.

Abbreviations

AFLP	:	amplified fragment length polymorphisms
DH	:	doubled haploids
EST	:	expressed sequence tags
GxE	:	genotype by environment
mRNAs	:	messenger RNA
NILs	:	near isogenic lines
NILs	:	Near isogenic lines
PCR	:	polymerase chain reaction
PIC	:	polymorphism information content
QTL	:	quantitative trait loci
RAPD	:	random amplified polymorphic DNAs
RFLP	:	restriction fragment length polymorphisms
RILs	:	recombinant inbred lines
SNP	:	single-nucleotide polymorphisms
SSR	:	simple sequence repeats

References

Botstein D, White RL, Skolnick M, Davis RW (1980) Construction of a genetic linkage map in man using restriction fragment length polymorphisms. Am J Hum Genet 32: 314–331.

Dudbridge F (2003) A survey of current software for linkage analysis. Human Genomics 1: 63–65.

Jones CJ, Edwards KJ, Castaglione S, Winfield MO, Sala F, van de Wiel C, Bredemeijer G, Vosman B, Matthes M, Daly A, Brettschneider R, Bettini P, Buiatti M, Maestri E, Malcevschi A, Marmiroli N, Aert R, Volckaert G, Rueda J, Linacero R, Vazquez A, Karp A (1997) Reproducibility testing of RAPD, AFLP and SSR markers in plants by a network of European laboratories. Mol Breed 3: 381–390.

Kristensen VN, Kelefiotis D, Kristensen T, Barresen-Dale AL (2001) High-throughput methods for detection of genetic variation. BioTechniques 30: 318–322.

Lai E (2001) Application of SNP Technologies in Medicine: Lessons Learned and Future Challenges. Genome Res 11: 927–929.

Lincoln SE, Daly MJ, Lander ES (1992) Mapping Genes Controlling Quantitative Traits: Using MAPMAKER/QTL Version 1.1, A Tutorial and Reference Manual. 2nd edn. A Whitehead Institute for Biomedical, Research Technical Report.

Panaud O, Chen X, McCouch SR (1995) Frequency of microsatellite sequences in rice (Oryza sativa L.). Genome 38: 1170–1176.

Santana Q, Coetzee M, Steenkamp E, Mlonyeni O, Hammond G, Wingfield M, Wingfield B (2009) Microsatellite discovery by deep sequencing of enriched genomic libraries. BioTechniques 46: 217–223.

Senior ML, Heun M (1993) Mapping maize microsatellites and polymerase chain reaction confirmation of the targeted repeats using a CT primer. Genome 36: 884–889.

Stam P (1993) Construction of integrated genetic linkage maps by means of a new computer package: Join Map. Plant J 3: 739–744.

Sturtevant AH (1913) The linear arrangement of six sex-linked factors in *Drosophila*, as shown by their mode of association. J Expt Zool 14: 43–59.

Suiter KA, Wendel JF, Case JS (1983) LINKAGE-1: a PASCAL computer program for the detection and analysis of genetic linkage. J Hered 74: 203–204.

Vos P, Hogers R, Bleeker M, Reijans M, van de Lee T, Hornes M, Frijters A, Pot J, Peleman J, Kuiper M (1995) AFLP: a new technique for DNA fingerprinting. Nucl Acid Res 23: 4407–4414.

Williams JG, Kubelik AR, Livak KJ, Rafalski JA, Tingey SV (1990) DNA polymorphisms amplified by arbitrary primers are useful as genetic markers. Nucleic Acids Research 18: 6531–6535.

6

Mapping and Tagging of Simply Inherited Traits in *Musa*

Rodomiro Ortiz

ABSTRACT

Few genetic markers are available in banana and plantain due to the lack of research in transmission genetics. The major obstacles for *Musa* genetic analysis include: triploidy in most popular cultivars, the very low reproductive fertility, and the few hybrids recovered after *in vitro* embryo germination leading to small population sizes. Nonetheless, transmission genetic research was undertaken in *Musa*, particularly in the 1990s. The genetic importance of some target traits including albinism, apical dominance, dwarfism, fruit parthenocarpy are discussed. The chapter includes a discussion of genes that are mapped.

Keywords: target traits in *Musa*, mapping of traits

6.1 Introduction

Despite the importance of *Musa* as a crop species, few genetic markers are available in banana and plantain due to the lack of research in transmission genetics (Ortiz and Vuylsteke 1994a). The major obstacles for *Musa* genetic analysis include: triploidy in most popular cultivars, the very low reproductive fertility, and the few hybrids recovered after *in vitro* embryo germination leading to small population sizes. Nonetheless, transmission genetic research was undertaken in *Musa*, particularly in the 1990s (Ortiz 2000). One of the first *Musa* genetic research results was about fruit parthenocarpy and persistence of bracts and flowers in the male

Department of Plant Breeding and Biotechnology, Swedish University of Agricultural Sciences, Box 101, SE-230 53, Alnarp, Sweden; e-mail: *rodomiroortiz@gmail.com*

List of abbreviations after the text.

part of the inflorescence axis. Three independent dominant P_i genes with complementary gene action control for the former whereas complementary gene control was suggested for the latter (Simmonds 1952). The members of the *Musa* Genomics Consortium (http://www.musagenomics.org) are developing segregating populations to identify genetic markers for host plant resistance to pathogens and pests, fruit parthenocarpy, linkage disequilibrium analysis and comparative genetic analysis with rice (http://www.musagenomics.org/index.php?id=49). The mapping populations are F_1 (segregating either as a test-cross if one parent is dominant heterozygous and the other is recessive, or as an F_2 if both parents share a di-allelic genotype) and F_2 generations. The crosses involved diploid species such as *M. acuminata* and *M. balbisiana* as well as some triploid cultigens. The first *Musa* genetic map was based on restriction fragment length polymorphism (RFLP) marker and isozymes in diploid populations (Faure et al. 1993), to which microsatellites (or simple sequence repeat, SSR) and amplified fragment length polymorphism (AFLP) markers were added.

6.2 Importance of the Target Traits

Albinism (a_i, Ortiz and Vuylsteke 1994a): Seedlings lacking chlorophyll can be seen in plantlets ensuing from *in vitro* germination of hybrid embryos. Segregation ratios observed in the hybrid offspring suggest that albinism may be at least under the control of two independent recessive alleles (a_1 and a_2). Albinism could be an interesting trait for studying genetic load in the crop and by monitoring the frequency of such deleterious recessive alleles to understand the gains brought by recurrent selection in diploid breeding populations.

Apical Dominance [*ad*, Ortiz and Vuylsteke 1994b]: The inhibition of lateral bud growth due to substances released by the terminal bud has been considered as a limiting factor for the perennial productivity of plantains. Segregation ratios in plantain-banana breeding populations suggest that inheritance of apical dominance is controlled by a major recessive gene *ad*. The dominant *Ad* allele improves the suckering of the crop as measured by the height of the tallest sucker in plantain-banana hybrids. The *Ad* gene may regulate gibberellic acid production but this allele shows incomplete penetrance and variable expressivity.

Dwarfism [*dw*, Ortiz and Vuylsteke 1995a]: The height of the pseudostem, that is, the distance between the soil and the petioles of the highest leaves, is used for sub-grouping plantain cultivars into giant, medium, and small stature. *Musa* clones with short internodes are called "dwarf" cultivars (Ortiz and Vuylsteke 1998). Such dwarfism is controlled by a single recessive gene *dw*, which appears to be located close to the centromere and shows a dosage effect at the tetraploid level. Dwarf cultivars are less-prone to wind damage and may show increased yield owing to high plant density and rapid sucker development.

Fruit parthenocarpy [P_i, Simmonds 1952]: *Musa* cultivars develop fruit by vegetative parthenocarpy, which occurs when a non-pollinated ovary automatically grows into a fruit that has its loculi filled with edible pulp. The autonomous stimulus giving rise to parthenocarpic fruits results from the interactions of at least three independent, complementary, dominant genes: P_1, P_2 and P_3. The P_1 locus appears to be far from the centromere and after the first crossing over. Ortiz and Vuylsteke (1995b) showed that fruit size and weight was partially associated with the effect of allele substitution in this P_1 locus.

Host plant resistance: Gaining further knowledge on the genetics of host plant resistance will enhance the development of resistant cultivars, which is considered the most appropriate means to achieve stable production. The genetic control of resistance to banana streak virus (Ortiz 1996), black Sigatoka (bs_1 +2 bsr_i, Ortiz and Vuylsteke 1994c) and burrowing nematode (2-gene system, Dochez et al. 2009) was elucidated using segregating populations at diploid and polyploid levels. The above listed pathogens and pests significantly affect bunch weight, thereby producing yield losses in the *Musa* crop, e.g., black Sigatoka due to *Mycosphaerella fijiensis* is the major constraint to plantain and banana production worldwide.

Segregation ratios to host plant resistance to black Sigatoka in diploid breeding populations fitted a genetic model having one major recessive resistance allele (bs_1), located between the centromere and first crossing-over, and two independent alleles with additive effects (bsr_2 and bsr_3). A similar model explains the results at the tetraploid level assuming that the favorable resistance alleles have a dosage effect when four copies of them are present at the respective loci (bs_i^4). A homozygous bs_1 genotype at the diploid level, alone or in combination with only one bsr_i genotype does not result in a partially resistant but rather in a less susceptible response. Only diploids with all three favorable alleles in the homozygous state will have a partially-resistant phenotype. Due to dosage effects, tetraploids with a homozygous bs_1 genotype will show partial resistance. Similarly, homozygous genotypes for any bsr_i loci have less susceptible phenotypes, even in the presence of an unfavorable Bs_1 susceptible allele, which means that an increased number of favorable alleles at the tetraploid level can overcome the action of the host plant's susceptibility alleles and the virulence genes of the pathogen. Further genetic research (Craenen and Ortiz 1997) showed that intra-locus interaction in the bs_1 locus apparently regulates the appearance of symptoms on the leaf surface, whereas the additive effect and the intra-locus interaction of the bs_1 locus affect disease development in the host plant. Hence, the gene action(s) at the bs_1 locus may provide durable resistance to black Sigatoka by slowing down disease development in the host plant.

6.3 Employment of Bulked Segregant Analysis for Gene Tagging

Individuals from segregating populations can be grouped according to their phenotypic expression for a specific trait and tested for differences in allele frequency between the population bulks; i.e., bulked segregant analysis (BSA; Michelmore et al. 1991). Molecular markers used for developing a genetic map can be used for BSA. Molecular polymorphism between the parents and respective markers closely-linked to a major gene regulating a particular trait will mainly co-segregate with that gene. The frequency of the two marker alleles present within each of the two bulks should deviate significantly from the expected 1:1 ratio for most populations after grouping plants as per their trait expression and testing such extreme groups with a polymorphic marker. The map location of closely linked DNA marker-trait can be deduced without having to genotype every individual in segregating populations when chromosomal locations of many molecular markers are known.

Banana streak virus (BSV) expression: Lheureux et al. (2003) undertook the analysis of banana streak disease (BSD) incidence in an F_1 triploid population ensuing from an interspecific cross between virus and pathogen-free diploid *M. balbisiana* and tetraploid *M. acuminata* parents, and with 50% of its offspring expressing BSV particles. Their aim was to elucidate whether BSD occurs in these hybrids as a result of the activation of BSV-O1 endogenous para-retrovirus sequences (EPVR) integrated into the *Musa* genome rather than from external sources of infection, and that hybridization during the breeding process may be one of the factors involved in triggering episomal expression of BSV integrants. Their research used polymerase chain reaction (PCR) amplification to determine the presence or absence of BSV-Ol EPRV. They determined that this endogenous sequence was specific to the *M. balbisiana* genome and occurred in a homozygous state. This research was able to identify through BSA, 10 AFLP markers co-segregating with the absence and presence of BSV infection in *M. balbisiana*, but absent in the *M. acuminata* genome. Seven AFLPs segregated with the presence of a BSV particle and three with the absence of BSV particles. These results suggest that the genetic control of BSV appearance is due to a monogenic allelic system that confers a carrier role to *M. balbisiana*.

Host plant resistance to Fusarium *wilt*: Resistance and susceptible seedlings and cultivars to *Fusarium* wilt coupled with the random amplified polymorphic DNA (RAPD) technique were used to identify markers linked to host plant resistance to this pathogen. The segregation ratio to *F. oxysporum* f. sp. *cubense* race 4 fitted to an 1:3 ratio in four seedling populations of wild *M. acuminata* ssp. *malaccensis*: a single recessive factor was advocated in three out of four seedling populations, whereas a single dominant factor for host plant resistance was suggested for the remaining seedling population (Javed et al. 2004). A 13:3 ratio, which suggests two loci controlling host plant resistance to *Fusarium* wilt, was also observed in a fifth seedling population. Such results among five seedling populations indicate that their wild parents are genetically distinct for their host plant resistance loci to *Fusarium* wilt race 4 and perhaps as a result of the co-evolution between the pathogen and its host(s). A few polymorphic RAPD markers showed banding patterns that were specific to resistant or susceptible seedlings: a 400 bp major band of primer-03 was shared by all susceptible seedlings and was absent in the resistant offspring, whereas a 300 bp of primer-24 was only observed in most susceptible seedlings, and a 1 kbp band derived from primer-21 was found only in the resistant seedlings to *Fusarium* wilt.

Zambrano et al. (2007) used 10 resistant and eight susceptible cultivars and wild accessions to *Fusarium* wilt, with various ploidy levels, to screen RAPD markers that could be associated with their host plant resistance to this pathogen. One primer, OPK-03, detected a major fragment of 485 bp, which was only found in the susceptible cultivars but absent in the resistant cultivars or accessions.

Dwarfism: Engelborghs et al. (2000) used AFLP and several variants of this DNA marker technique to characterize four pairs of the dwarf plantain "Curare enano" and its normal-size *in vitro* off-type. Three endonuclease-AFLP (TE-ALP) generated fewer and shorter fragments, thereby leading to fewer polymorphisms between the dwarf cultivar and its normal size off-type. cDNA-AFLP revealed a normal-specific fragment whereas cDNA-TE-AFLP was able to identify a dwarf-specific fragment, which shows the value of AFLP and its variants for distinguishing very similar *Musa* genotypes.

6.4 Genes Mapped

Banana Streak Virus (BSV) Expression: Segregation of AFLP markers, considering a test-cross configuration, allowed the construction of a maternal genetic map of the linkage group containing the locus associated with BSV infection in a F_1 hybrid population (Lheureux et al. 2003). For BSA analysis, two pools of AFLP templates composed of five healthy individuals and five diseased F_1 offspring, respectively, were used. Both template pools were fingerprinted with AFLP using 64 combinations of *Eco*RI and 3/*Mse*I plus 3 primer extensions. The marker-span was 69.8 cM and the genetic distances between markers varied from 0.6 to 39.2 cM. The BSV expressed locus (*BEL*) was 1.2 cM apart from the nearest marker (ACACAC).

Host plant resistance to black Sigatoka: Carreel et al. (1999) used a F_2 segregating population of 153 plants, ensuing from a cross between *M. acuminata* spp. *burmannicoides* ("Calcutta 4") and *M. acuminata* spp. *banksii* ("Madang"), to develop a linkage map with 110 AFLP markers in association with 39 codominant RFLP and SSR markers to anchor the respective linkage groups to the *Musa* core map. One RFLP was strongly associated to host plant resistance to black Sigatoka and a second locus accounting for this host plant resistance at a lower significance level was also mapped onto another linkage group.

Abbreviations

AFLP	:	amplified fragment length polymorphisms
BSA	:	bulk segregant analysis
BSD	:	banana streak disease
BSV	:	banana streak virus expression
EPVR	:	endogenous para-retrovirus sequences
RAPD	:	random amplified polymorphic DNA
RFLP	:	restriction fragment length polymorphism
SSR	:	micro-satellites

References

Carreel F, Abadie C, Carlier J, Tomekpe K, Lagoda P, Bakry F (1999) Genome mapping and genetic analysis of the black leaf streak resistance in bananas. Infomusa 8: II–III.

Craenen K, Ortiz R (1997) Effect of the *bs 1* gene in plantain-banana hybrids on response to black sigatoka. Theor Appl Genet 95: 497–505.

Dochez C, Tenkouano A, Ortiz R, Whyte J, De Waele D (2009) Host plant resistance to *Radopholus similis* in a diploid banana hybrid population. Nematology 11: 329–335.

Engelborghs I, Swennen R, Sagi L (2000) Fluorescent AFLP analysis on azacytidine and gibberellin treated banana (*Musa* spp.) plants to assess differences in cytosine methylation and the mechanism of dwarfism. Med Fac Landbouww Univ Gent 65/3b: 387–396.

Faure S, Noyer JL, Horry JP, Bakry F, Lanaud C, Goazalez de Lean D (1993) A molecular marker-based linkage map of diploid bananas (*Musa acuminata*). Theor Appl Genet 87: 517–526.

Javed MA, Chai M, Othman YR (2004) Study of resistance of *Musa acuminata* to *Fusarium oxysporum* using RAPD markers. Biologia Plantarum 48: 93–99.

Lheureux F, Carreel F, Jenny C, Lockhart BEL, Iskra C (2003) Identification of genetic markers linked to banana streak disease expression in inter-specific *Musa* hybrids. Theor Appl Genet 106: 594–598.

Michelmore RW, Paran I, Kesseli RV (1991) Identification of markers linked to disease-resistance genes by bulked segregant analysis: A rapid method to detect markers in specific genomic regions by using segregating populations. Proc Natl Acad Sci USA 88: 9828–9832

Ortiz R (1996) The potential of AMMI analysis for field assessment of *Musa* genotypes to virus infection. HortScience 31: 742–762.

Ortiz R (2000) Understanding the *Musa* genome: An update. Acta Hort 540: 157–168.
Ortiz R, Vuylsteke D (1994a) Genetic analysis of apical dominance and improvement of suckering behaviour in plantain. HortScience 119: 1050–1053.
Ortiz R, Vuylsteke D (1994b) Inheritance of albinism in banana and plantain (*Musa* spp.). HortScience 29: 903–905.
Ortiz R, Vuylsteke D (1994c) Inheritance of black sigatoka disease resistance in plantain-banana (*Musa* spp.) hybrids. Theor Appl Genet 89: 146–152.
Ortiz R, Vuylsteke D (1995a) Effect of the parthenocarpy gene *P1* and ploidy on fruit and bunch traits of plantain-banana hybrids. Heredity 75: 460–465.
Ortiz R, Vuylsteke D (1995b) Inheritance of dwarfism in plantain (*Musa* spp., AAB group). Plant Breeding 114: 466–468.
Ortiz R, Vuylsteke D (1998) Quantitative variation and phenotypic correlations in banana and plantain. Sci Hort 72: 239–253.
Simmonds NW (1952) Segregations in some diploid bananas. J Genet 51: 458–469.
Zambrano AY, Martinez G, Gutierrez Z, Manzanilla E, Villardon JLV, Demey RY (2007) Marcador RAPD asociado a la resistencia a *Fusarium oxysporum en Musa*. Interciencia 32: 775–779.

7

Molecular Mapping of Complex Traits

Rodomiro Ortiz

ABSTRACT

Mendelian genetics has been useful for elucidating qualitative trait variation in *Musa*. Recent advances in high throughput "omics" technology will provide a means for mapping and gaining insights into complex traits, enabling the translation of genetic information into biological function. Host plant resistance to pests and pathogens, and fruit parthenocarpy are among the most important target traits among the mapping populations available from the *Musa* Genomics Consortium. In spite of the advances on developing molecular markers and their further use in *Musa* genetic maps, genetics and breeding research, reports on quantitative trait loci (QTL) mapping are scanty in banana and plantain. Advances made in mapping quantitative traits in *Musa* are discussed.

Keywords: quantitative traits, mapping populations

7.1 Introduction

Mendelian genetics has been useful for elucidating qualitative trait variation in *Musa* (Ortiz 2000). Likewise, biometrical approaches have facilitated the understanding of host plant resistance genetics to pathogens and pests affecting crop yield, or attributing to ploidy or gene action to bunch and fruit traits. Quantitative trait variation may be also affected by the environment and the genotype-by-environment interaction (Ortiz and Tenkouano 2011). Recent advances in high-throughput omics technology will further provide

---

Department of Plant Breeding and Biotechnology, Swedish University of Agricultural Sciences, Box 101, SE-230 53, Alnarp, Sweden; e-mail: *rodomiroortiz@gmail.com*

List of abbreviations after the text.

a means for mapping and gaining insights into complex traits, enabling to translate genetic information into biological function.

7.2 Target Traits and Populations

Host plant resistance to pests and pathogens, and fruit parthenocarpy are among the most important target traits among the mapping populations available from the *Musa* Genomics Consortium (http://www.musagenomics.org/index.php?id=49). Most of the F_1 or F_2 segregating populations come from crosses between diploid accessions, e.g., *M. acuminata* ssp. *malaccencis*, spp. *banksii*, or more recently *M. balbisiana*. The molecular markers that have been used are restriction fragment length polymorphism (RFLP), microsatellites or simple sequence repeat (SSR), random amplified polymorphic DNA (RAPD) amplified fragment length polymorphism (AFLP) or resistance gene analog (RGA) by research teams in Brazil (UCB, EMBRAPA), Cameroon (CRBP), France (CIRAD), India (IIHR), Nigeria (IITA) and Uganda (IITA and NARO).

Although bacteria, fungi, insects, nematodes, viruses affect banana and plantain yields and their stability across environments significantly, there are other target traits for *Musa* breeders (Fig. 7-1), viz., dwarfism, earliness, short cycling, suckering behavior, improved root systems, high bunch weight, large fruits sizes and quality that meets end-users' demands in both diploid

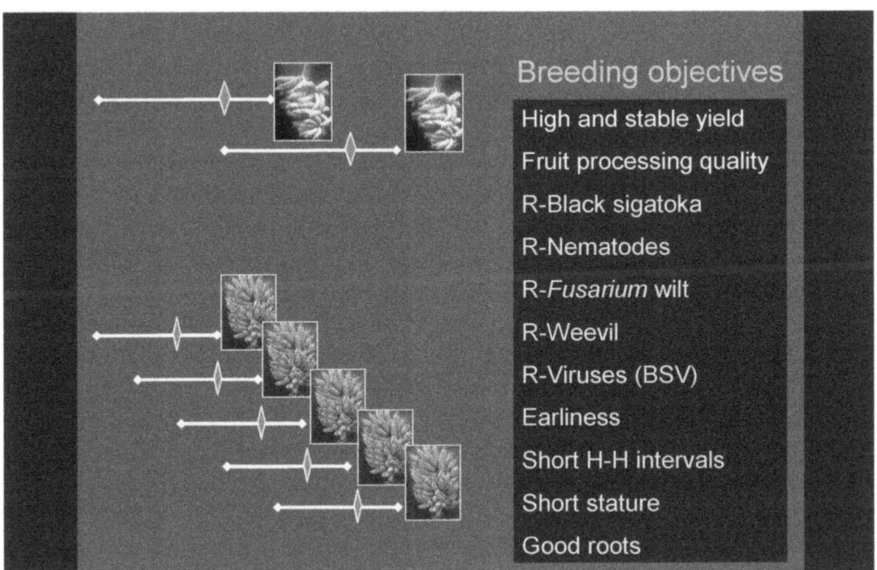

Figure 7-1 Breeding objectives in plantain and banana breeding H-H: harvest to harvest, R = host plant resistance. [Drawing courtesy of Dr. Abdou Tenkouano, AVRDC, Tanzania].

and polyploid breeding populations. Post-harvest losses are also a serious deterrent to expanded production in some areas of the developing world, particularly in tropical sub-Saharan Africa (Vuylsteke et al. 1997).

7.3 Framework Maps and Markers

The first *Musa* genetic map was based on 58 RFLP, 28 RAPD and four isozyme markers that segregated in a F_2 population of 92 diploid individuals. A total of 90 loci were detected, 77 of which were on 15 linkage groups while 13 segregated independently. Segregation distortions were evident in 36% of the loci, mostly favoring the male parent and are likely due to chromosome structural rearrangements (Fauré et al. 1993). The first saturated map using 89 diploid individuals from selfing the highly fertile diploid banana cultivar "M53" included more than 300 markers dispersed in 11 linkage groups representing the genome of *Musa acuminata*: 100 were co-dominant RFLP, 30 were SSR, and remaining markers were dominant AFLPs. Several markers showed high segregation distortion levels due to *M. acuminata* subspecies-specific translocations (Noyer et al. 1997). Carreel et al. (1999) used 153 diploid individuals from the cross "Calcutta" 4 x "Madang" to develop a linkage map with 110 AFLPs, which were associated with 39 RFLPs and SSRs that allowed to anchor the linkage groups to the banana genetic core map. Significant allelic distorted segregations were observed for 58 markers in the F_2 progeny. Recently, two parental maps composed of 167 SSRs and 322 DArT markers covering 1,197 cM were established from 180 F_1 progeny from a cross between two genetically distant accessions of *M. acuminata*, "Borneo" and "Pisang Lilin" (P. Lilin) (Hippolyte et al. 2010). Based on the gametic recombination of each parent, a significant proportion of the markers (21.7%) deviated ($P < 0.05$) from the expected Mendelian ratios. These skewed markers were distributed in different linkage groups for each parent. This first saturated map was proposed as a "reference *Musa* map" for further analyses (Hippolyte et al. 2010).

Microsatellites and other DNA marker systems have been used for characterizing *Musa* breeding populations. For example, SSR markers provided insights into the segregation of $2n$ eggs in triploid plantains and heterozigosity levels in tetraploid hybrids and their parents used in $3x$ x $2x$ crosses. About 75% of the alleles donated by the maternal plantain genotype ("Obino l'Ewai") segregated, thereby confirming the occurrence of recombination during the formation of $2n$ megaspores in triploid fertile plantain (Crouch et al. 1998). Analysis of wild diploid accession "Calcutta 4" and its selfed-progeny suggested that this accession was highly heterozygous. Variable numbers of tandem repeat (VNTR) markers were also used to assess the variation among diploid and tetraploid plantain-banana hybrids and their parents (Crouch et al. 1999a). The $4x$ hybrids were generally more closely related to their triploid plantain parent than their diploid full-sibs. VNTR analysis was able to identify hybrids that may be most useful in subsequent breeding of secondary triploid hybrids. A negative association was observed between the VNTR-based genetic similarity of hybrids to the triploid plantain parent and a phenotypic distance index based on eight agronomic descriptors, whereas there was no association between the molecular genetic similarity of hybrids to the diploid wild parent and the respective phenotypic distance index. Many VNTRs generated an unexpectedly high number of amplification products from diploid and triploid parents plus their progeny that suggested the presence of a high frequency of loci duplication in both A and B genomes. Further research by Crouch et al. (1999b) showed that AFLP, RAPD and VNTR assays detected a high level of polymorphism between parental genotypes and within progeny populations. Nonetheless, AFLP assays had the highest multiplex ratio whereas VNTR analysis detected the highest levels of polymorphism. AFLP analysis of a full-sib tetraploid hybrid population confirmed previous reports based on VNTR analysis, of a high frequency of recombination during 2*n* gamete formation by triploid plantain landraces. Both VNTR and RAPD analyses of a full-sib triploid hybrid population suggested a high frequency of homoeologous recombination during "*n*" gamete formation by tetraploid hybrids. Low correlations were observed between estimates of genetic similarity based on different types of markers used. Microsatellites were also used for DNA-fingerprinting to distinguish among secondary triploid full-sib hybrids (Ortiz et al. 1998). These research results show the importance of genetic characterization of *Musa* breeding stocks, and the value of microsatellites and other DNA markers for marker-aided genetic research in *Musa*.

Inspite of the advances on developing molecular markers and their further use in *Musa* genetic maps, genetics and breeding research, reports on quantitative trait loci (QTL) mapping are scanty in banana and plantain (see Chapter 6). Hence, this chapter does not include references regarding software used for QTL mapping, or details on trait-wise QTL and chromosomal locations.

7.4 Mendelization of QTL

In the 1990s there were some research attempts for using ploidy polymorphism and the few available genetic markers to account for quantitative trait variation (QTV) for traits such as plant height (Ortiz and Vuylsteke 1995a), host plant resistance to black Sigatoka (Craenen and Ortiz 1997), and bunch weight and fruits sizes (Craenen and Ortiz 1996; Ortiz et al. 1997). The amount of QTV accounted by ploidy and genetic markers varied according to the characteristic and cross in which the markers were examined. Table 7-1 summarizes the genetic models of QTV for bunch and fruit traits as explained individually by the major locus controlling host plant resistance to black sigatoka (*bs_1*) and one of the fruit parthenocarpy (*P_1*) genes, plus their epistatic interactions in the most successful cross for plantain improvement.

These results were not surprising because ploidy, fruit parthenocarpy, and host-plant response to black Sigatoka significantly influence the phenotype for bunch and fruit traits. This research showed the feasibility for dissecting QTV using segregating ploidy and genetic markers in hybrid offspring derived from a triploid crop such as plantain.

Table 7-1 Multiple regression models based on ploidy (X), marker loci that account for quantitative trait variation for bunch and fruit traits in the most successful breeding plantain-banana cross ("Obino l'ewai" x "Calcutta-4") for producing promising tetraploid offspring (After Ortiz et al. 1997).

Trait	Regression model	Standard error	R^2 (%)
Bunch weight (BW, kg plant^{-1})	$BW = 4.4\,X + 3.5\,A_{P1} + 4.2\,D_{bs1} - 5.5\,A_{bs1} + 6.3$	2.4	88
Number of hands (H)	$H = 3.61\,D_{P1} + 3.1$	1.2	75
Fruit number (F)	$F = 55.3\,D_{P1} + 44$	31.3	82
Average fruit weight (AFW, g)	$AFW = 58\,X + 33.2\,A_{P1} + 49.4\,D_{P1} - 7.3\,A_{bs1} - 65.5$	22.4	94
Average fruit length (FL, cm)	$FL = 4.2 - 10.2\,A_{P1} - 7\,D_{bs1} + 6.5\,A_{bs1} - 0.01\,(A_{P1} \times D_{bs1}) + 9.3\,(A_{bs1} \times D_{P1})$	2.4	94
Average fruit width (FW, cm)	$FC = 2.9 - 8\,A_{P1} - 4.7\,D_{bs1} + 4.4\,A_{bs1} + 0.4\,(A_{P1} \times A_{bs1}) - 0.2\,(A_{P1} \times D_{bs1}) + 6.2\,(A_{bs1} \times D_{P1})$	1.2	94

A_{P1} = effect of allele substitution in *P_1* locus, A_{bs1} = effect of allele substitution in *bs_1* locus, D_{P1} = effect of dominance (i.e., intra-locus interaction) in *P_1* locus, D_{bs1} = net effect of intra-locus interaction in *bs_1* locus

7.5 Marker-aided Breeding

The first attempts for using DNA markers as indirect selection tools in plantain and banana breeding were undertaken in the late 1990s. For example, microsatellites revealed that hybrid performance was associated but not significantly correlated with genetic similarity indices based on both pedigree and molecular data (Tenkouano et al. 1998). This research showed that the available genetic models were not appropriate for populations with intergeneration genome size polymorphism. Further research by Tenkouano et al. (1999a, b) showed that pedigree-based estimates of parent-offspring relationships were substantially different from molecular data. The marker-based contribution of triploid maternal accessions to their diploid offspring was greater than expected from available models of meiosis in

Musa. Conversely, the maternal contribution to their tetraploid offspring was less than expected. Pedigree-based similarity was smallest for clones with no common parent and greatest for full-sibs. There was no association between marker-based similarity and pedigree relationships. DNA markers may therefore provide a more accurate description of genetic relatedness but pedigree-based analysis may prove useful for selecting promising crosses in *Musa* breeding. Likewise, prediction of progeny bunch weight in secondary triploids was best revealed by genealogical distance and equal parental contribution. Predicted fruit size was most accurate when DNA marker data were used with the assumption of an unequal parental contribution. Genetic models considering polyploidy and population structure, as determined by coefficients of parentage and an independent set of DNA markers, may be therefore sought for further association mapping research in the *Musa* crop.

7.6 Conclusion

Quantitative traits in *Musa* such as bunch mass, its components (hands and fruits number) or fruit weight and sizes, depend on genetics, ploidy and the environment where and when the crop grows (plant or ratoon cycle, season, location, year, crop husbandry). Although *Musa* landraces are available, hybrids and breeding materials vary on yield and other quantitative traits, their genetic factors are not easily dissected because of the large number of genes involved and the interactions among each other (i.e., epistasis) and with the environment. However, there are examples of complex traits in other crops in which individual genes and their contribution to the specific trait could be determined. Although such complex traits are not controlled by as many genes as for example bunch weight *per se*, its component traits (fruit number or average fruit weight) may involve simpler pathways that can more easily be measured phenotypically and characterized at the DNA sequence level. By determining the genes involved in component traits, one can find the alleles for these genes that are responsible for improved performance and then introduce these alleles through marker-assisted breeding into advanced breeding materials. In this regard, association mapping should allow finding associations between simple changes in gene sequence with distinct phenotypes. Some techniques identify one or two changes per genes and the analysis may expand by completely resequencing such genes across the larger diverse panel of genetic resources available in a *Musa* breeding program. The exact sequencing of all the naturally occurring alleles will be able to ascertain the function of alleles improving multi-genic traits. High quality phenotyping of allele effects under different environments will be needed to succeed in this endeavor.

Abbreviations

AFLP	:	amplified fragment length polymorphisms
CRBP	:	Centre de Recherches Regionales sur Bananiers et Plantains
CIRAD	:	Centre de Coopération Internationale en Recherche Agronomique pour le Dévelopement
EMBRAPA	:	Empresa Brasileira de Pesquisas Agropecuarias
IIHR	:	Indian Institute of Horticultural Research
IITA	:	International Institute of Tropical Agriculture
NARO	:	National Agricultural Research Organization
QTV	:	quantitative trait variation
RAPD	:	random amplified polymorphic DNA
RFLP	:	restriction fragment length polymorphism
RGA	:	resistance gene analogs
SSR	:	micro-satellites
UCB	:	Universidade Católica de Brasília
VNTR	:	variable number of tandem repeats

References

Carreel F, Abadie C, Carlier J, Tomepke K, Lagoda PJL, Bakry F (1999) Genome mapping and genetic analysis of the black leaf streak resistance in bananas. InfoMusa (ProMusa) 8: III–IV.

Craenen K, Ortiz R (1996) Effect of the black sigatoka resistance gene *bs1* and ploidy level in fruit and bunch traits of plantain-banana hybrids. Euphytica 87: 97–101.

Craenen K, Ortiz R (1997) Effect of the bs_1 gene in plantain-banana hybrids on response to black sigatoka. Theor Appl Genet 95: 497–505.

Crouch HK, Crouch JH, Jarret RL, Cregan PB, Ortiz R (1998) Segregation at microsatellite loci in haploid and diploid gametes of *Musa*. Crop Sci 38: 211–217.

Crouch JH, Crouch HK, Tenkouano A, Ortiz R (1999a) VNTR-based diversity analysis of 2x and 4x full-sib *Musa* hybrids. Elec J Biotechnol 2: 99–108: http://www.ejb.org/content/vol2/issue3/full/1/

Crouch JH, Crouch HK, Constand H, Van Gysel A, Breyne P, Van Montagu M, Jarret RL, Ortiz R (1999b) Comparison of PCR-based molecular marker analyses of *Musa* breeding populations. Mol Breed 5: 233–244.

Fauré S, Noyer JL, Horry JP, Bakry F, Lanaud C, Goñzalez de León D (1993) A molecular marker-based linkage map of diploid bananas (*Musa acuminata*) Theor Appl Genet 87: 517–526.

Hippolyte I, Bakry F, Seguin M, Gardes L, Rivallan R, Risterucci A-M, Jenny C, Perrier X, Carreel F, Argout X, Piffanelli P, Khan I, Miller R, Pappas G, Mbeguie-A-Mbeguie D, Matsumoto T, De Bernardinis V, Huttner E, Kilian A, Baurens F-C, D'Hont A, Cote F, Courtois B, Glaszmann J-C (2010) A saturated SSR/DArT linkage map of *Musa acuminata* addressing genome rearrangements among bananas. BMC Plant Biol 10: 65–65.

Noyer JL, Dambier D, Lanaud C, Lagoda P (1997) The saturated map of diploid banana (*Musa acuminata*). In: Plant and Animal Genomes V Conference Abstracts, 12–16 January San Diego, USA.

Ortiz R (2000) Understanding the *Musa* genome: an update. Acta Hort 540: 157–168.

Ortiz R, Vuylsteke D (1995a) Inheritance of dwarfism in AAB plantains. Plant Breeding 114: 466–468.
Ortiz R, Vuylsteke D (1995b) Effect of the parthenocarpy gene P_1 and ploidy in bunch and fruit traits of plantain and banana hybrids. Heredity 75: 460–465.
Ortiz R, Tenkouano A (2011) Genotype by environment interaction and *Musa* improvement. In: M Pillay, A Tenkouano (eds) Banana Breeding: Progress and Challenges. CRC Press, Boca Raton, FL, USA, pp 235–247.
Ortiz R, Craenen K, Vuylsteke D (1997) Ploidy manipulations and genetic markers as tools for analysis of quantitative trait variation in progeny derived from triploid plantains. Hereditas 126: 255–259.
Ortiz R, Vuylsteke D, Crouch HK, Crouch J (1998) TM3x: triploid black sigatoka resistant *Musa* hybrid germplasm. HortScience 33: 362–365.
Tenkouano A, Crouch JH, Crouch HK, Ortiz R (1998) Genetic diversity, hybrid performance and combining ability for yield in *Musa* germplasm. Euphytica 102: 281–288.
Tenkouano A, Crouch JH, Crouch HK, Vuylsteke D, Ortiz R (1999a) A comparison of DNA marker and pedigree methods for genetic analysis in plantain and banana (*Musa* spp.) clones. I. Estimation of genetic relationships. Theor Appl Genet 98: 62–68.
Tenkouano A, Crouch JH, Crouch HK, Vuylsteke D, Ortiz R (1999b) A comparison of DNA marker and pedigree methods for genetic analysis in plantain and banana (*Musa* spp.) clones. II. Predicting hybrid performance. Theor Appl Genet 98: 69–75.
Vuylsteke D, Ortiz R, Ferris RSB, Crouch JH (1997) Plantain improvement. Plant Breed Rev 14: 267–320.

8

Map-Based Cloning in *Musa* spp.

Andrew James,[1,*] *Rodomiro Ortiz*[2] and *Robert Miller*[3]

ABSTRACT

Map based cloning (MBC) is a forward genetics approach which aims to identify the nucleotide sequences that underlies a specific trait which may be the product of either induced or naturally occurring variation. Quantitative Trait Loci (QTL), the basis of many important agronomic traits, are defined by position and, unless a corresponding mutant can be identified, MBC or association mapping are the only ways to obtain the sequence of genes underlying a quantitative trait. Quantitative trait loci (QTL) are controlled by many loci and environmental factors and therefore the phenotype does not provide reliable genotypic information. However, QTL cloning by MBC has been successful in model plants.

The most common problem in MBC is inconsistencies in mapping data. The small population size of segregating offspring from *Musa* interspecific, interploidy crosses has been regarded as the major obstacle to genetic analysis. However, some progress has been achieved in developing genetic markers and quantitative trait variation in *Musa*. Genomic resources available for map-based cloning in *Musa* include, resistance genes, resistance gene analogs (RGA), and RGA genetic markers. Bacterial artificial chromosome (BAC) resources are being used for physical mapping of RGAs in *Musa*.

[1]Centro de Investigacion Cientifica de Yucatan, Calle 43 No. 130, Col. Churburna de Hidalgo, Merida, C.P. 97200, Yucatan, Mexico; e-mail: *andyj007@cicy.mx*
[2]Department of Plant Breeding and Biotechnology, Swedish University of Agricultural Sciences, Box 101, SE-230 53, Alnarp, Sweden; e-mail: *rodomiroortiz@gmail.com*
[3]Laboratório de Microbiologia: Interação Planta-Praga, Instituto de Ciências Biológicas, Departamento de Biologia Celular, Universidade de Brasília, Campus Universitário Darcy Ribeiro, Asa Norte, 70910-900, Brasília, D.F., Brasil; and Universidade Católica de Brasília, SGAN Qd. 916 Módulo B, Av. W5 Norte, CEP: 70.790-160, Brasília/DF, Brazil; e-mail: *rmiller@pos.ucb.br or bruna@tba.com.br*
*Corresponding author

List of abbreviations after the text.

When compared to other crops of equal or lesser economic importance, platforms (viz.: high resolution genetic maps, mapping populations, ESTs and the whole genome sequence) for map based cloning in banana and plantain, are lagging in development. However, a relatively high density reference genetic map of *M. acuminata* was recently made public. As this genetic map includes many SSR markers, it should be useful for researchers who do not have access to expensive high throughput technology such as microarray technology.

Keywords: map based cloning, genetic markers, resistant gene analogues, genomic libraries

8.1 Introduction

Many traits of agronomic interest are controlled by genes whose products, function, sequence and position in the genome are unknown. To isolate these genes it is necessary to use methods, which do not require prior knowledge of the gene product(s). One such method is map-based cloning (MBC), also referred to as positional cloning. MBC is a forward genetics approach, which aims to identify the nucleotide sequence(s) that underlie a specific trait, which may be the product of either induced or naturally occurring variation. Quantitative trait loci (QTL), the basis of many important agronomic traits, are defined by position and, unless a corresponding mutant can be identified, MBC or association mapping are the only ways to obtain the sequence of genes underlying a quantitative trait. MBC is also essential for understanding the genetic basis underlying phenotypes amongst ecotypes of model species such as *Arabidopsis* (Jander et al. 2002).

The basis of MBC is that recombination and physical distance are directly related; the greater the physical distance between markers/genes, the more recombination events (RE) occur between homologous chromosomes. Linkage analysis to identify chromosomal regions containing trait loci in model organisms depends on following the inheritance of traits in pedigrees and aims to find co- inheritance of segregating traits and genetic markers. Trait loci are inferred to map in the vicinity of marker loci. For MBC two conditions must be met: (1) segregation for the trait must be based on genetic differences, and (2) the genes responsible can be mapped to a chromosomal position linked to a molecular marker. MBC in plants is most efficient when one has the resources available including facultative autogamy in the plant; high quality phenotyping data available across environments; availability of large F_2 segregating populations or chromosome deficiency lines; high density molecular marker genetic maps, ideally derived from isogenic lines (because this maximizes linkage disequilibrium between markers and trait loci); short generation times; large insert genomic libraries maintained in yeast or bacterial artificial chromosome (BAC) vectors; several well-characterized expressed sequence tag (EST) or cDNA libraries; and, ideally, a physical map (for example, of contiguous clones assembled from a fluorescence-based high information content fingerprinted BAC library) or the complete genome sequence.

For detecting the maximum number of DNA polymorphisms, the sequences of two closely related genomes are required. For example the two genomes of *Oryza sativa* and *O. indica* could be distinguished by 80,127 polymorphic sites (International Rice Genome Sequencing Project 2005) and the two ecotypes of *Arabidopsis*, Col-0 and L*er*, by 56,670 polymorphisms (Jander et al. 2002).

MBC follows these steps: Identify a phenotype/mutant of interest (for example, for disease resistance) in a large F_2 mapping population that segregates for the desired phenotype. Resolution is primarily determined by size of the mapping population; in *Arabidopsis* a resolution of 10–40 kb requires 1,000 plants or 2,000 chromosomes (Lukowitz et al. 2000).

- Identify a molecular marker/s that co-segregates with the locus/loci of interest in a F_2 mapping population. Markers should be co-dominant so that both homologous chromosomes of a plant are genotyped; thus giving the maximum amount of information.
- Identify large insert genomic clone/s to which the marker probe hybridizes; this may be in a COSMID, FOSMID, bacterial (BAC), yeast (YAC) or a transformation-competent (TAC or BIBAC) artificial chromosome vector. If a plant transformation protocol exists then a TAC or BIBAC clone may be directly used for a complementation test.
- Once a marker is found, which co-segregates with the trait, a "chromosome walk" is initiated from the marker using hybridization between overlapping clones; this continues until the "walk" encounters a marker known to be on the opposite side of the gene. The major difficulty with this technique in plants is the abundance of repetitive sequence. If the walk went through such a region, the walk could be diverted to a different chromosome, and the gene would not be found. This laborious, time-consuming approach has now been superseded by "chromosome landing" (Tanksley et al. 1995). This technique aims to reduce the "walk" by identifying molecular markers with a physical distance from the targeted gene that is less than the average insert size of the genomic library being used for clone isolation. This can be achieved by using higher density molecular marker maps and/or by using orthologous probes derived from related species/families (comparative maps). The DNA marker is then used to screen the library and "land" on the clone containing the gene.

- For genetic complementation (transformation) to demonstrate functionality of the candidate gene, DNA fragments between flanking markers are cloned and introduced into a plant lacking the gene.
- To sequence the candidate gene and determine if the function is known.

8.2 Bulked Segregant Analysis

Bulked segregant analysis (BSA) is a rapid mapping strategy suitable for monogenic qualitative traits to identify markers that are genetically linked to a mutation (Michelmore et al. 1991). BSA compares two bulked DNA samples of individuals from an F_2 or RI segregating population originating from a single cross. Within each pool, the individuals are identical for the trait or gene of interest but arbitrary for other genes. Two pools contrasting for a trait are analyzed to identify markers that distinguish them. Two bulked pools of segregants, differing for one trait, will differ only at the locus harboring that trait. Markers, polymorphic between two pools, will be genetically linked to loci determining the trait used to construct the pools. Because pools of DNA are examined rather than individuals, the number of PCR reactions, or marker arrays such as DArT is reduced when compared to building a genetic map for the purpose of identifying markers associated with a phenotype. As linkage maps become more saturated with random markers the mapping process becomes less efficient; therefore BSA provides a method to focus on regions of interest, areas sparsely populated with markers, or for locating genes that do not segregate in populations initially used to generate the genetic map.

8.3 Map-based cloning of Quantitative Trait Loci (QTL)

QTL are controlled by an unknown number of loci and environmental factors, therefore phenotypes of individuals do not provide reliable genotypic information; however QTL cloning by MBC has been successful in tomato (*Fw*2.2, *Brix*9-2-5), *Arabidopsis* (*Frigida*) and rice (*Hd*1 and *Hd*6); all were found to be controlled by single genes (reviewed by Remington et al. 2001). Four of these had major effects and the fifth a minor effect; therefore it is possible to clone those with minor effects also. These QTL were converted to single Mendelian factors by minimizing residual genetic and environmental variation using near isogenic lines (NILs), which were developed by repeated backcrossing of one F_1 to one parent. Fine scale QTL mapping was then carried out on F_2 or F_3 progeny. Problems arise if plants are self-incompatible, have inbreeding depression, have long generation times and if there is no transformation protocol available for complementation. To overcome these problems chromosome deficiency lines have been used in *Drosophila*; however, deficiency lines are not yet available in most plants. The other possibility is to use association mapping although there may be problems of population admixture and inconsistent relationships between linkage disequilibrium (LD) and physical distance and the LD decay rates may be too slow or too rapid. If the genome is well characterized, e.g., maize, then candidate genes may be identified in QTL regions, although a QTL may harbor hundreds of genes.

8.4 Application of MBC in Model Plants

One of the first plant genes to be cloned on the basis of its position was the disease resistance gene *Pto* of tomato (Martin et al. 1993), although it took several years of work. Similarly, it took several years to clone the first genes by chromosome walking in *Arabidopsis*. However, with the completion of sequencing of two *Arabidopsis* genomes and use of high-throughput methods for single nucleotide polymorphisms (SNP) and indel detection, over 50,000 polymorphic markers are now available and any *Arabidopsis* gene that has a mutant phenotype can be cloned with less than one person/year of work (Lukowitz et al. 2000; Jander et al. 2002), assuming that five life cycles can be completed within 12 months. Knowledge of gene function in *Arabidopsis* also aids in the identification of orthologous genes in other plants.

In *Zea mays*, mapping populations are easily created because of separate male and female flowers. The first step of MBC in maize is to identify the chromosome arm on which the mutation is located. There are two ways to do this: (1) use of translocation stocks (1 or 2 crosses needed) or (2) BSA analysis (which is more efficient). Chromosomes are divided into several "bins" and mutants are mapped to bins (10 normal plant DNA and 10 mutants DNA, plus two markers per chromosome arm). DNA preparations are made from individuals to map a mutation to a defined interval. As in *Arabidopsis*, this requires 1,000 plants to map to a single BAC because although the maize genome is 20 times larger, it probably has a similar number of genes and recombination mostly occurs within genes (Bortiri et al. 2006). The genetic map of maize for 2006 included 3,149 gene–based indel polymorphisms (IDPs) giving a total of 5,179 polymorphic markers available from two mapping projects. To enable the anchoring of genes to the physical map, BAC contigs are searched with gene-specific oligonucleotide-based ("overgo") probes derived from ESTs. Problems may arise when overgos detect members of gene families or there is inaccurate BAC placement; these problems may be avoided by eliminating those positive BACs that are detected in many contigs. Another source of markers is BAC end sequences (BES); 154,293 BES have been generated from one maize BAC library.

Due to the conservation in gene order and number between the genomes of cereals, maize sequences from which markers have been developed can be used to blast rice sequences using the Gramene website. Once a syntenous rice region is identified, rice genes can be used to find orthologous maize genes by blast or the comparative map feature of Gramene. The linking of the physical to the genetic map of maize uses fluorescence-based high information content fingerprinting techniques for assembling BAC clones into contigs. The mapping interval between two markers is reduced so that it spans several maize BACs (0.1 cM may be contained in one or two BAC clones) and several genes in the orthologous rice sequence. The rice sequence is then searched for likely candidate genes to identify candidate maize equivalents. If several candidate genes are found linked to a marker, it may be necessary to look at RNA levels for comparison. Candidate genes are verified by recombination, comparing and sequencing alleles, and by transformation. The first published positional cloning paper in maize was a QTL region that distinguished teozinte and maize, identified using a molecular marker tightly linked to a BAC contig, flanking markers, and a syntenic region of the rice genome. The QTL was then narrowed down to a gene of 1,042 bp by recombination using 3,106 individuals (Doebly and Stec 1993).

8.4.1 *Problems Likely to be Encountered with MBC (Examples from* **Arabidopsis)**

1) The most common problem is inconsistencies in mapping data; therefore it is important to make more than one mapping population. Errors in phenotype scoring may occur because many phenotypes are not fully penetrant or are affected by growth conditions; these have to be re-examined in F_3 or F_4 populations.
2) Analysis of naturally occurring variation has problems when a given trait is influenced by more than one locus. Fine mapping of these loci requires reduction of genetic complexity of the mapping population; for example, by creating recombinant inbred lines in which only one of the loci remains polymorphic. Examples in *Arabidopsis* are circadian rhythm, flowering time, seed size, dormancy, secondary metabolism and trichome density.
3) Epigenetic mutations; these are heritable changes in expression and function of a gene that are not due to changes in the DNA sequence; for example, the *SUPERMAN* flower development gene of *Arabidopsis*, this is possibly due to cytosine methylation; therefore it is not possible to map these loci.

4) The ratio of genetic distance to physical distance may differ with respect to location on a chromosome; for example, the *GURKE* gene shows little recombination because it is located very close to the centromere and the presence of repetitive DNA has a strong effect on recombination; an 1% recombination rate corresponds to 100 to 400 kb but close to the centromere 1% recombination corresponds to 1,000 to 2,500 kb. Similar problems arise if the gene is located close to the telomere.
5) The presence of DNA rearrangements; for example, an inversion of 500 kb between two accessions means that all recombination events within the inversion will create non-viable meiotic products which are impossible to map. Irradiation and T-DNA insertions can also cause DNA rearrangements, which are difficult to map.

8.5 Traits, Genes of Interest and Genetic Knowledge for *Musa* Improvement

8.5.1 *State of the art of* Musa *genetics*

In spite of being a staple food in the tropics, scanty results on trait inheritance in plantain and bananas were not available until the 1990s (for a summary see Ortiz 1995 and references therein). Although the breeding system, ploidy series, and asexual propagation of the *Musa* crop make genetic analysis difficult in banana and plantain, genetic research was undertaken throughout the 1990s mainly in West Africa at the High Rainfall Station of the International Institute of Tropical Agriculture (IITA, Onne, Nigeria) and in East Africa at IITA Eastern and Southern Africa Regional Centre (ESARC, Kampala, Uganda). Other genetic research results (especially facilitated by molecular biology analysis; Crouch et al. 1998, 1999a, 1999b) also became available mainly from the Centre de Coopération Internationale en Recherche Agronomique pour le Dévelopement (CIRAD, France) and the Empresa Brasileira de Pesquisas Agropecuarias (EMBRAPA, Brazil).

The small population size of segregating offspring from *Musa* interspecific, interploidy crosses has been regarded as the major obstacle to genetic analysis, which explains the absence of inheritance studies on nearly or quite sterile triploid plantain and banana. Nonetheless, as a result of research at IITA, the inheritance of some traits was unraveled and some genetic markers became available (Table 8-1). Furthermore, the availability of segregating offspring in test crosses from triploid heterozygous plantain-diploid true breeding wild banana (Ortiz et al. 1998) crosses and of diploid-diploid plantain-banana hybrids paved the way for further *Musa* genetic research (Ortiz and Vuylsteke 1996; Vuylsteke et al. 1997; Ortiz 2000). Through such means, insights into the triploid plantain genome, which was thought to be inaccessible for genetic analysis, were revealed.

Table 8-1 Type of gene action of morphological, reproductive and agronomic traits.

Trait	Type of gene action	Reference
Albinism	Two independent recessive alleles (a_i) with complementary gene action	Ortiz and Vuylsteke 1994a
2*n* pollen	At least one dominant gene	Ortiz 1997a
Dwarfism	One major recessive (*dw*) for short false internodes with modifiers affecting plant height	Ortiz and Vuylsteke 1995a Ortiz 2000
Apical dominance and suckering behavior	One major recessive gene (*ad*). The dominant allele (*Ad*) shows incomplete penetrance and variable expressivity	Ortiz and Vuylsteke 1994b Ortiz 2000
Persistent male bracts and hermaphrodite flowers	Two independent 2-locus systems of complementary dominant genes	Simmonds 1952 Ortiz 1996a
Fruit parthenocarpy	Three independent complementary dominant genes (P_i), with at least one allele (P_1) segregating locus in plantain-Calcutta 4 offspring	Simmonds 1952 Ortiz and Vuylsteke 1995b
Bunch orientation	Oligogenic system regulated by the epistatic effects of at least three dominant loci	Ortiz and Vuylsteke 1998
Pseudostem waxiness	One recessive gene (*wx*)	Ortiz et al. 1995b

8.5.2 Trait Inheritance

Fruit parthenocarpy inheritance in diploid banana (Simmonds 1952) was the most well known genetic research on an important descriptor until the 1990s. Albinism—complete lack of chlorophyll in any plant tissue, in diploid plantain-banana hybrid seedlings (Ortiz and Vuylsteke 1994a) was among the first research results of the 1990s, which provided a means to re-investigate the almost unapproachable *Musa* genome. More than 40 years after previous *Musa* genetic research, this report established that albinism is under the genetic control of two independent recessive alleles. This finding also showed that deleterious recessive alleles are in French plantains and in both cultivars and breeding populations of diploid bananas. The latter suggested therefore that phenotypic recurrent selection for agronomic traits might be based on the elimination of such deleterious genes.

Further genetic research in plantain-banana offspring led to the discovery of alleles inherited in a Mendelian fashion –in either monogenic or polygenic epistatic systems, controlling 2*n* pollen, dwarfism, apical dominance, persistency of male bracts and hermaphrodite flowers, bunch orientation and pseudostem waxiness (Table 8-1). Furthermore, the underpinnings of host plant resistance to black Sigatoka, banana weevil, banana streak virus and burrowing nematodes were elucidated and added to the previous knowledge on genetics for resistance to yellow Sigatoka, Moko and Panama diseases (Table 8-2).

Table 8-2 Inheritance of host plant resistance to major *Musa* pests.

Pest	Host plant resistance inheritance	Reference
Black Sigatoka	One major recessive resistance allele (bs_1) and two independent alleles with additive effects (bsr_2 and bsr_3) with a dosage effect of favorable alleles	Ortiz and Vuylsteke 1994c
	Intralocus interaction in the bs_1 locus apparently regulates the appearance of symptoms on the leaf surface, whereas the additive effect and the intralocus interaction of the bs_1 locus affect disease development in the host plant	Craenen and Ortiz 1997 Ortiz 2000
Yellow Sigatoka	Multiple genes with dosage effects in *M. acuminata* spp. *microcarpa* and spp. *errans*	Vakili 1968
	Dominant genes in *M. acuminata* spp. *Malaccensis*	Rowe 1984
	Recessive genes in *M. acuminata* spp. *burmannica*	Shepherd 1990
Banana streak virus (BSV)	Epistatic interactions arising from a specific cross may induce susceptibility to BSV and symptom expression	Ortiz 1996b
	A genetic mechanism involved in BSV appearance, and a monogenic allelic system confers the role of carrier to the *M. balbisiana* parent	Lheureux et al. 2003
Burrowing nematode	Two dominant genes with additive and interactive effects, whereby a recessive homozygous genotype of one locus suppresses dominant allele in other locus	Dochez et al. 2009
Banana weevil	Gene(s) with incomplete or partial dominance towards resistance parent	Ortiz et al. 1995a
Fusarium wilt (or Panama disease)	One major dominant gene for race 1	Vakili 1965a
Bacterial wilt (or Moko disease)	Several recessive genes	Vakili 1965b

8.5.3 Phenotyping of Host Plant Resistance

Precise phenotyping remains the main factor for an appropriate assessment of quantitative trait variation, and particularly for partial resistance to a pest. Field screening should be in areas where the specific pathogens or pests are ubiquitous. Comparisons are always made with respect to a known susceptible check. In this regard a scale for rating host-plant resistance based on biometrics offers a powerful tool for avoiding wrong groupings (Table 8-3). The uniform infection by a pathogen or pest may be enhanced by using "spread rows" of a known susceptible host. Spatial adjustments due to non-uniform variation, should be pursued after a significant analysis of variance for the "spread rows" (Table 8-4). Optimum lot size (or number of plants per plot), and appropriate replication numbers are available for field screening of host plant resistance to black Sigatoka in plantain-banana hybrids, desert banana and East African highland bananas (Ortiz and Vuylsteke 1994a; Okoro et al. 1997; Nokoe and Ortiz 1998), and root health in plantains (Blomme and Ortiz 1996). Likewise there are screening protocols for host plant resistance to banana weevil (Ortiz et al. 1995a; Kassim et al. 2010) and burrowing nematodes (Dochez et al. 2009).

Table 8-3 Grouping host response to a pathogen or pest as per comparison with a known host response to a susceptible and a resistant reference cultivar.

Comparison with known susceptible host*	Comparison with known resistant host*	Host response of hybrid
Significantly different	Not significantly different	Resistant
Not significantly different	Significantly different	Susceptible
Significantly different	Significantly different	Partial resistant
Not significantly different	Not significantly different	Inconclusive

According to the Dunnett's test

Table 8-4 Analysis of variance for host response to a pathogen using *b* blocks, *r* reps and *p* plants.

Source of variation	Degrees of freedom	*P* in F test
Blocks (B)	$b-1$	Non significant = uniform spread
Reps/B	$(r-1)b$	Non significant = uniform spread
Residual *	$(p-1)br$	
Total	$pbr-1$	

*Minimize this source of variation to get a low coefficient of variation

8.5.4 Genetic Markers and Quantitative Trait Variation

Major recessive genes control apical dominance, black Sigatoka resistance and dwarfism in plantain-banana hybrids (Tables 8-1 and 8-2). However, continuous variation was observed in their segregating offspring. Moreover, the dominant gene for fruit parthenocarpy (P_1), the recessive gene for black Sigatoka resistance (bs_1) and ploidy significantly affected the quantitative variation observed in plantain-banana tetraploid, triploid and diploid hybrids (Table 8-5; Vandenhout et al. 1995). Such results showed that in multigenic systems, much of the observed quantitative trait variation can be accounted by alleles with large phenotypic effects in a few loci. The results also confirm the influence of black Sigatoka in yield of plantain-banana

Table 8-5 Gene effects on host plant resistance to black Sigatoka, plant height, bunch weight and fruit sizes in plantain-banana hybrids.

Gene	Phenotypic effect	Reference
dw in Bobby Tannap plantain-Calcutta 4 hybrids	Homozygous *dw* hybrids showing shorter false-internodes had smaller plant height than normal plants	Ortiz and Vuylsteke 1995a
P_1 in plantain-Calcutta 4 hybrids	Higher ploidy and change of recessive to dominant allele in P_1 locus improve bunch and fruit weights, though both factors cannot account for all quantitative trait variation for such traits	Ortiz and Vuylsteke 1995b
bs_1 in plantain-Calcutta 4 hybrids	Combined effect of ploidy and *bs1* partially responsible for quantitative trait variation of bunch weight and fruit size: partially resistant and less susceptible hybrids showed higher yield than their more susceptible full sibs	Craenen and Ortiz 1996
P_1 and bs_1 in plantain-Calcutta 4 hybrids	Both loci account up to 80% of the quantitative variation in plantain yield	Ortiz et al. 1997

hybrids. For example, the bunch weight of tetraploid-banana hybrids appears to be significantly correlated with the disease's development time (Craenen and Ortiz 1998).

Epistasis can further enhance fruit sizes of high yielding plantain-banana hybrids. Both additive x additive and additive x intralocus interactions in of bs_1 and P_1 loci may also increase fruit mass. This finding was not surprising since non-additive gene action controls yield in asexual crops (Peloquin and Ortiz 1992). This type of gene action was further corroborated by comparing half-sib hybrid offspring derived from French plantains and the wild banana Calcutta 4 (Ortiz 1997c). Hybrid offspring from "Obino l'Ewai" showed on average higher bunch weight than "Bobby Tannap"-derived hybrid offspring, whereas "Bobby Tannap" had higher bunch weight *per se* than "Obino l'Ewai". Such a result can be accounted for by non-additive gene action (heterozygosity and epistasis) controlling bunch weight in plantain. Hence the potential bunch weight cannot be predicted based on parental performance but instead by using specific combining ability tests (Tenkouano et al. 1998a).

Hybrid performance was also associated but not significantly correlated with genetic similarity indices based on both pedigree and molecular data (Tenkouano et al. 1998b). Such research suggests that genetic models, assuming either disomic or polysomic inheritance, may not be adequate for populations with intergeneration genome size polymorphism. In this regard, Tenkouano et al. (1999b) reported that prediction of progeny bunch weight was best when based on genealogical distance and equal parental contribution. Predicted fruit size was most accurate when DNA marker data were used and the assumption of an unequal parental contribution was made. Consideration of parental heterozygosity produced larger residuals for all traits. Tenkouano et al. (1999a) further indicated that DNA markers may provide a more accurate description of genetic relatedness but pedigree-based analysis may prove useful for the selection of prospective parental combinations in *Musa* breeding.

8.5.5 Quantitative Genetics

Variance components are calculated based on plot means per cycle for genotypes, environments, and genotype-by-environment (GE) interactions, which allows the estimation of repeatability, broad-sense heritability and intraclass correlation coefficients (Ortiz 1995). Table 8-6 provides some estimates for morphological and agronomic traits as well as for measurements of host plant resistance to black Sigatoka. Low heritability values result from either lack of genetic variation for such trait(s) or significant G×E. Conversely, high trait heritability suggests either a non-significant G×E interaction or a high degree of genetic variation. A low repeatability indicates a significant influence of the environment. The magnitudes of intraclass correlations may tell how much variation among parental sources, ploidy or hybrids, accounts for the total genetic variance. This genetic information allows the use of distinct breeding schemes. For example, phenotypic recurrent selection for traits with high heritability and among hybrid intra-class correlation due to additive genetic variation, and recurrent selection with progeny testing for traits with low intra-class correlation due to non-additive genetic variation among hybrids, or reciprocal recurrent selection for traits in which both types of genetic variation are significant.

8.5.6 Genetic Knowledge Facilitated by Marker Analysis in Breeding Populations

Although there is no information on linkage disequilibrium in *Musa*, one can infer from other research reports how important it could be in the origin of banana and plantain or in the breeding of both arising from 2*n* gametes (or gametes with the sporophytic chromosome number). For example, Crouch et al. (1998, 1999b) analyzed with microsatellites the genetic constitution of tetraploid plantain-banana hybrids to characterize the nature of the gametes from which they were derived. During analysis of

Table 8-6 Broad-sense heritability (H^2), intraclass correlation among hybrids at same ploidy level within same cross (r_x) and (R) for host plant resistance to black Sigatoka, agronomic traits, yield and components in *Musa* hybrids (PB) and other germplasm (M).

Trait	H^2	r_x	R	Reference
Plant height (cm)	0.75 M		1.75 M	Ortiz 1997b
		0.59 P-B		Ortiz 1997c
Plant girth (cm)	0.86 M		3.08 M	Ortiz 1997b
Number of standing leaves at flowering	0.64 P-B	0.78 P-B	0.56 P-B	Craenen and Ortiz 1997
Youngest leaf spotted with black Sigatoka at flowering	0.57 P-B	0.69 P-B	0.59 P-B	Craenen and Ortiz 1997
Youngest leaf with black Sigatoka symptoms at flowering	0.57 P-B	0.72 P-B	0.49 P-B	Craenen and Ortiz 1997
Total leaf area affected by black Sigatoka (%) at flowering	0.45 P-B	0.58 P-B	0.33 P-B	Craenen and Ortiz 1997
Leaf index with necrotic spots due to black Sigatoka at flowering	0.37 P-B	0.47 P-B	0.29 P-B	Craenen and Ortiz 1997
Tallest sucker height (cm) at	0.82 M		2.20	
bunch harvest		0.57 P-B		Ortiz 1997c
Bunch weight (kg)	0.68 M		1.05 M	Ortiz 1997b
	0.88 P-B		2.35 P-B	Ortiz and Vuylsteke 1995b
		0.57 P-B		Ortiz 1997c
Hands per bunch (#)	0.78 M		1.70 M	Ortiz 1997b
	0.75 P-B		0.33 P-B	Ortiz and Vuylsteke 1995b
		0.43 P-B		Ortiz 1997c
Fruit per bunch (#)	0.80 M		1.99 M	Ortiz 1997b
	0.82 P-B		0.33 P-B	Ortiz and Vuylsteke 1995b
		0.54 P-B		Ortiz 1997c
Time to fruit filling (days)	0.14 P-B		0.01 P-B	Ortiz and Vuylsteke 1995b
		0.01 P-B		Ortiz 1997c
Fruit weight (g)	0.94 P-B		1.32 P-B	Ortiz and Vuylsteke 1995b
		0.54 P-B		Ortiz 1997c
Fruit length (cm)	0.90 – 0.93 M		4.87 – 6.42 M	Ortiz 1997b
			7.84 P-B	
	0.94 P-B			Ortiz and Vuylsteke 1995b
		0.65 P-B		Ortiz 1997c
Fruit circumference (cm)	0.86 – 0.90 M		3.48 – 4.24 M	Ortiz 1997b
	0.94 P-B		2.02 P-B	Ortiz and Vuylsteke 1995b
		0.67 P-B		Ortiz 1997c

the tetraploid hybrid population, 75% of the alleles donated by the maternal plantain genotype were observed to segregate. These data demonstrate the occurrence of recombination during the formation of 2*n* megaspores in triploid plantain.

Crouch et al. (1999a) undertook a comparative analysis of 2*x* and 4*x* full-sib plantain-banana hybrids with their parental genotypes and the results indicated that tetraploid hybrids are generally more closely related to their triploid plantain parent than their diploid full-sibs. There was a significant negative association between the variable number of tandem repeat (VNTR)-based genetic similarity of hybrids to their plantain parent and a phenotypic distance index based on eight agronomic descriptors. However, there was no association between the molecular genetic similarity of the hybrids to their diploid wild banana parent and the respective phenotypic distance index. Many microsatellite markers generated an unexpectedly high number of amplification products from the wild banana and the plantain genotypes plus their progeny, which may suggest the presence of a high frequency of loci duplication in both A and B genomes, which may pose another challenge for determining the extent of linkage disequilibrium in polyploid *Musa* with the aid of DNA markers.

Along with sexual polyploidization and segregation of 2*n* gametes, trisomic inheritance may affect the analysis of linkage disequilibrium in *Musa*. In this regard, Ortiz and Vuylsteke (1994d) reported trisomic pattern of segregation for some genetic markers in diploid hybrids ensuing from triploid plantain-diploid banana crosses. Their results suggest that there could not be preferential pairing among the homologous chromosomes of the so-called "A genome" but that random distribution may occur of the paired chromosomes to the cell poles during anaphase I of the first meiotic division, which may account for the high percentage of fertile diploid hybrids since AB hybrids should be sterile. Such a finding challenges the genome differentiation hypothesis between the ancestors of banana and plantain: *M. acuminata* and *M. balbisiana* –the donor of the so-called "B genome".

8.6 Genomic Resources Available for Map-Based Cloning in *Musa*

8.6.1 Plant Resistance Genes

Plant "gene for gene" race-specific resistance is known to be initiated following activation of genetically dominant, allele specific constitutive disease resistance (R) gene protein products, or receptors, by a dominant allele specific pathogen avirulence (*Avr*) gene products, or elicitors (Flor 1971). Conversely, Avr proteins, in the absence of their R-protein counterparts, can act as virulence determinants (Dangl 1994; Chisholm et al. 2006). Pathogen recognition, also known as incompatibility, activates a large number of inducible defense mechanisms, resulting in diverse defense responses. A number of key signaling molecules and responses appear in common in plants. These include, at the site of infection, a rapid hypersensitive cell death (HR), to limit pathogen spread, downstream production of signaling molecules such as reactive oxygen species (ROS), nitric oxide (NO) and phenolic accumulation, tissue and cell wall reinforcement (Dixon et al. 1994), vessel blockage, synthesis of pathogenesis-related (PR) proteins, accumulation of phytohormone signaling molecules salycilic acid (SA), jasmonic acid (JA) and ethylene, and formation of antimicrobial proteins such as phytoalexins (Hammond-Kosack and Jones 1996). Cellular release of ROS (Lamb and Dixon 1997) can increase levels of superoxide anion radicals (O_2^-), hydroxyl radicals (OH^-) and hydrogen peroxide (H_2O_2). This latter component is believed to play an important role in oxidation of pathogens, in phytoalexin biosynthesis, activation of defense genes, and triggering systemic acquired resistance (SAR), a priming of plant tissues far from the site of invasion against pathogen attack (Dong 2001; Métraux 2001).

Specific alleles from one or few resistance genes likely determine pathogen recognition (termed monogenic or oligogenic), initiating incompatibility. Although a direct interaction between R-proteins and their corresponding Avr proteins has been hypothesized, such one-to-one interactions have rarely been proved to occur. In the guard hypothesis, put forward to explain why Pto protein kinase requires the NBS-LRR protein Prf to activate defense following Avrpto recognition (Van der Biezen and Jones 1998), an R-gene may interact with a plant protein known as a guardee, which is typically targeted by an Avr protein. Modification of the guardee protein by the attacking virulence protein is therefore monitored by the R-gene, such that a complex of R-protein-modified guardee-Avr protein induces signal transduction and defense responses. Evidence for the guard hypothesis has been observed with an *Arabidopsis* R-protein (Mackey et al. 2002). Given that R-genes can provide resistance to pathogens, with minimal cost to the plant in terms of fitness reduction, introgression of such sources of resistance in conventional breeding programs is common (Pink 2002). Although such race-specific resistance can limit pathogen reproduction, limiting build up of inoculum potential, and slowing development of epidemics, resistance can often be ephemeral, as often seen in monocultures. This may arise as a result of mutation in pathogen Avr genes, which can occur without fitness reduction.

Known to be abundant in every plant species (Michelmore and Meyers 1998), over 40 R-genes from model plants and diverse crop species have now been characterized (Martin et al. 2003). Despite the broad range of recognized pathogens, R-genes share significant homologies in amino acid sequences and structural protein motifs, suggesting common mechanisms for protein-protein interactions as receptor systems and common roles in transduction of pathogen signals in plant defense (Ellis et al. 2000). A number of resistance gene structural classes have been identified based upon conserved protein domains.

The first R-protein class contains four subfamilies, all with cytoplasmic nucleotide binding site (NBS) and leucine-rich repeat (LRR) domains. In a second R-protein class, extracellular LRRs can be anchored by a short transmembrane domain (receptor like proteins) and, in a third class, can be linked to a cytoplasmic serine-threonine kinase domain. A fourth class comprises serine-threonine kinases alone, without accompanied LRRs, whilst a fifth class groups R-proteins with a coiled-coil domain anchored to the membrane. A sixth class contains diverse proteins.

The most prevalent class of R-genes encode proteins with NBS and LRR domains (Meyers et al. 1999). To date, their only known function in plants is in conferring disease resistance (Nimchuk et al. 2003), with over 20 proven R-genes originating from numerous monocotylendon and dicotyledon plant species. In *Arabidopsis*, of the 207 genes coding for domains characterisitic of disease resistance proteins (TAGI 2000), 149 possess NBS-LRR domains (Meyers et al. 2003). Genes of this class have been proven to confer resistance to diverse pathogens and pests, including fungi, bactéria, viruses, nematodes and aphids (Dangl and Jones 2001). Specificity is encoded, at least in part, in the fast-evolving C-terminal part of the LRR domain, which consists of a varying number of LRR repeats. Subsequent signaling is thought to involve the N-terminal region of the NBS domain (Belkhadir et al. 2004).

8.6.2 Resistance Gene Analogs

Resistance gene analogs (RGAs) have been characterized across diverse plant taxa such as *A. thaliana* (Aarts et al. 1998) and rice (Leister and Katagiri 2000), amongst others (for review see Chelkowski and Koczyk 2003). In *Musa*, non-TIR NBS-LRR RGAs have recently been characterized in wild and cultivated *M. acuminata* and *M. balbisiana* accessions (Pei et al. 2007). Miller et al. (2008) described a large scale analysis of non-TIR NBS-LRR RGA diversity in *M. acuminata* Calcutta 4, employed as a disease resistant donor species in banana breeding programs. From 52 distinct NBS-encoding protein sequence contigs, 33 contained contiguous open-reading frames (ORFs). This is a considerable number, given that 30% of the 161 putative genes in the *Arabidopsis* genome that code for NBS-type resistance proteins are of the non-TIR class (TAGI 2000). Non-TIR NBS-LRR RGA characterization has also been described in *M. acuminata* ssp. *malaccensis* (Peraza-Echeverria et al. 2008), together with diverse A and B genome cultivars and *Musa* species *M. ornata, M. schizocarpa, M. textilis,* and *M. velutina* (Mohamad and Heslop-Harrison 2008). Exhaustive analysis of diversity of NBS-LRR sequences in *M. acuminata* is likely to require further input, considering the abundance of such gene sequences reported in rice (Goff et al. 2002; Zhou et al. 2004). RGA characterization from additional R-gene families has also been reported, with receptor-like protein *Cf* orthologs isolated in Zebrina GF (Wiame et al. 2000), and *Pto*-type receptor like serine-threonine kinases in *M. acuminata* cv Tuu Gia (Peraza-Echeverria et al. 2007).

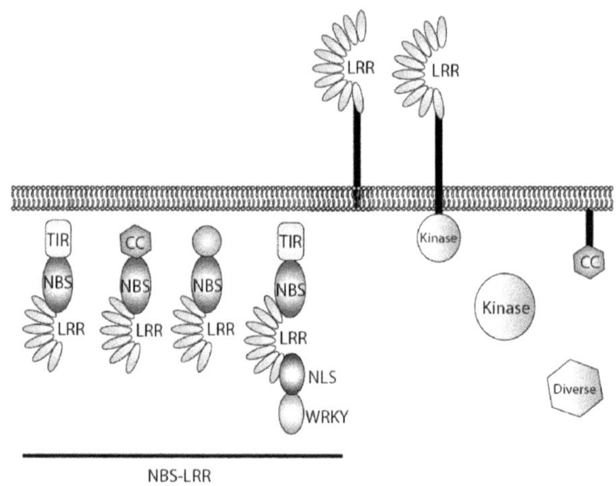

Figure 8-1 Representation of known R protein classes.

8.6.3 RGA Genetic Markers

RGAs have been mapped in diverse plant species, to genomic regions for quantitative trait loci (QTL) for disease resistance (e.g., Irigoyen et al. 2004, Xu et al. 2005; Xiao et al. 2007). Such co-localization has led to the hypothesis of an involvement of these genes in the effects of resistance QTL (Pflieger et al. 2001). Miller and colleagues (2008) reported 30 RGAs displaying single locus or multiple loci polymorphisms on restriction digested DNA from *M. acuminata* parentals for the reference genetic map in *Musa* under construction (Table 8-7; Fig. 8-2) (Hippolyte et al. 2008). Such polymorphism in RGA markers has also been reported in rice (e.g., Ramalingam et al. 2003). Identification of RGA markers co-segregating with QTL for resistance on segregating populations for disease resistance could offer potential for marker-assisted trait selection. Plans are underway to place *Musa* RGA markers onto future maps developed at Embrapa derived from crosses

Table 8-7 Polymorphisms in *M. acuminata* RFLP-RGA markers.

RGA contig	Genbank accesion number	*Polymorphisms observed on* M Table 7. Polymorphisms in *M. acuminata* RFLP-RGA markers **. *acuminata* parentals Borneo and Pisang Lilin**[a]		
		*Dra*I	*Hind*III	*Eco*RV
MaRGA03	ER935975	Monomorphic	monomorphic	polymorphic (multiple loci)
MaRGA04	ER935976	polymorphic (multiple loci)	polymorphic (multiple loci)	polymorphic (multiple loci)
MaRGA07	ER935979	polymorphic (multiple loci)	polymorphic (multiple loci)	polymorphic (multiple loci)
MaRGA08	ER935980	polymorphic (multiple loci)	polymorphic (multiple loci)	polymorphic (multiple loci)
MaRGA09	ER935981	polymorphic (multiple loci)	polymorphic (multiple loci)	polymorphic (multiple loci)
MaRGA10	ER935982	polymorphic (multiple loci)	polymorphic (multiple loci)	polymorphic (multiple loci)
MaRGA11	ER935983	polymorphic (multiple loci)	polymorphic (multiple loci)	polymorphic (multiple loci)
MaRGA12	ER935984	polymorphic (multiple loci)	polymorphic (multiple loci)	polymorphic (multiple loci)
MaRGA13	ER935985	Monomorphic	polymorphic (multiple loci)	Monomorphic
MaRGA14	ER935986	polymorphic (multiple loci)	polymorphic (multiple loci)	polymorphic (multiple loci)
MaRGA15	ER935987	polymorphic (multiple loci)	polymorphic (multiple loci)	polymorphic (multiple loci)
MaRGA16	ER935988	polymorphic (single locus)	Monomorphic	Monomorphic
MaRGA17	ER935989	polymorphic (multiple loci)	polymorphic (multiple loci)	polymorphic (multiple loci)
MaRGA18	ER935990	Monomorphic	polymorphic (multiple loci)	polymorphic (multiple loci)
MaRGA19	ER935991	polymorphic (multiple loci)	polymorphic (multiple loci)	polymorphic (multiple loci)
MaRGA20	ER935992	polymorphic (multiple loci)	polymorphic (multiple loci)	polymorphic (multiple loci)
MaRGA22	ER935994	polymorphic (multiple loci)	polymorphic (multiple loci)	polymorphic (multiple loci)
MaRGA23	ER935995	polymorphic (multiple loci)	polymorphic (multiple loci)	polymorphic (multiple loci)
MaRGA24	ER935996	Monomorphic	polymorphic (multiple loci)	polymorphic (multiple loci)
MaRGA25	ER935997	Monomorphic	polymorphic (multiple loci)	polymorphic (multiple loci)
MaRGA27	ER935999	polymorphic (multiple loci)	polymorphic (multiple loci)	polymorphic (multiple loci)
MaRGA28	ER936000	polymorphic (multiple loci)	polymorphic (multiple loci)	polymorphic (multiple loci)

Table 8-7 contd....

Table 8-7 contd....

RGA contig	Genbank accesion number	*Polymorphisms observed* on *M. acuminata* parentals Borneo and Pisang Lilin[a]		
		*Dra*I	*Hind*III	*Eco*RV
MaRGA29	ER936001	polymorphic (multiple loci)	polymorphic (multiple loci)	polymorphic (multiple loci)
MaRGA37	ER936009	polymorphic (multiple loci)	polymorphic (multiple loci)	polymorphic (multiple loci)
MaRGA38	ER936010	monomorphic	monomorphic	Monomorphic
MaRGA39	ER936011	monomorphic	monomorphic	Monomorphic
MaRGA40	ER936012	polymorphic (multiple loci)	polymorphic (multiple loci)	polymorphic (multiple loci)
MaRGA41	ER936013	polymorphic (multiple loci)	monomorphic	monomorphic
MaRGA42	ER936014	no hybridization	no hybridization	no hybridization
MaRGA43	ER936015	polymorphic (multiple loci)	polymorphic (multiple loci)	polymorphic (multiple loci)
MaRGA44	ER936016	polymorphic (multiple loci)	polymorphic (multiple loci)	polymorphic (multiple loci)
MaRGA45	ER936017	polymorphic (multiple loci)	polymorphic (multiple loci)	polymorphic (multiple loci)
MaRGA46	ER936018	monomorphic	polymorphic (multiple loci)	Monomorphic

Adapted from Miller et al. 2008

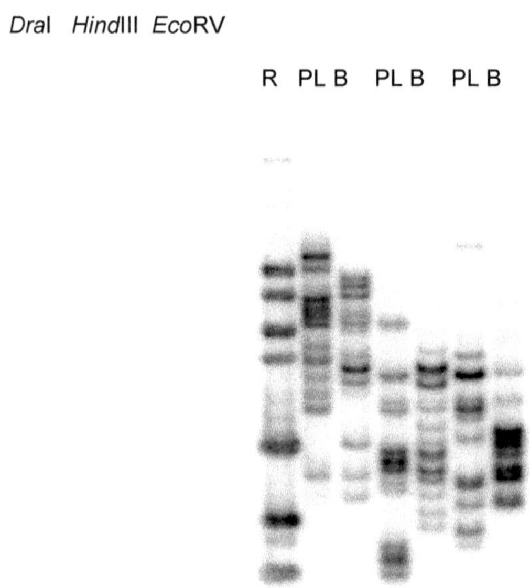

Figure 8-2 Polymorphisms observed in *M. acuminata* parentals with an RGA genetic marker.

contrasting in resistance to Sigatoka diseases. RGA marker application in high resolution genetic mapping and subsequent map-based cloning will also facilitate *Musa* R-gene discovery.

Examples of polymorphisms observed in restriction-digested genomic DNA from *M. acuminata* spp. *microcarpa* genetic map parentals "Borneo" (B) and "Pisang Lilin" (PL), following hybridization of Southern blots with RGA probe MaRGA43. R represents size marker Raoul (Appligene, Illkirch, France; from Miller et al. 2008).

8.6.4 Physical Mapping of RGAs Using BAC Resources

A total of 86 unique RGA-positive BAC clones have been identified to date in *M. acuminata* Calcutta 4 (AA), *M. acuminata* "Grande Naine" (AAA), and *M. balbisiana* "Pisang Klutuk Wulung" (PKW) (BB), using two distinct P^{32} labeled RGA probes (Miller et al. 2008). Such high numbers of hits may be expected for R-genes, given the frequent presence of these large gene families across multiple genomic loci. High-resolution fluorescence in situ hybridization (FISH) has also been used to physically map one such RGA BAC clone (MBP_32/N20) to pachytene chromosomal positions (de Capdeville et al. 2008), with the technique applicable for both cytogenetic map development and marker position validation on genetic maps.

Individual BAC clones were also often reported by Miller and colleagues (2008) to contain tandem copies of RGAs. Such clustering of R-gene families and RGAs is common in plant genomes (Hulbert et al. 2001), as a result of tandem duplications of paralogs (Meyers et al. 2003). Within such clusters, R-genes may be conferring resistance to different strains of a pathogen or to different taxa (van der Vossen et al. 2000). A clustered organization may also represent a variation reservoir for R-gene evolution. Given that Miller et al. (2008) described 33 contiguous *Musa* RGAs, continued application as probes provides an efficient approach for identification of unique putative resistance loci in *Musa*. The sequence and analysis of *Musa* BAC clones containing RGA loci is in progress (Baurens et al. 2007) and will shed light on organization and mechanisms governing the evolution of these gene regions. Genetic marker development from such data may then enable comparisons between physical and genetic maps.

8.7 Genetic Markers Derived from cDNA and EST Libraries

In an ongoing collaborative study between UCB, Embrapa, CIRAD and NIAS, a large scale characterization has been conducted for analysis of gene expression in resistant and susceptible *M. acuminata* cultivar leaf material inoculated *in vitro* with *Mycosphaerella fijiensis* (Passos et al. 2007). To date, unidirectional 5'-end sequencing has been conducted on approximately 8,000 randomly selected clones per library, with chromatogram data deposited in the *Musa* genome database Data*Musa* (http://genoma.embrapa.br/musa/index.html/DATA_musa.html). Using an in-house EST processing pipeline, in addition to gene discovery and expression profiling, data has been used to create molecular markers. A computational search over consensi was performed to locate SSRs with at least two repeating units spanning more than 10 bases. Advantages exist in deriving such markers from ESTs, given that genetic maps can be enriched for genic regions. To date, primers have been designed for over 300 consensi with SSRs, with numerous validated for polymorphism on parental diploid *M. acuminata* cultivars contrasting in resistance to Sigatoka diseases.

8.8 Genetic Mapping and Creation of Segregating Populations

Several segregating populations have been generated to identify the location of traits of interest, however the only peer- reviewed publication of a mapping population is that generated by CIRAD (Fauré et al. 1993). A brief summary of ongoing work on segregating populations is given in Table 8-8 and the work of CIRAD is reported in more detail below and in Table 8-9.

Mapping populations were studied by Fauré et al. 1993, with the first map involving a cross between the wild diploid *M. acuminata* spp. *banksii* and a diploid AA cultivated type (SF265) segregating for parthenocarpy. The map was based on an F_2 population of 92 indivduals using 58 restriction fragment length polymorphisms (RFLPs), four isozyme and 28 random amplified polymorphic DNAs (RAPDs) markers; of the 90 loci detected, 77 were placed on 15 linkage groups and 13 segregated independently. The total length of the map was 606 cM. Segregation distortions were shown by 36% of all loci. This initial attempt to map the banana genome highlighted that translocation events are a major concern in *Musa* genetics. A second genetic map was developed using selfed progeny of two diploid AA cultivars (M53 and SFB5), and showed 11 linkage groups putatively corresponding to the 11 pairs of chromosome of the banana (Noyer et al. 1997; Lagoda et al. 1998). This was based on 100 co-dominant RFLPs, 30 microsatellites and 170 AFLPs; of which 17 were considered to be co-dominant. The total length of the map was 1200 cM. Segregation distortions were again shown by 36% of all markers (Table 8-9).

In *M. acuminata*, seven groups of translocations, comprising accessions having the same chromosome structure, have been identified (Standard, North Malaysia, Malaysia Mountains, North A, North B, Indonesia, and East Africa) by workers at CIRAD. A map comparison was made from a segregating F_2 population (AFCAM; Table 8-9) and the two previous genetic maps in order to further study translocation events in banana (Vilarinhos

Table 8-8 Segregating populations of *Musa* species and cultivars.

Parents	Institute	Purpose
M. acuminata burmannicoides Calcutta 1V (AA) x *M. acuminata banksii madang* (AA) *M. acuminata truncata (AA)* x *M. a. madang* (AA) *M. a. pahang* (AA) *x* Pisang jari (cv. AA) *M. a. pahang* (AA) x Yangambi Km5 (AAA)	Centre de Coopération Internationale en Recherche Agronomique pour le Developpement (CIRAD)	Resistance/susceptibility to *M. fijiensis* Nematode resistance (www.musagenomics.org)
Musa acuminata X wild *M. balbisiana* AA x AA BB x BB	Indian Institute of Horticultural Research	For mapping using molecular markers (Pers. comm. R. Kundapura, 2007)
Contrasting cultivars/wild accessions	Indian Institute of Horticultural Research	For fusarium wilt and nematode resistance (www.musagenomics.org)
Calcutta 1V and Pisang Lillin	National Agricultural Research Organisation (Uganda)	Population segregating for parthenocarpy (www.musagenomics.org)
SF247 (AA) x *M.a. microcarpa* Borneo (AA) 6142 (AA) x 8075 (AA)	International Institute for Tropical Agriculture	Populations segregating for parthenocarpy and black sigatoka resistance.(Vroh-Bi 2008)
M. acuminata ssp. *malaccencis Foc- (Fusarium oxysporum* f.sp. *cubense*) Tropical Race 4 resistant (male) x *Foc*-TR4 susceptible (female)	University of Malaysia, Kuala Lumpur, Malaysia	Population segregating for resistance to *Foc*- TR4. (Kayat et al. 2009)

et al. 2004, 2006). This comparison confirmed that linkage groups are grossly conserved but translocation events complicate gamete formation and segregation. Moreover, these attempts at generation of genetic maps revealed the complexity of gamete segregation in *Musa*, even at the diploid level. Studies involving polyploid donors have been reported (Crouch et al. 1998, 1999) and show high levels of skewed segregation ratios. The populations that were used for genetic mapping have also diminished in size due to the weakness of the progeny derived from translocation-rich materials.

Given that most mapping programs have faced problems with production and maintenance of large populations, mainly because of translocations, one of the latest attempts to construct a genetic map in banana involved a cross between two wild diploid *M. acuminata* accessions *M. acuminata* spp. *microcarpa* "Borneo" and *M. acuminata* spp. *malaccensis* "Pisang lilin", that was reported to carry only a single translocation event. This map is saturated with SSR markers and DArTs (Table 8-9) (Hippolyte

Table 8-9 Segregating populations of *Musa* created for the purpose of genetic mapping using molecular markers.

Species/cultivar	Pop. size	F_1/F_2	No. loci	Linkage groups	Total cM	cM between markers	LOD score	% distorted segregation	Markers	Unlinked loci	Phenotypic traits	Source
M. acuminata spp. *banksii* x AA cultivated type (SF265) Pop. Name :'SFB5'	92	F_2	90	15	606	10 cM	3.0	36%	58 RFLPs; 4 isozymes; 28 RAPDs	13	Parthenocarpy	Faure et al. 1993
Selfed M53 (*M. acuminata malaccencis* x *M. acuminata banksii* and *M. acuminata banksii* x AA cv `Paka)	89	F_1	185	18			4.0		120 AFLPs 40 RFLPs 24 SSRs 1 isozyme	9	Bunchy leaf disease	Noyer et al. 1997
Composite map of M53 and SFB5	181	F_1	373	11	1227	3 cM	4.75	36%	100 co-dominant RFLPs; 30 SSRs; 170 AFLPs	9		Lagoda et al. 1998
AAAA IDN 110 x *Musa balbisiana* `Pisang Klutuk Wulung (BB)	249	F_1	n/a	n/a			3.0	n/a	Bulked segregant analyis using AFLPs	n/a	Banana streak disease release factor	Lheureux et al. 2003
M. *acuminata* AA `Calcutta 1V` (trans. group North) x *M. acuminata* AA `Madang' (trans. group Standard) Pop. Name: `AFCAM`	244	F_2	244	13	n/a	n/a	n/a	n/a	19 RFLP; 47 SSR; 178 AFLP	n/a	Resistance/ suceptability Black Sigatoka	Vilarinhos et al. 2006
M. acuminata microcarpa Borneo x *M. acuminata* Pisang Lilin Pop. Name: `BORLI'	180	F_1	489	11	1197	2.8 cM	5.0	n/a	167 SSR 322 DArT	n/a		Hippolyte et al. 2010

et al. 2010). In the frame of the current banana genome sequencing project CIRAD has developed a new mapping population derived from a selfed wild *M. acuminata* spp. *malaccencis* "Pahang" through embryo rescues. The objective is to obtain a map with SSR and DArT markers including two markers every 1Mb. This mapping project will help assemble the sequences of the genome of "Pahang" and will serve as a base for a core set of microsatellite markers for uptake in future mapping projects in banana.

The application of bulked segregant analyis to locate specific phenotypic traits in banana has been demonstrated by Lheureux et al. (2003) who investigated the activation of banana streak endogeneous pararetrovirus (BSV) in interspecific hybrids created in a allotriploid mapping population (AAB) between a AAAA tetraploid and a BB diploid (Table 8-9). This analysis used 52 pools of five diseased and five healthy individuals using 64 primer AFLP combinations; markers specific for each condition were then used to test the rest of the progeny, permitting the construction of a genetic map of the BSV expressed locus (BEL). The distance of the nearest marker to the BEL was 1.7 cM. Although the locus was not cloned, this work demonstrates the potential of BSA and genetic mapping for positional cloning in *Musa*.

8.9 Development of Large Insert Genomic Libraries

Table 8-10 summarizes the large insert BAC and BIBAC genomic libraries available to the *Musa* Genomics Consortium (www.musagenomics.org). Both the Calcutta *Hind*lll and the *M. balbisiana* libraries are also available as three dimensional pools, which reduces the cost and the time required to find candidate genes in BACs. The wild diploid, "Calcutta 1V" has been used in breeding programs, principally for genetic analysis and for generation of new cultivars resistant to black Sigatoka and was considered for whole genome sequencing. However, molecular analysis indicted that this clone is highly heterozygous; consequently a dihaploid has been made from the *M. acuminata* spp. *malaccencis* Pahang. The BAC libraries of "Pahang" are now contributing to the whole genome sequencing being undertaken by Genoscope, France.

8.10 Prospects for MBC in *Musa*

It is evident that, in comparison to other crops of equal or lesser economic importance, platforms (viz., high resolution genetic maps, mapping populations, ESTs and the whole genome sequence) for map-based cloning in banana and plantain, are lagging in development. However, a relatively high density reference genetic map of *M. acuminata* was recently been made public (Hippolyte et al. 2010). As this genetic map includes many SSR markers, it should be useful for researchers who do not have

Table 8-10 Large insert genomic libraries available to the *Musa* Genomics Consortium.

Species	Clone name	Cloning vector	Cloning site	Mean insert size	Number of clones	Genome coverage	Reference
Musa acuminata	Calcutta 1V	pECBAC-1	*Bam*H1	110	17,280	3x	INIBAP Annual Report 2003 (www.musagenomics.org)
Musa acuminata	Calcutta 1V	pIndigoBAC-5	*Hind*111	100	55,152	9x	Vilarinhos et al. 2003
Musa acuminata	Grand Naine	pIndigoBAC-5	*Hind*111	145	55,296	4.5x	Vilarinhos et al. (www.musagenomics.org)
Musa acuminata	Tuu Gia	pCLD04541	*Hind*111	100	30,700	5.1x	Ortiz-Vàzquez et al. 2005
Musa acuminata	Pahang, dihaploid	pCC1BAC	*Bam*H1	140	23,040	5x	www.musagenomics.org
Musa acuminata	Pahang dihaploid	pIndigoBAC-5	*Hind*111	110	31,104	5.7x	www.musagenomics.org
Musa balbisiana	Pisang Klutuk Wulung	pIndigoBAC-5	*Hind*111	135	36,864	9x	Safàr et al. 2004

access to expensive high throughput technology (for example, microarray technology). More importantly, the genome of the doubled haploid *M. acuminata malaccensis* "Pahang" is currently being sequenced by Genoscope and CIRAD (funded by the French National Research Agency) and the sequence will be made available to the public. Together with a higher density genetic map this will enable the beginning of positional cloning in *Musa*. It will also serve as a template for resequencing the A genome and for sequencing the B genome which should lead to the discovery of large numbers of single nucleotide polymorphisms, facilitating both genetic mapping, positional cloning, and diversity analysis of the *Musa* genome.

Abbreviations

AFLP	:	amplified fragment length polymorphism
BAC	:	bacterial artificial chromosome
BES	:	BAC End Sequences
BIBAC	:	binary bacterial artificial chromosome
BSA	:	Bulked segregant analysis
cDNA	:	complementary DNA
CIRAD	:	Centre de Coopération Internationale en Recherche Agronomique pour le Dévelopement
cM	:	centimorgan
DArT	:	diversity array technology
EMBRAPA	:	Empresa Brasileira de Pesquisas Agropecuarias
ESARC	:	Eastern and Southern Africa Regional Centre
EST	:	expressed sequence tags
FISH	:	fluorescent *in situ* hybridization
HR	:	hypersensitive response
IDPs	:	InDel polymorphisms
JA	:	jasmonic acid
LD	:	linkage disequilibrium
MBC	:	map based cloning
NBS-LRR	:	nucleotide-binding- site-leucine- rich- repeat
NILs	:	near isogenic lines
NO	:	nitric oxide
PCR	:	polymerase chain reaction
PR	:	pathogeneis related
QTL	:	quantitative Trait Loci
RAPDs	:	random amplified polymorphic DNAs
RE	:	recombination events
RFLPs	:	restriction fragment length polymorphisms
RGAs	:	resistance gene analogs
ROS	:	reactive oxygen species

SA	:	salycilic acid
SAR	:	systemic acquired resistance
SNP	:	Single Nucleotide Polymorphisms
TAC	:	Transformation-Competent
UCB	:	Universidade Católica de Brasília
YAC	:	yeast artificial chromosome

Acknowledgments

We are grateful to Drs. Franc Baurens, Angelique d'Hont and François Carreel of CIRAD, France, for revision of the section dealing with genetic mapping of *Musa*.

References

Aarts MG, Lintel TE, Hekkert B, Holub BT, Beynon JL, Stiekema WJ, Pereira A (1998) Identification of R-gene homologous DNA fragments genetically linked to disease resistance loci in *Arabidopsis thaliana*. Mol Plant Microbe Interact 11: 251–258.

Baurens FC, Miller RNG, Lescot M, Pappas Jr. MJ , Matsumoto T, Roux N, Sasaki T, Sidibe Bocs S (2007) Comparative genomics of *Musa* and rice genomes: insight into an RGA cluster. 6th Plant Genomics European Meeting (Plant GEM), October 3–6, Tenerife.

Belkhadir Y, Subramaniam R, Dangl JL (2004) Plant disease resistance protein signalling: NBS-LRR proteins and their partners. Curr Opin Plant Biol 7: 391–399.

Blomme G, Ortiz R (1996) Field techniques for root health assessment in plantains. MusAfrica 9: 6–7.

Bortiri E, Dave J, Sarah H (2006) Advances in maize genomics: the emergence of positional cloning. Curr Opin Plant Biol 9 (2): 164–71.

Bradeen JM, Simon PW (1998) Conversion of an AFLP fragment linked to the carrot Y_2locus to a simple, codominant, PCR-based marker form. Theor Appl Genet 97: 960–967.

Chelkowski J, Koczyk G (2003) Resistance gene analogues of *Arabidopsis thaliana*: recognition by structure. J Appl Genet 44: 311–321.

Cheung F, Town CD (2007) A BAC end view of the *Musa accuminata* genome. BMC Plant Biol 7: 29.

Chisholm ST, Coaker G, Day B, Staskawicz BJ (2006) Host-microbe interactions: shaping the evolution of the plant immune response. Cell 124: 803–814.

Craenen K, Ortiz R (1996) Effect of the black sigatoka resistance gene bs_1 and ploidy level in fruit and bunch traits of plantain-banana hybrids. Euphytica 87: 97–101.

Craenen K, Ortiz R (1997) Effect of the bs_1 gene in plantain-banana hybrids on response to black sigatoka. Theor Appl Genet 95: 497–505.

Craenen K, Ortiz R (1998) Influence of black Sigatoka disease on the growth and yield of diploid and tetraploid hybrid plantains. Crop Protec 17: 13–18.

Crouch HK, Crouch JH, Jarret RL, Cregan PB, Ortiz R (1998) Segregation at microsatellite loci in haploid and diploid gametes of *Musa*. Crop Sci 38: 211–217.

Crouch JH, Crouch HK, Constandt H, Van Gysel A, Breyne P, van Montagu M, Jarret RL, Ortiz R (1999a) Comparison of PCR-based molecular marker analyses of *Musa* breeding populations. Mol Breed 5: 233–244.

Crouch JH, Vuylsteke D, Ortiz R (1998b) Perspectives on the application of biotechnology to assist the genetic enhancement of plantain and banana (*Musa* spp.). Elec J Biotechnol 1(1): 1–12 http://www.ejb.org/content/vol1/issue1/full/2/

Crouch JH, Crouch HK, Tenkouano A, Ortiz R (2003) VNTR-based diversity analysis of 2x and 4x full-sib *Musa* hybrids. Electronic J Biotechnol 2:130–139.

De Capdeville G, Souza Jr. MY, Szinay D, Diniz L, Wijnker E, de Jong H, Swennen R, Kema GHJ, De Jong H (2008) The potential of high-resolution BAC-FISH in banana breeding. Euphytica 166: 431–443.

Dangl JL (1994) The enigmatic avirulence genes of phytopathogenic bacteria. Curr Topics Microbiol Immunol 192: 99–118.

Dangl L, Jones J (2001) Plant pathogens and integrated defence responses to infection. Nature 411: 826–833.

Dixon RA, Harrison MJ, Lamb CJ (1994) Early events in the activation of plant defense responses. Annu Rev Phytopathol 32: 479–501.

Dochez C, Tenkouano A, Ortiz R, Whyte J, De Waele D (2009) Host plant resistance to *Radopholus similis* in a diploid banana hybrid population. Nematology 6: 215–221.

Doebly J, Stec A (1993) Inheritance of the morphological differences between maize and teosinte: comparison of the results for two F_2 populations. Genetics 134: 559–570.

Dong X (2001) Genetic dissection of systemic acquired resistance. Curr Opin Plant Biol 4: 309–314.

Ellis J, Dodds P, Pryor T (2000) Structure, function and evolution of plant disease resistance genes. Curr Opin Plant Biol 3: 278–284.

Faure S, Noyer JL, Horry JP, Bakry F, Lanaud C, Gonzalez de Leon D (1993). A molecular marker-based linkage map of diploid bananas (*Musa acuminata*). Theor Appl Genet 87: 517–526.

Flor HH (1971) Current status of the gene-for-gene concept. Annu Rev Phytopathol 9: 275–276.

Goff SA, Ricke D, Lan TH, Presting G, Wang R, Dunn M, Glazebrook J, Sessions A, Oeller P, Varma H, Hadley D, Hutchison D, Martin C, Katagiri F, Lange BM, Moughamer T, Xia Y, Budworth P, Zhong J, Miguel T, Paszkowski U, Zhang S, Colbert M, Sun WL, Chen L, Cooper B, Park S, Wood TC, Mao L, Quail P, Wing R, Dean R, Yu Y, Zharkikh A, Shen R, Sahasrabudhe A, Thomas A, Cannings R, Gutin A, Pruss D, Reid J, Tavtigian S, Mitchell J, Eldredge G, Scholl T, Miller RM, Bhatnagar S, Adey N, Rubano T, Tusneem N, Robinson R, Feldhaus J, Macalma T, Oliphant A, Briggs S (2002) A draft sequence of the rice genome (*Oryza sativa* L. ssp. *japonica*). Science 296: 92–100.

Hammond-Kosack KE, Jones JDG (1996) Inducible plant defense mechanisms and resistance gene function. Plant Cell 8: 1773–1791.

Hippolyte I, Seguin M, Bakry F, Gardes L, Baurens FC, Miller RN, Khan IA, Jenny C, Carreel F, Huttner E, Perrier X, Kilian A, Risterucci A-M, Glaszmann J-C (2008) *Musa* Genetic Mapping. XVI International Plant and Animal genome conference (PAG), January, San Diego, CA, USA.

Hippolyte I, Bakry F, Seguin M, Gardes L, Rivallan R, Risterucci A-M, Jenny C, Perrier X, Carreel F, Argout X, Piffanelli P, Khan I, Miller R, Pappas G, Mbeguie-A-Mbeguie D, Matsumoto T, De Bernardinis V, Huttner E, Kilian A, Baurens F-C, D'Hont A, Cote F, Courtois B, Glaszmann J-C (2010) A saturated SSR/DArT linkage map of *Musa acuminata* addressing genome rearrangements among bananas. BMC Plant Biol 10: 65–65.

Hulbert SH, Webb CA, Smith CM, Sun Q (2001) Resistance gene complexes: evolution and utilization. Annu Rev Phytopathol 39: 285–312.

INIBAP Annual Report (2003) The Musa Genomics Consortium, pp 29. International Network for Improvement of Banana and Plantain, Montpellier, France.

Irigoyen ML, Loarce Y, Fominaya A, Ferrer E (2004) Isolation and mapping of resistance gene analogs from the *Avena strigosa* genome. Theor Appl Genet 109: 713–724.

Jander G, Norris SR, Rounsley SD, Bush DF, Levin IM, Last RL (2002) Arabidopsis map-based cloning in the post-genome era. Plant Physiol 129: 440–450.

Kassim S, Nyine M, Pillay M (2010) A screening method for banana weevil (*Cosmopolites sordidus* Germar) resistance using reference genotypes. Afr J Biotechnol 9: 4725–4750.

Kayat F, Bonar N, Waugh R, Rajinder S, Rahimah AR, Rashid KA, Othman RY 2009. Development of a genetic linkage map for genes associated with resistance and susceptibility to *Fusarium oxysporum* f. sp. *cubense* from an F_1 hybrid population of *Musa acuminata* ssp. *malaccensis*. Acta Hort 828: 333–340.

Lagoda PJL, Noyer JL, Baurens FC, Grivet L, Raboin LM, Amson C, Aubert G, Costet L, Lanaud C 1998. Results of a *Musa* mapping project. In: Use of novel DNA fingerprinting techniques for the detection and characterization of genetic variation in vegetatively propagated crops, IAEA Technical Doc 1047: 35–43.

Lamb C, Dixon RA (1997) The oxidative burst in plant disease resistance. Annu Rev Plant Physiol Plant Mol Biol 48: 251–275.

Leister RT, Katagiri F (2000) A resistance gene product of the nucleotide binding site—leucine rich repeats class can form a complex with bacterial avirulence proteins *in vivo*. Plant J 22: 345–354.

Lheureux F, Carreel F, Jenny C, Lockhart BEL, Iskra-Caruana ML (2003) Identification of genetic markers linked to banana streak disease expression in inter-specific *Musa* hybrids. Theor Appl Genet 106: 594–598.

Lukowitz W, Gillmor CS, Scheible WR (2000) Positional cloning in Arabidopsis. Why it feels good to have a genome initiative working for you. Plant Physiol 123: 795–805.

Mackey D, Holt BF, Wiig A, Dangl JL (2002) RIN4 interacts with *Pseudomonas syringae* type III effector molecules and is required for RPM1-mediated resistance in *Arabidopsis*. Cell 108: 743–754.

Martin GB, Brommonschenkel SH, Chunwongse J (1993) Map-based cloning of a protein kinase gene conferring disease resistance in tomato. Science 262: 1432–1436.

Martin GB, Bogdanove AJ, Sessa G (2003) Understanding the functions of plant disease resistance proteins. Annu Rev Plant Biol 54: 23–61.

Meyers BC, Dickerman AW, Michelmore RW, Sivaramakrishnan S, Sobral BW, Young ND (1999) Plant disease resistance genes encode members of an ancient and diverse protein family within the nucleotide-binding superfamily. Plant J 20: 317–332.

Meyers BC, Kozik A, Griego A, Kuang H, Michelmore RW (2003) Genome-wide analysis of NBS-LRR-encoding genes in Arabidopsis. Plant Cell 15: 809–834.

Métraux JP (2001) Systemic acquired resistance and salicylic acid: current state of knowledge. Eur J Plant Pathol 107: 13–18.

Michelmore RW, Meyers BC (1998) Clusters of resistance genes in plants evolve by divergent selection and a birth-and-death process. Genome Res 8: 1113–1130.

Michelmore RW, Paran I, Kesseli RV (1991) Identification of markers linked to disease-resistance genes by bulked segregant analysis: A rapid method to detect markers in specific genomic regions by using segregating populations. Proc Natl Acad Sci USA, 88: 9828–9832.

Miller RNG, Bertioli DJ, Baurens FC, Santos CM, Alves PC, Martins NF, Togawa RC, Souza Jr. MT , Pappas Jr. GJ (2008) Analysis of non-TIR NBS-LRR resistance gene analogs in *Musa acuminata* Colla: Isolation, RFLP marker development, and physical mapping. BMC Plant Biol 8: 15.

Miller RNG, Passos MAN, Menezes NNP, Souza Jr. MT, Mota M, Azevedo VCR, Amorim EP, Pappas Jr. GJ, Ciampi AY (2010) Characterization of novel microsatellite markers in *Musa acuminata* subsp. *burmannicoides*, var Calcutta 4. BMC Res. Notes 3:149.

Mohamad A, Heslop-Harrison JS (2008) Genomes, diversity and resistance gene analogs in *Musa* species. Cytogenet. Genome Res 121: 59–66.

Nimchuk Z, Eulgem T, Holt BF, Dangl JL 2003. Recognition and response in the plant immune system. Annu Rev Genet 37: 579–609.

Nokoe S, Ortiz R (1998) Optimum plot sizes for banana trials. HortScience 33: 130–132.

Noyer JL, Dambier D, Lanaud C, Lagoda P (1997) The saturated map of diploid banana (*Musa acuminata*). In : Plant and Animal Genomes V Conference, January 12–16, San Diego, CA, USA.

Okoro J, Ortiz R, Vuylsteke D (1997) Optimum plot size for black sigatoka evaluation in East African highland bananas. Tropicultura 15: 186–189.

Ortiz R (1995) *Musa* genetics. In: S Gowen (ed) Bananas and Plantains. Chapman and Hall, London, UK, pp 84–109.

Ortiz R (1996a) Segregation for persistent neutral (hermaphrodite) flowers and male bracts in plantain-banana hybrids. InfoMusa 5 (2): 19–20.

Ortiz R (1996b) The potential of AMMI analysis for field assessment of *Musa* genotypes to virus infection. HortScience 31: 829–832.

Ortiz R (1997a) Occurrence and inheritance of 2n pollen in *Musa*. Ann Bot 79: 449–453.

Ortiz R (1997b) Morphological variation in *Musa* germplasm. Genet esour Crop Evol 44: 393–404.

Ortiz R (1997c) Genetic and phenotypic correlations in plantain-banana euploid hybrids. Plant Breed 116: 487–491.

Ortiz R (2000) Understanding the *Musa* genome: an update. Acta Hort 540: 157–168.

Ortiz R, Vuylsteke D (1994a) Inheritance of albinism in banana and plantain (*Musa* spp.). HortScience 29: 903–905.

Ortiz R, Vuylsteke D (1994b) Genetic analysis of apical dominance and improvement of suckering behaviour in plantain. HortScience 119: 1050–1053.

Ortiz R, Vuylsteke D (1994c) Inheritance of black sigatoka disease resistance in plantain-banana (*Musa* spp.) hybrids. Theor Appl Genet 89: 146–152.

Ortiz R, Vuylsteke D (1994d) Trisomic segregation ratios and genome differentiation in AAB plantains. InfoMusa 3 (1): 21.

Ortiz R, Vuylsteke D (1995a) Inheritance of dwarfism in AAB plantains. Plant Breed 114: 466–468.

Ortiz R, Vuylsteke D (1995b) Effect of the parthenocarpy gene P_1 and ploidy in bunch and fruit traits of plantain and banana hybrids. Heredity 75: 460–465.

Ortiz R, Vuylsteke D 1996. Recent advances in *Musa* genetics, breeding and biotechnology. Plant Breed Abst 66: 1355–1363.

Ortiz R, Vuylsteke D (1998) Segregation for bunch orientation in banana and plantain hybrids. Euphytica 101: 79–82.

Ortiz R, Vuylsteke D, Dumpe B, Ferris RSB (1995a) Banana weevil resistance and corm hardness in *Musa* germplasm. Euphytica 86: 95–102.

Ortiz R, Vuylsteke D, Ogburia NM (1995b) Inheritance of waxiness in the pseudostem of banana and plantain. J Hered 86: 297–299.

Ortiz R, Craenen K, Vuylsteke D (1997) Ploidy manipulations and genetic markers as tools for analysis of quantitative trait variation in progeny derived from triploid plantains. Hereditas 126: 255–259.

Ortiz R, Vuylsteke D, Crouch JH (1998) *Musa* genetics, 'Calcutta-4', and scientific ethics: reply to Shepherd's letter. InfoMusa 7 (2): 31–32.

Ortiz-Vàzquez E, Kaemmer D, Zhang HB, Muth J, Rodríguez Mendiola M, Arias Castro C, James A (2005) Construction and characterization of a plant transformation-competent BIBAC library of the black Sigatoka-resistant banana *Musa acuminata* cv. 'Tuu Gia' (AA). Theor Appl Genet 110(4): 706–713.

Passos MAN, Emediato FL, Nunes FAC, Pappas Jr. GJ, Miller RNG (2007) Analysis of ESTs in the interaction between *Musa acuminata* and *Mycosphaerella fijiensis*. In: 53rd Brazilian Genetics Congress. Águas de Lindóia /SP, Sociedade Brasileira de Genética.

Pei X, Li S, Jiang Y, Zhang Y, Wang Z, Jia S (2007) Isolation, characterization and phylogenetic analysis of the resistance gene analogues (RGAs) in banana (*Musa* spp.). Plant Sci 172: 1166–1174.

Peloquin SJ, Ortiz R (1992) Techniques for introgressing unadapted germplasm to breeding populations. In: HT Stalker, JP Murphy (eds) Plant Breeding in the 1990s. CAB International, Wallingford, UK, pp 485–507.

Peraza-Echeverria S, James-Kay A, Canto-Canche B, Castillo-Castro E (2007) Structural and phylogenetic analysis of Pto-type disease resistance gene candidates in banana. Mol Genet Genom 278: 443–453.

Peraza-Echeverria S, Dale JL, Harding RM, Smith MK, Collet C (2008) Characterization of disease resistance gene candidates of the nucleotide binding site (NBS) type from banana and correlation of a transcriptional polymorphism with resistance to *Fusarium oxysporum* f.sp. *cubense* race 4. Mol Breed 22 : 565–579.

Pink DAC (2002) Strategies using genes for non-durable resistance. Euphytica 1: 227–236.

Pflieger S, Palloix A, Caranta C, Blattes A, Lefebvre V (2001) Defense response genes co-localize with quantitative disease resistance loci in pepper. Theor Appl Genet 103: 920–929.

Ramalingam J, Vera-Cruz CM, Kukreja K, Chittoor JM, Wu JL, Lee SW, Baraoidan M, George ML, Cohen MB, Hulbert SH, Leach JE, Leung H (2003) Candidate defense genes from rice, barley and maize and their association with qualitative and quantitative resistance in rice. Mol Plant Microbe Interact 16: 14–24.

Remington DL, Ungerer MC, Purugganan MD (2001) Map-based cloning of quantitative trait loci: progress and prospects. Genet Res Camb 78: 213–218.

Rowe P (1984) Breeding bananas and plantains. Plant Breed Rev 2: 135–155.

Safár J, Noa-Carrazana JC, Vrána J, Bartos J, Alkhimova O, Sabau X, Simková H, Lheureux F, Caruana ML, Dolezel J, Piffanelli P (2004) Creation of a BAC resource to study the structure and evolution of the banana (*Musa balbisiana*) genome. Genome; 47(6): 1182–91.

Shepherd K (1990) Genetic improvement of bananas in Brazil: aspects related to resistance to the genus *Mycosphaerella*. In: RA Fullerton, RH Stover (eds) Sigatoka Leaf Spot Diseases of Bananas. Proceedings of International Workshop, San José, Costa Rica, 28 March–1 April 1989. INIBAP, Montpellier, France, pp 237–242.

Simmonds NW (1952) Segregations in some diploid bananas. J Genet 51: 458–469.

Tanksley SD, Ganal MW, Martin GB (1995) Trends in Genetics 11(2): 63–68.

TAGI: The Arabidopsis Genome Initiative (2000) Analysis of the genome sequence of the flowering plant *Arabidopsis thaliana*. Nature 6814: 796–815.

Tenkouano A, Ortiz R, Vuylsteke D (1998a) Combining ability for yield and plant phenology in plantain-derived populations. Euphytica 104:151–158.

Tenkouano A, Crouch JH, Crouch HK, Ortiz R (1998b) Genetic diversity, hybrid performance and combining ability for yield in *Musa* germplasm. Euphytica 102: 281–288.

Tenkouano A, Crouch JH, Crouch HK, Vuylsteke D, Ortiz R (1999a) A comparison of DNA marker and pedigree methods for genetic analysis in plantain and banana (*Musa* spp.) clones. I. Estimation of genetic relationships. Theor Appl Genet 98: 62–68.

Tenkouano A, Crouch JH, Crouch HK, Vuylsteke D, Ortiz R (1999b) A comparison of DNA marker and pedigree methods for genetic analysis in plantain and banana (*Musa* spp.) clones. II. Predicting hybrid performance. Theor Appl Genet 98: 69–75.

Vakili NG (1965a) Fusarium wilt resistance in seedlings and mature plants of *Musa* species. Phytopathology 55: 135–140.

Vakili NG (1965b) Inheritance of resistance in *M. acuminata* in bacterial wilt caused by the tomato race of *Pseudomonas solanacearum*. Phytopathology 55: 1206–1209.

Vakili NG (1968) Responses of *Musa acuminata* species and edible cultivars to infection by *Mycosphaerella musicola*. Trop Agric (Trinidad) 45: 13–22.

Van der Biezen EA, Jones JDG (1998) Plant disease-resistance proteins and the gene-for-gene concept. Trends Biochem Sci 12: 454–456.

Van der Vossen EA, van der Voort JN, Kanyuka K, Bendahmane A, Snedbrink H, Baulcombe DC, Bakker J, Stiekema WJ, Klein-Lankhorst RM (2000) Homologues of a single resistance-gene cluster in potato confer resistance to distinct pathogens: a virus and a nematode. Plant J 23: 567–576.

Vandenhout H, Ortiz R, Vuylsteke D, Swennen R, Bai KV (1995) Effect of ploidy on stomatal and other quantitative traits in plantain and banana hybrids. Euphytica 83: 117–122.

Vilarinhos AD, Piffanelli P, Lagoda P, Thibivilliers S, Sabau X, Carreel F, D'Hont A (2003) Construction and characterization of a bacterial artificial chromosome library of banana (*Musa acuminata* Colla). Theor Appl Genet 106: 1102–1106.

Vilarinhos AD, Benabdelmouna A, Bakry F, Piffanelli P, Triaire D, Lagoda P, Noyer JL, Courtois B, Carreel F, D'Hont A (2004) In: C Picq, A Vézina (eds) First International Congress on *Musa*: Harnessing Research for Improved Livelihoods, Penang, Malaysia.

Vilarinhos A, Carreel F, Rodier M, Hippolyte I, Benabdelmouna A, Triaire D, Bakry F, Courtois B, D'Hont A (2006) Characterization of translocations in banana by FISH of BAC clones anchored to a genetic map. In: Abstracts of the International Conference Plant and Animal Genome XIV. January, San Diego, CA, USA.

Vos R, Hogers R, Bleeker M, Reijans M, Van de Lee T, Hornes M, Friters A, Pot J, Paleman J, Kuiper M, Zabeau M (1995) AFLP: a new technique for DNA fingerprinting. Nucl Acids Res 23 (21): 4407–4414.

Vroh-Bi I (2008) Mapping Agronomic Traits In Bananas And Plantains (*Musa* sp.) In: Abstract W47, Musa Genomics. Plant and Animal Genomes Conference, San Diego, California, USA.

Vuylsteke D, Ortiz R, Ferris RSB, Crouch JH (1997) Plantain improvement. Plant Breed Rev 14: 267–320.

Wiame I, Swennen R, Sági L (2000) PCR-based cloning of candidate disease resistance genes from banana (*Musa acuminata*). Acta Hort 521: 51–57.

Xiao W, Zhao J, Fan S, Li L, Dai J, Xu M (2007) Mapping of genome-wide resistance gene analogs (RGAs) in maize (*Zea mays* L.). Appl Theor Genet 115: 501–8.

Xu Q, Wen X, Deng X (2005) Isolation of TIR and nonTIR NBS–LRR resistance gene analogues and identification of molecular markers linked to a powdery mildew resistance locus in chestnut rose (*Rosa roxburghii* Tratt). Theor Appl Genet 111: 819–830.

Zhou T, Wang T, Chen JQ, Araki H, Jing Z, Jiang K, Shen J, Tian D (2004) Genome-wide identification of NBS genes in japonica rice reveals significant expansion of divergent non-TIR NBS-LRR genes. Mol Genet Genom 271: 402–415.

9

Functional Genomics and Transcriptomics in *Musa*

Michael Pillay,[1,*] *Sivalingam Elayabalan,*[2] *Kaliyaperumal Ashokkumar*[3] and *Silvas Jabekumar Prince Kirubarkaran*[4]

ABSTRACT

Genome analysis of multicellular organisms reveal a large number of genes for which no function is known or can be predicted. Many genetic tools are now available for investigating gene function. In general they are divided into two broad categories-forward and reverse genetics. This chapter reviews some of the techniques for gene functional analysis with special emphasis on insertional mutagenesis, targeted gene disruption by homologous recombination, RNA interference, virus-induced gene silencing (VIGS) technology, and Targeting Induced Local Lesions IN Genomes (TILLING). In addition progress in some areas of transcriptomics research in *Musa* is reviewed. Insertional mutagenesis has been widely used for cloning genes, promoters, enhancers and other regulatory sequences in the model plant *Arabidopsis*. Insertional mutagenesis provides a direct route to determining function. Most other approaches are correlative and do not necessarily prove a causal relationship between gene sequence and function. Gene targeting by homologous recombination, although cumbersome, is now feasible in rice. RNA interference (RNAi) is based on sequence-specific RNA degradation that follows the formation of double-stranded RNA (dsRNA) homologous in sequence to the targeted gene. RNAi

[1]Vaal University of Technology, Private Bag X021, Vanderbijlpark 1900 South Africa; e-mail: *mpillay@vut.ac.za*

[2]Crop Tech Ltd., P.O. Box 1367, Arusha, Tanzania; e-mail: *balabiotech@gmail.com*

[3]Department of Plant Sciences, University of Saskatchewan, 51 Campus Drive, Saskatoon, SK, S7N 5A8, Canada; e-mail: *ashok.kumar@usask.ca*

[4]Department of Plant Molecular Biology and Biotechnology, Center for Plant Molecular Biology, Tamil Nadu Agricultural University, Coimbatore-641003-India; e-mail: *symsil07@gmail.com*

*Corresponding author

List of abbreviations after the text.

allows silencing one, several, or all members of a multigene family or homoeologous gene copies in polyploids. Virus-induced gene silencing (VIGS) is a powerful and rapid technique for analysis of gene function in plant growth and development. To overcome the limitations of knocking out an entire gene and to expand knowledge of active gene mutations, a new process called Targeting Induced Local Lesions in Genomes or TILLING was developed. TILLING uses chemical mutagenesis to yield a traditional allelic series of point mutations for virtually all genes. Serial Analysis of Gene Expression (SAGE) is based on the massive sequential analysis of short cDNA sequence tags. The chapter also summarizes the role of Bacterial Artificial chromosomes (BACs), Expressed Sequence Tags (EST) libraries and Microarray technology in the functional genomics of *Musa*. Finally recent studies on comparative functional genomics to study drought stress in *Musa* is also addressed.

Keywords: gene knockout, insertional mutagenesis, RNA interference, TILLING, transcriptomics

9.1 Introduction

One of the most significant findings revealed by analysis of genomes of multicellular organisms is the large number of genes for which no function is known or can be predicted (Alonso et al. 2003). Functional genomics is a relatively new technology that uses high throughput methods to analyze a large number of genes at a time. It studies the function of individual or groups of genes. Functional genomics allows researchers to gain an understanding of the control of genes during normal growth and development, and in response to environmental stresses. Researchers can further investigate the complex interactions that occur in genetic and cellular networks and ideally, the technology will allow the entire genetic make-up of an organism to be studied.

Many genetic tools are now available for investigating gene function. In general, they are divided into two broad categories—forward and reverse genetics. Forward genetics refers to a process where studies are initiated to determine the genetics underpinnings of observable phenotypic variation (Tierney and Lamour 2005). In many cases, the observable variation has been induced using a DNA damaging agent (mutagen) but also may be naturally occurring. The investigator eventually ends up sequencing the gene or genes thought to be involved. Instead of going from phenotype to sequence as in forward genetics, reverse genetics works in the opposite direction—a gene sequence is known, but its exact function is uncertain. In reverse genetics, a specific gene or gene product is disrupted or modified and then the phenotype is measured (Tierney and Lamour 2005).

A number of strategies are now available to disrupt or alter a gene and observe its phenotypic effect on an organism. These include methods that target genes specifically or those that bring about non-targeted random disruptions and include:

(i) Targeted gene disruption (Oliver 1996; Kempin et al. 1997). Although this method has been used in *Arabidopsis*, it is laborious, involving the generation of hundreds or thousands of transgenic plants for every gene assayed (Kempin et al. 1997).
(ii) Gene silencing via sense or antisense suppression has been a popular method in the past (Baulcombe 1999). However, this method also requires that several independent transgenic lines to be generated for every gene.
(iii) Chemical, radiation, and transposon mutagenesis is regarded as the most versatile method for assessing gene function. Transposon and T-DNA [portion of the Ti (tumor-inducing) plasmid that is transferred to plant cells] insertions offer the additional advantage of tagging the target gene molecularly and in many cases genetically via reporter and selectable marker genes carried by the insertion (Martienssen 1998).

Transcriptomics, which is the high-throughput analysis of differential gene expression is also a powerful tool for discovering novel genes or for gaining additional information about certain biological processes on a genomic scale (Mochida and Shinozaki 2010). Genome-wide expression analysis of almost all genes has become possible, even though the profiling techniques are technically rather demanding and expensive, e.g., serial analysis of gene expression (SAGE) (Velculescu et al. 1995) and massively parallel signature sequencing (MPSS) (Brenner et al. 2000).

This chapter reviews some of the techniques for gene functional analysis with special emphasis on insertional mutagenesis, targeted gene disruption by homologous recombination, RNA interference, virus-induced gene silencing (VIGS) technology, and targeting induced local lesions in genomes (TILLING). In addition progress in some areas of transcriptomics research in *Musa* is reviewed.

9.2 Insertional Mutagenesis and Gene Functional Analysis

Gene suppression (deletion/mutation knockout) or over-expression permits the gene sequence to be linked to a phenotype from which the function of the gene can be deduced (Matzke and Matzke 1995).

Gene disruption is a powerful way to obtain knockout mutants that help in ascertaining biological function of the numerous uncharacterized open reading frames (ORFs) from genome sequences or from expressed sequence tags (Radhamony et al. 2005). Insertional mutagenesis has been extensively used for cloning genes, promoters, enhancers and other regulatory sequences from *Arabidopsis* (Radhamony et al. 2005). Gene knockout systems provide a direct route to determining function. Most other approaches to dissect gene function are correlative and do not necessarily prove a causal relationship between gene sequence and function. For example, DNA chips provide a means to discover conditions under which gene expression is regulated on a genome wide scale (Becker et al. 2003; Birnbaum et al. 2003).

However, factors other than mRNA level may also determine the activity of a gene product *in situ*. Therefore, expression studies cannot prove a causal relationship. By contrast, the availability of a null mutation for the gene of interest allows one to directly monitor the effect of this deficiency on the organism's ability to function (Radhamony et al. 2005). Although methods such as gene replacement through homologous recombination work well in yeast and mice, this technique has not been easy for targeted insertion in plants. However, efficient gene targeting by homologous recombination has been reported in rice (Terada et al. 2002; Yang et al. 2010; Yu et al. 2010).

The use of insertional mutagenesis, in principle, provides a more rapid way to clone a mutated gene. DNA elements that are able to insert at random within chromosomes such as transposons (Sundaresan et al. 1995; Martienssen 1998) or the T-DNA of *Agrobacterium tumefacians* (Azpiroz-Leehan and Feldmann 1997), can be used as mutagens to create loss of function mutations in plants. Transposons are mobile genetic elements that can relocate from one genomic location to another (Hayes 2003). They are DNA sequences that can insert themselves at a new location in the genome without having any sequence relationship with the target locus (Lewin 2004). Transposon-based signature-tagged mutagenesis has been successful in identifying essential genes as well as genes involved in infectivity of a variety of pathogens. Strategies for insertional mutagenesis using transposons have been developed for a number of animal and plant models (Hayes 2003). Transposon-based methods have been used in *Arabidopsis*, maize and other plants (Stemple 2004).

In T-DNA insertional mutagenesis, use is made of the segment of the Ti plasmid of *A. tumefaciens* known as T-DNA that carries genes to transform the plant cell has also been utilized for this purpose. Since the sequence of the inserted element is known, the gene in which the insertion has occurred can be recovered, using various cloning or PCR-based strategies. One drawback of insertional mutagenesis is the low frequency of mutations, necessitating the screening of large numbers of individuals to find mutations in any given gene (Gilchrist and Haughn 2005). Insertions in essential genes will usually cause lethality, and less severe mutations must be generated in these genes in order to understand gene function (Till et al. 2003).

An advantage of using T-DNA as the insertional mutagen as compared to transposons is that the T-DNA insertions do not transpose subsequent to insertion and are chemically and physically stable through multiple generations (Radhamony et al. 2005). The T-DNA not only disrupts the expression of the gene into which it is inserted, but also acts as a marker for subsequent identification of the mutation. It is important to establish linkage between the mutant phenotype and the insertion element. Even if some linkage can be established, this does not exclude the possibility that the phenotype is caused by a nearby but independent mutation. Consequently it is necessary to identify the gene disrupted by the insertion and test for complementation of the mutant phenotype by the wild type allele in transgenic plants.

A highly efficient procedure for obtaining mutants in genes identified in sequencing programs takes advantage of the availability of large collections of plants mutagenized by an insertion element.

Mutational approaches have been successfully used for the study of genetic and molecular bases for many traits in plant biology. Access to the mutation is obtained using positional cloning strategies. This strategy is facilitated in model species such as *Arabidopsis*, for which dense genetic maps with many visible and molecular genetic markers exist, and for which the complete physical map consisting of a collection of overlapping cloned DNA fragments and total genome sequence are already available. The limiting factors for this approach are the time and effort required for creating the mapping population and the fine mapping of the mutant locus.

Besides the model plant *Arabidopsis*, insertional mutagenesis for gene functional analysis has been applied to other plants such as rice, tomato and *Brassica*. The maize Ac/Ds transposon system has been shown to be efficient for gene tagging and functional analysis in rice (Hiroyuki et al. 1999). T-DNA tagging has also been used successfully for gene discovery in rice. In tomato new tools for functional analysis based on insertional mutagenesis, with Ac/Ds system in the background of the miniature cultivar Micro-Tom has been established (Meissner et al. 2000). In addition, promoter trapping using firefly luciferase reporter gene and enhancer trapping using GUS was also developed. T-DNA tagging strategy for isolation of new promoters has been tried in *Brassica napus* employing hypocotyl transformation with a promoterless *gus*: *npt*II tagging construct.

The use of mutagenesis to find and study plant genes is increasingly being used in functional genomics. Thus T-DNA tagging in conjunction with other mutation based techniques like transposon insertion, TILLING, etc., would not only continue to provide useful information in *Arabidopsis* but is likely to prove an efficient tool for functional genomics in other plants also. Genome sequences of banana are rapidly becoming available (Cheung and Town 2007) and these techniques may be used for its gene functional analysis by these methods. Santos et al. (2009) conducted a genome-wide T-DNA tagging strategy for the identification and characterization of novel banana promoters. The method proved valuable for the identification of novel promoters and genes in banana and for monitoring expression patterns throughout *in vitro* development and low temperature treatment.

9.3 Targeted Gene Disruption by Homologous Recombination

Homologous recombination is a reciprocal exchange of DNA sequences, as in between two chromosomes that carry the same genetic loci (Lewin 2004). Just as homologous recombination has been found to be mainly initiated with a double-strand break, gene targeting by homologous recombination is associated with the repair of double strand breaks. The double-strand break repair and synthesis-dependent strand-annealing models are the most generally accepted models to explain gene targeting (Iida and Terada 2004).

A reverse-genetic system using homologous recombination has recently been developed for *Drosophila*. It is promising, but is a lengthy procedure and requires generation of specific transgenic flies (Stemple 2004). Reproducible gene targeting by homologous recombination is now also feasible in rice. With the combination of site-specific recombination systems (such as Cre-lox), the future of gene targeting by homologous recombination as a routine procedure for engineering the genome of rice and presumably other plants is bright (Iida and Terada 2004).

9.4 Gene Silencing by RNA Interference

In contrast to insertional mutagenesis, RNA interference (RNAi) is based on sequence-specific RNA degradation that follows the formation of double-stranded RNA (dsRNA) homologous in sequence to the targeted gene (Marx 2000; Baulcombe 2004). RNAi allows silencing one, several, or all members of a multigene family or homoeologous gene copies in polyploids by targeting sequences that are unique or shared by several genes (Lawrence and Pikaard 2003; Miki et al. 2005) and may be useful in banana, which is a polyploid crop.

RNA interference (RNAi) is the process by which expression of a target gene is inhibited by antisense and sense RNAs. It works based on the ability of double-stranded sequences to recognize and degrade sequences that are complementary to them (Lewin 2004). RNA interference has recently become a powerful tool to silence the expression of genes and analyze their loss-of-function phenotype, allowing analysis of gene function when mutant alleles are not available.

RNAi has already become a useful method for gene targeting in fungi (Nakayashiki 2005), viruses (Baulcombe 2004; Wani et al. 2010), bacteria (Escobar et al. 2001) and plants (Brodersen and Voinnet 2006) as it allows the study of the function of hundreds of thousands of genes to be tested (Godge et al. 2008). It can silence a gene throughout an organism or in specific tissues (Islam et al. 2005), offer the versatility to partially silence or completely turn off genes, work in both cultured cells and whole organisms and can selectively silence genes at particular stages of the organism's life cycle (Milhavet et al. 2003). One interesting feature of RNA silencing in plants is that once it is triggered in a certain cell, a mobile signal is produced and spreads through the whole plant causing the entire plant to be silenced (Dunoyer et al. 2007).

The greatest difficulty in using RNAi as a reverse genetic technique in plants is that throughput is limited by the ability to deliver siRNAs to target loci (Henikoff et al. 2004). A number of methods for delivering double stranded (dsRNA) or siRNA into different cells and tissue has been tested and includes:

(i) transformation with dsRNA forming vectors for selected gene(s) by an *Agrobacterium*- mediated transformation (Waterhouse et al. 2001; Chuang and Meyerowtiz 2000);
(ii) delivery of cognate dsRNA of *uidA* GUS (β-glucuronidase) and TaGLP2a: GFP (green fluorescent protein) reporter genes into single epidermal cells by particle bombardment (Schweizer et al. 2000);
(iii) introducing a vector in plants by infiltration (Liu et al. 2002a);
(iv) delivery of dsRNA into suspension cells by cationic oligopeptide polyarginine-siRNA complex;
(v) infecting plants with viral vectors that produce dsRNA (Dalmay et al. 2000);
(vi) delivery of siRNA into cultured plant cells; and
(vi) gene silencing by nanosense pulsed laser-induced stress wave (LISW) (Tang et al. 2006).

RNA-mediated gene silencing (RNA silencing) has been used as a tool for gene targeting in fungi. Homology based gene silencing induced by transgenes (co-suppression), antisense, or dsRNA has been demonstrated in many plant pathogenic fungi, including *Cladosporium fulvum* (Hamada and Spanu 1998), *Magnaporthae oryzae* (Kadotani et al. 2003; Kim et al. 2009; Chen et al. 2010), *Venturia inaequalis* (Fitzgerald et al. 2004), *Neurospora crassa* (Goldoni et al. 2004), *Aspergillus nidulans* (Hammond and Keller 2005), and *Fusarium graminearum* (Nakayashiki 2005) whether it is suitable for large scale mutagenesis in fungal pathogens remains to be tested.

Antiviral RNAi technology has been used for viral disease management in human cell lines (Bitko and Barik 2001; Gitlin et al. 2002; Jacque et al.

2002; Novina et al. 2002). Such silencing mechanisms (RNAi) have been exploited to protect and manage viral infections in plants (Waterhouse et al. 2001; Ullu et al. 2002). The effectiveness of the technology in generating virus resistant plants was demonstrated in potato (Waterhouse et al. 1998), black gram (Pooggin et al. 2003), cassava (Vanitharani et al. 2003), and tobacco (Qu et al. 2007).

The origin of RNAi technology from the classical studies with the nematode *C. elegans* has shown ways and means to explore the possibilities of this mechanism for protecting plants from nematode damage (Wani et al. 2010). Two approaches have been advocated. One that relies on targeting plant genes that are involved with the infection process and the other targets the essential genes within the nematode. RNAi can be induced in *C. elegans* by feeding it dsRNA, hence it was reasoned that expressing hpRNAs containing sequences of vital nematode genes in the host plant might deliver dsRNA to a feeding nematode to incapacitate or kill it (Wani et al. 2010).

RNAi-mediated suppression of a gene plays an indispensable role in hampering the nematode development and may adversely affect the progression of pathogenesis in direct or indirect ways. There are accumulating evidences for the efficacy of RNAi in plant parasitic nematode management and a wide range of genes have been targeted for silencing in cyst and rootknot nematode species.

Ongoing research demonstrates the relevance of the targeted parasitism genes during the nematode life cycle and more importantly, suggest that a viable level of resistance in crop plants may be accomplished in the future by using RNAi technology against cyst nematodes.

RNAi and miRNA technologies of gene silencing are newly developed genomics tools that have great advantages over antisense and co-suppression due to their higher silencing efficiency and shorter time requirements for screening (Wani et al. 2010).

Future directions will focus on developing finely tuned RNAi-based gene silencing vectors that are able to operate in a temporally and spatially controlled manner. However, a better and comprehensive understanding of RNAi would allow researchers to work effectively and efficiently in order to improve crop plants nutritionally and manage various intruders of crop plants (Wani et al. 2010).

9.5 Virus-induced Gene Silencing (VIGS): A Tool for Plant Functional Genomics

Virus-induced gene silencing is a powerful and rapid technique for analysis of gene function in plant growth and development (Baulcombe 1999). VIGS is a RNA-mediated post-transcriptional gene silencing mechanism that protects plants against foreign gene invasion. In addition to allowing a better understanding of how plants defend themselves against plant viruses, VIGS is also an extremely powerful functional genomics tool for knocking out gene expression of target plant genes in some plants. If the virus is used as a vector incorporating sequence from the host plant, the enzymatic pathway induced will silence the host gene (gene knock-out). Numerous methods have been employed to initiate this homology-based RNA degradation process, but all rely on the activity of double-stranded RNAs (dsRNAs) corresponding to the gene of interest. It involves creation of engineered viruses carrying sequences corresponding to the host gene to be silenced. dsRNA can be introduced into plants either as stable hairpin-RNA-synthesizing transgenes or by infection with recombinant viruses that carry segments of plant genes (Baulcombe 1999; Wang and Waterhouse 2002). This was first noticed when infection of *Nicotiana benthamiana* with tobacco rattle virus (TRV) containing a fragment of the gene encoding phytoene desaturase, an essential enzyme in carotene synthesis, resulted in a photo-bleached phenotype in photosynthetic tissues. Although many plant viruses have been used to develop VIGS vectors (Ruiz et al. 1998; Ratcliff et al. 2001; Gossele et al. 2002; Turnage et al. 2002), the TRV-based vectors have been suggested to have potential for use with a wide range of plant species due to the wide host range of TRV (Ratcliff et al. 2001). TRV provides the most robust results in terms of efficiency, ease of application, and absence of disease symptoms.

VIGS offers several advantages over existing functional genomics approaches because it does not require genetic transformation to silence a target plant gene, and phenotypes can be identified within a few weeks of virus inoculation. It is possible to silence multiple classes of closely related genes, as well as genes where true knock-outs are embryo-lethal. It is also more amenable to high-throughput studies and has been used successfully in model and non-model host systems (Liu et al. 2002c; Lu et al. 2003; Burch-Smith et al. 2004, 2006; Constantin et al. 2004; Fofana et al. 2004; Ding et al. 2006).

To date there are no research reports available for VIGS in banana and plantain. Renner et al. (2009) demonstrated that barley stripe mosaic virus (BSMV) can infect two species within the Zingiberaceae, and that BSMV–VIGS can be applied to specifically down-regulate phytoene desaturase in ginger *Zingiber officinale*. These results suggest that extension of BSMV–VIGS to other monocots such as banana many be possible in the future.

9.6 Tilling and Ecotilling

Traditional reverse genetic methods, such as the use of transposons to "knock out" a specific gene, can accurately determine phenotype but require time consuming transgenic or sophisticated tissue culture methodologies (Colbert et al. 2001). Such "knockout" methods are limiting because the entire gene is knocked out—the effects of partial loss of function of an active gene cannot be observed.

To overcome the limitations of knocking out an entire gene and to expand knowledge of active gene mutations, a new process called targeting induced local lesions in genomes or TILLING was developed (McCallum et al. 2000). TILLING uses chemical mutagenesis to yield a traditional allelic series of point mutations for virtually all genes. The TILLING process is of particular value for essential genes where sublethal alleles are required for phenotypic analysis. Furthermore, because TILLING does not involve transgenic modifications, it is attractive not only for functional genomics but also for agricultural applications (Henikof et al. 2004). The value of TILLING for genetic research is enhanced by its proven viability for a rapidly growing range of organisms including *Drosophila, Arabidopsis,* zebrafish, *maize, Medicago truncatula* and others. A full description of TILLING for *Arabidopsis* is described in Henikof et al. (2004).

The high densities of EMS (ethylmethane sulfonate) mutagenesis in TILLING raise the concern about background mutations being mistaken for mutations in target genes during phenotypic analysis. However, EMS-generated mutations at densities comparable to those in TILLING lines continue to be a basic learning tool for genetics, where background mutations have not been a problem. On the one hand, mutations in genes expected to impact a phenotypic trait controlled by many genes, such as plant height or size or leaf shape, may be subject to epistatic interactions, and outcrossing to the wild type may be necessary. On the other hand, mutations in genes expected to impact a phenotype that is controlled by few genes are unlikely to produce phenotypes perturbed by background mutations, and outcrossing is not a prerequisite for analysis (Henikoff and Comai 2003).

Ecotilling an adaptation of TILLING (Colbert et al. 2001; Comai et al. 2004) is a high-throughput method for the discovery and characterization of single nucleotide polymorphisms (SNPs) and small insertions/deletions (indels). It has been shown to be an accurate, lowcost and high-throughput method for the discovery and evaluation of nucleotide diversity in humans, plants and other organisms (Gilchrist et al. 2006; Till et al. 2006a, b; Nieto et al. 2007; Weil 2009).

Till et al. (2008) were the first to apply TILLING techniques in banana. EMS mutagenized populations of the Grand Naine variety are being developed. An evaluation of utilizing gamma irradiation for TILLING different varieties of banana including "Calcutta 4" is being investigated. Mutagenized populations of diploid "Calcutta 4" AA for TILLING are currently being developed at INIVIT in Cuba as part of an FAO/IAEA

Coordinated Research Project, where mutagenesis of cell cultures is also being investigated (Jain et al. 2011).

Till et al. (2010) used EcoTILLING to identify polymorphisms in the *Musa* genepool. They discovered over 800 novel alleles in 80 *Musa* accessions. In addition two single nucleotide polymorphisms that may be deleterious for the function of a gene putatively important for phototropism were identified. Evaluation of heterozygous polymorphism and haplotype blocks revealed a high level of nucleotide diversity in *Musa* accessions. Till et al. (2010) applied this strategy for the simultaneous discovery of heterozygous and homozygous polymorphisms in diploid accessions to rapidly evaluate nucleotide diversity in accessions of the same genome type. They also concluded that Ecotilling is suitable for diversity studies in *Musa*, that it can be considered for functional genomics studies and as a tool in selecting germplasm for traditional and mutation breeding approaches. Many challenges arise when TILLING is considered for banana. Widely consumed triploid varieties are largely sterile, infertile and/or parthenocarpic, thus requiring vegetative propagation, making seed mutagenesis impractical. Mutation induction using *in vitro* material can therefore be considered. Embryogenic cell suspensions are now available in banana (Strosse et al. 2003), although the technique appears to be cultivar specific. The problem of chimerism can be reduced through successive rounds of meristem isolation, followed by cutting meristems and allowing plantlets to regenerate (Roux 2004). The development of mutagenized populations for reverse genetic screens, and the optimization of TILLING and EcoTILLING in banana, will become useful resources for *Musa* research.

9.7 Functional Genomics Tools for Banana Improvement

Functional genomics and systems biology research are facilitating the identification of gene networks that are involved in controlling genetic variation for agronomically valuable traits (Varshney and Tuberosa 2007). Furthermore, combining new knowledge from functional genomic research with conventional breeding methods is essential for enhancing response to banana improvement. Currently, a number of functional genomics tools have used in banana.

In this section Serial analysis of gene expression (SAGE), massively parallel signature sequencing (MPSS), bacterial artificial chromosomes (BACs) expressed sequence tags (ESTs), and microarrays and their possible use for banana improvement is presented.

9.7.1 Serial Analysis of Gene Expression (SAGE)

Serial analysis of gene expression (SAGE) is based on the massive sequential analysis of short cDNA sequence tags. Each tag is derived from a defined position within a transcript. Its size (14 bp) is sufficient to identify the corresponding gene and the number of times each tag is observed provides an accurate measurement of its expression level. SAGE allows an accurate quantitative analysis of gene expression for thousands of genes at a time (Velculescu et al. 1995). SAGE is also a powerful tool for the identification of novel candidate genes and the characterization of specific metabolic or regulatory pathways (Coemans et al. 2005). SAGE in plants has mainly been limited to model organisms such as *Arabidopsis* and rice since the annotation of the tags depends mainly on the availability of cDNA libraries or EST collections (Matsumura et al. 1999, 2003a,b; Fizames et al. 2004). Various ways to increase the tag length, thereby enhancing annotation frequency has been made. These improvements include modified SAGE (18-bp tag; Ryo et al. 2000), Long SAGE (21-bp tag; Saha et al. 2002) and SuperSAGE (26-bp tag; Matsumura et al. 2003a). In banana and plantain with limited or no genomic DNA and cDNA/EST sequences available, classical SAGE yielding short tags would not be practical due to the very low chance and reliability of annotating the sequenced tags. However, non-model plant species possess numerous important traits not available for study in model plants, which emphasizes the need for high-throughput transcript profiling generally applicable to all crop plants. These traits may include different organs (e.g., fleshy fruits), special developmental processes (e.g., apomixis or parthenocarpy) and distinct quality traits such as flavor, nutrient or medicinal substances. Certain plant–pathogen interactions with serious economical consequences may also justify a large-scale functional analysis in the target organisms directly.

In banana, leaf material from the wild diploid banana *Musa acuminata* was used to generate a SuperSAGE library (Coemans et al. 2005). A total of 10,196 SuperSAGE tags were analyzed, which corresponded to 5,292 expressed genes. Although the most abundant tag was represented more than 300 times, which is close to 3% of the total transcripts, the vast majority of transcripts occurred only once. The number of unique tags steeply decreased with increasing tag copy number. For example, 4,409 tags, which correspond to 83.3% of all unique tags, occurred once and constituted 43.2% of the transcripts present in the analyzed tissue. In contrast, only 214 tags, representative of 2.1% of all unique tags, had a tag copy number ≥5. These 214 tags, however, represented 40.9% of the transcript pool analyzed

(Coemans et al. 2005). These findings are similar to a SuperSAGE study in rice leaves, where the sequencing of 12,119 tags identified 7,546 (62.3%) as being unique (Matsumura et al. 2003b). Furthermore, the 26-bp SuperSAGE tags were successfully applied as primers in 3'RACE, thus allowing the identification of unknown transcripts. Hence, SuperSAGE combined with 3'RACE and TAIL-PCR provides a powerful tool for functional genomics in non-model organisms for which supporting sequence resources are less extensive (Coemans et al. 2005).

Further studies by Sagi et al. (2005) reported SuperSAGE libraries with more than 5,000 unique 26-bp long tags generated from banana leaves, allowing a quantitative analysis of the transcriptome. Besides the identification of extremely low-abundance transcripts putative differentially expressed genes were identified after infection with *Mycosphaerella fijiensis*, the most important fungal pathogen of banana.

Extension of both types of tags to full-length genes is in progress and the functional analysis of these genes will contribute to the understanding of the banana genome.

9.7.2 *Massively Parallel Signature Sequencing (MPSS)*

Massively parallel signature sequencing (MPSS) stands for high-throughput sequencing of millions of cDNAs conjugated to 32-mer capture oligonucleotide tags on the surface of 5 μm diameter microbeads, where each microbead harbors some 100,000 identical copies of a particular cDNA ("microbead library"), that circumvents separate cDNA isolation, template processing and robotic procedures (Brenner et al. 2000). In brief, the abundance of each and every mRNA of a cell can be estimated by counting the number of clones with identical signatures. However, the multitude of steps (e.g., combinatorial synthesis of capture oligonucleotides, cDNA synthesis and restriction, addition of capture oligos to the 3'-end of each cDNA, cloning with PCR handles, PCR, denaturing of the product, ligation of annealed sequences, fluorescence-activated cell sorting, sequencing of about 16–20 bases at the free ends of the cloned templates, flow-cell management, adaptor ligation, etc.) makes this technique cumbersome for expression profiling. Apart from these sophisticated technologies, a number of less complex techniques are available that produce enough information for an initial transcriptome analysis (e.g., random activation of gene expression, RAGE; Harrington 2001). Other variations of MPSS are described in Mochida and Shinozaki (2010).

9.7.3 Bacterial Artificial Chromosomes (BACs) and Expressed Sequence Tags (EST) Libraries

ESTs are created by partial "one-pass" sequencing of randomly picked gene transcripts that have been converted into cDNA (Adams et al. 1993). Since cDNA and EST collections can be acquired regardless of genomic complexity, this approach has been applied not only to model species but also to a number of applied species with large genome sizes due to polyploidy and/or to their number of repetitive sequences (Mochida and Shinozaki 2010). The rapid accumulation of ESTs and cDNA clones have become significant resources for function al genomics.

Several thousand ESTs (gene sequences analyzed by isolation of mRNA), important for examination of gene expression, responses and differentiation of the plants, and examination of diversity, have been published (Santos et al. 2005) and many tens of thousands more are becoming available. Comparisons of EST libraries will be very valuable for identification of genes that are differentially expressed under stress conditions. Santos et al. (2005) made libraries from banana plants grown in cold (5°C) and hot (45°C) conditions, and found that about 30% of the genes in their library had been identified in other species as being involved in responses to environmental stress, and that there were substantial differences in the expression between the two libraries.

Several BAC libraries are already available in for *M. acuminata* and *M. balbisiana* (Vilarinhos et al. 2003; Safar et al. 2004; Ortiz-Vázquez et al. 2005; Piffanelli et al. 2008). Several BAC clones have been sequenced and show that a coding gene occurs every 6·4 to 6·9 kb in genomic DNA (Aert et al. 2004; Cheung and Town 2007) except in regions with large numbers of transposable elements where there are fewer genes. The application of low-depth 454 sequencing provided the largest amount of DNA sequence data, and enabled a detailed analysis of repetitive components of its nuclear genome (Hribova et al. 2010).

9.8 Transcriptomics of Fruit Ripening in Banana

The banana fruit undergoes several important physico-chemical changes during ripening (Godoy et al. 2010). The knowledge and understanding of the control points occurring during ripening can provide the scientific basis for the development of new post-harvest technologies, for improvement of shelf-life, but also for the enrichment of the nutritional and sensorial attributes of banana products (Lajolo 2004).

Many studies have examined fruit ripening aspects in banana, especially the genes involved in the production of ethylene. Analysis of gene expression would permit identification of important genes and regulatory elements involved in this process to also manipulate post-harvest banana ripening. The isolation of mRNA transcripts encoding proteins associated with the ripening process is a powerful tool for this purpose.

Differential screening of cDNA libraries from banana pulp at ripening stages 1 and 3 revealed that the following cDNAs were down-regulated in pulp during the ripening process: starch synthase, granule-bound starch synthase (GBSS), chitinase, lectin, and a type-2 metallothionein (Clendennen and May 1997). In contrast, the following cDNAs showed increased abundances early in ripening:

Endochitinase, [beta]-1,3-glucanase, a thaumatin-like protein, ascorbate peroxidase, metallothionein, and a putative senescence-related protein (Clendennen and May 1997).

In another study, cDNA libraries from banana pulp at different ripening stages were differentially screened using a novel microtiter plate method. The identified ripening-related cDNAs encoded enzymes involved in ethylene biosynthesis, respiration, starch metabolism, cell wall degradation, and several other key metabolic events (Medina-Suárez et al. 1997). Liu et al. (1999) investigated the characteristics of ethylene biosynthesis associated with ripening in banana. At the onset of the climacteric period in naturally ripened fruit, ethylene production increased greatly, with a sharp peak concomitant with an increase in the accumulation of *MA-ACS1* mRNA, and then decreased rapidly. At the onset of ripening, the *in vivo* ACC oxidase activity was enhanced greatly, followed by an immediate and rapid decrease. Expression of the *MA-ACO1* gene encoding banana ACC oxidase was detectable at the preclimacteric stage, increased when ripening commenced, and then remained high throughout the later ripening stage despite of a rapid reduction in the ACC oxidase activity. The results suggested that ethylene production in banana fruit is regulated by transcription of *MA-ACS1*.

Most of the cDNAs isolated in these studies are abundantly expressed in bananas, reflecting a limitation of differential screening. Therefore to isolate genes that are differentially expressed at lower levels, a more powerful tool is suppression subtractive hybridization (SSH; Diatchenko et al. 1996). Manrique-Trujillo et al. (2007) reported the first set of unigenes differentially expressed at late stages of banana fruit ripening, obtained from a pulp cDNA subtraction library, comparing the PCI 1 and PCI 5 ripening stages. The unigene set was produced by assembling the ESTs from the subtraction library. This allowed us to improve sequences in length and quality in addition to removing redundancy. The differential expressed groups of genes were involved in processes associated with fruit ripening, such as stress, detoxification, cytoskeleton and biosynthesis of volatile compounds. Kesari et al. (2007) used SSH approach, and identified 37 EST-unigenes from banana, which are expressed differentially during ripening. About 50% of these belonged to processes such as stress, defense and detoxification. In addition, genes, which are known to involve regulation of gene expression and other processes, were identified.

Four banana cDNAs encoding EIN3-like proteins were isolated from Cavendish banana (Mbeguie´-A-Mbeguie´ et al. 2008). Coinciding with the biochemical changes observed during banana fruit ripening, differentially expressed genes were isolated from the pulp (Clendennen and May 1997; Medina-Suarez et al. 1997) and peel (Drury et al. 1999; Liu et al. 2002b) of banana fruit after the initiation of the ripening process.

Jin et al. (2009) conducted a cDNA microarray analysis coupled with RT-PCR on banana fruit 10 days after harvest, and reported that 16 cDNAs were found to be up-regulated and six were down-regulated. The cDNAs identified are involved in signal transduction, amino acid metabolism, lipid metabolism, proteolysis, citrate biosynthesis and metabolism, and the uptake and transport of potassium. This information provides new clues that aid in understanding the regulation of ethylene biosynthesis initiation in postharvest banana ripening. Transcriptomics for fruit ripening appears to be one of the most active research areas in *Musa.*

To isolate differentially expressed genes at the early stage of postharvest banana ripening, a forward SSH cDNA library was constructed (Xu et al. 2007). SSH was performed with cDNA from banana fruit on the day of harvest as the "driver" and cDNA from banana fruit 2 days postharvest (DPH) as the "tester." A total of 289 clones in the SSH library were sequenced. BLASTX results revealed that 191 cDNAs had significant sequence homologies with known sequences in the NCBI database. Of the 191 cDNAs, 138 were singletons, and 53 belonged to divergent clusters containing 2–8 sequences. The identified cDNAs encoded proteins involved in cellular processes such as: metabolism; protein destination and storage; protein synthesis; signal transduction; transport and intracellular traffic; cell structure, growth, and division; transcription and post-transcription; and disease and defense. To characterize differentially expressed cDNAs in the SSH library, cDNA microarray analysis was conducted. A total of 26 cDNAs in the 2-DPH banana fruit were found to be up-regulated and these results were confirmed by using reverse transcriptase-PCR (RT-PCR).

Godoy et al. (2010) isolated 12 differentially expressed cDNAs, which were confirmed by dot-blots and Northern blots. Among the sequences identified were sequences homologous to plant aquaporins, adenine nucleotide translocator, immunophilin, legumin-like proteins, deoxyguanosine kinase and omega-3 fatty acid desaturase. Some of these cDNAs correspond to newly isolated genes involved in changes related to the respiratory climacteric, or stress-defense responses. The information generated in these studies provides new clues to aid in the understanding of banana ripening.

9.9 Microarray Technology

Microarray analysis has emerging technologies in the last few years as a flexible method for analyzing large number of nucleic acids fragments in parallel. Its origin can be traced to several different disciplines and techniques. Microarrays can be seen as a continued development of molecular biology hybridization methods, as an extension of the use of fluorescence microscopy in cell biology, as well as a diagnostic assay using capture to solid surface as a way to reduce the amount of analysis needed. With the recent and rapid increase in the number of sequenced species in whole-genome and/or large-scale cDNA clones, a number of DNA microarrays have also been developed for transcriptome analysis in various plant species. The convergence of ideas and principles utilized in these fields, together with technological advancements in preparing miniaturized collection of nucleic acids on solid supports, have all contributed to the emergence of microarray and microchip technologies (www.crownscientific.com). Systems-wide approaches such as microarray RNA-profiling are ideally suited to the study of the gene complex overlapping responses of plants to biotic and abiotic stresses. Tiling arrays, which are high-density oligonucleotide probes spanning the entire genome in a particular organism, are a platform for analyzing expressed regions throughout a whole genome; it is an effective method by which to discover novel genes and elucidate their structure. Transcriptome analysis in *Arabidopsis* under abiotic stress conditions using a whole-genome tiling array and discovered a number of antisense transcripts induced by abiotic stresses (Matsui et al. 2008). Similar analysis could be conducted in banana.

9.10 Comparative Functional Genomics to Study Drought Stress in *Musa*

Musa spp. is mainly cultivated in tropical and subtropical regions where rainfall is often erratic and dry seasons may exist. To date there have been few comparative studies on *Musa* drought stress tolerance, but field observations suggest that B-genome confers greater tolerance to drought than the A-genome (Thomas et al. 1998). Banana and plantains display a wide range of overlapping responses to both biotic and abiotic conditions and the diversity of physiological, biochemical and molecular strategies adopted during adaptation to unfavorable environmental conditions (including drought) create problems to study the molecular level and find out the stress responsible genes and expression profile (Dunoyer et al. 2007). However use of commercially available high-density oligonucleotide microarrays from closely related, heterologous species can be used to probe the transcriptomes of non-model plants.

Davey et al. (2009) evaluated the use of commercial, high-density Affymetrix rice and *Arabidopsis* ATH-1 Gene-Chip® microarrays to analyze complex plant responses in banana. They were able to rapidly profile the response of the *Musa* transcriptome to chronic drought stress, and to identify a range of structural and regulatory *Musa* gene-homologs previously found to be associated with the water-deficit response in other (model) plants. Further, despite the large phylogenetic difference between *Musa* and rice or *Arabidopsis* probes, comparisons to publically available transcriptome-profiling experiments identified a range of common drought-responsive genes, supporting the assigned *Musa* gene identities and descriptions. This approach outlines the potential of this strategy for the characterization of stress-resistance in banana and plantain varieties for which relatively little sequence information is currently available. Following cross-hybridization of *Musa* gDNA to the Rice GeneChip® Genome Array, ~33,700 gene-specific probe-sets had a sufficiently high degree of homology to be retained for transcriptomic analyses. In a proof-of-concept approach, pooled RNA representing a single biological replicate of control and drought stressed leaves of the *Musa* cultivar "Cachaco" was hybridized to the Affymetrix Rice Genome Array. A total of 2,910 *Musa* gene homologs with a >2-fold difference in expression levels were subsequently identified. These drought-responsive transcripts included many functional classes associated with plant biotic and abiotic stress responses, as well as a range of regulatory genes known to be involved in coordinating abiotic stress responses. This latter group included members of the ERF, DREB, MYB, bZIP and bHLH transcription factor families. Fifty-two of these drought-sensitive *Musa* transcripts were homologous to genes underlying quantitative trait loci (QTLs) for drought and cold tolerance in rice, including in two instances QTLs associated with a single underlying gene. The list of drought-responsive transcripts also included genes identified in publicly-available comparative transcriptomics experiments (Davey et al. 2009).

9.11 Conclusion

Banana is a good candidate for functional genomics because it has a relatively small genome size (600 Mbp) and important characteristics such as climacteric fruit, parthenocarpy or vegetative propagation, which are absent in model organisms (Sagi et al. 2005). Today, a number of advances in genetics and genomics have enhanced our understanding of structural and functional aspects of plant genomes and have included basic knowledge in ways that can enhance our ability to improve crop plants. Combining the new knowledge from genomic research with conventional breeding methods is vital for enhancing crop improvement. Current genomic techniques can contribute to banana and other crops improvement in two ways. First, a better understanding of the biological mechanisms can lead to improved screening methods for selecting superior genotypes more efficiently. Second, new knowledge can improve the decision-making process for more efficient breeding strategies (Varsheny et al. 2005).

Abbreviations

3'RACE	:	rapid amplification of 3' cDNA end
Ac/Ds	:	activator-dissociator
BACs	:	bacterial artificial chromosomes
BSMV	:	barley stripe mosaic virus
cDNA	:	complementary DNA
dsRNA	:	double-stranded RNA
EMS	:	ethylmethane sulphonate
ESTs	:	expressed sequence tags
GBSS	:	granule-bound starch synthase
GUS	:	Beta-glucuronidases
MPSS	:	massively parallel signature sequencing
PCR	:	polymerase chain reaction
RNAi	:	RNA interference
RT-PCR	:	reverse transcriptase
SAGE	:	serial analysis of gene expression
SSH	:	suppression subtractive hybridization
TAIL-PCR	:	thermal asymmetric interlaced PCR
T-DNA	:	tumor-inducing DNA
TILLING	:	Targeting Induced Local Lesions IN Genomes
TRV	:	Tobacco Rattle Virus
VIGS	:	virus-induced gene silencing

References

Adams M, Soares M, Kerlavage A, Fields C, Venter JC (1993) Rapid cDNA sequencing (expressed sequence tags) from a directionally cloned human infant brain cDNA library. Nat Genet 4: 373–380.

Aert R, Sagi L, Volckaert G (2004) Gene content and density in banana (*Musa acuminata*) as revealed by genomic sequencing of BAC clones. Theor Appl Genet 109: 129–139.

Alonso JM, Stepanova AN, Leisse TJ, Kim CJ, Chen H, Shinn P, Stevenson DK, Zimmerman J, Barajas P, Cheuk R, Gadrinab C, Heller C, Jeske A, Koesema E, Meyers CC, Parker H, Prednis L, Ansari Y, Choy N, Deen H, Geralt M, Hazari N, Home E, Karnes M, Mulholland C, Ndubaku R, Schmidt I, Guzman P, Aguilar-Henonin L, Schmid M, Weigel D, Carter

DE, Marchand T, Risseeuw E, Brogden D, Zeko A, Crosby WL, Berry CC, Ecker JR (2003) Genome-wide insertional mutagenesis of *Arabidopsis thaliana*. Science 301: 653–657.

Azpiroz-Leehan R, Feldmann KA (1997) T-DNA insertion mutagenesis in Arabidopsis: going back and forth. Trends Genet 13: 152–156.

Baulcombe DC (1999) Fast forward genetics based on virus induced gene silencing. Curr Opin Plant Biol 2: 109–13.

Baulcombe DC (2004) RNA silencing in Plants. Nature 431: 356–363.

Becker JD, Boavida LC, Carneiro J, Haury M, Feijo JA (2003) Transcriptional profiling of *Arabidopsis* tissues reveals the unique characteristics of the pollen transcriptome. Plant Physiol 133: 713–725.

Bitko V, Barik S (2001) Phenotypic Silencing of Cytoplasmic Genes with Sequence Specific Double Stranded Short Interfering RNA and Its Applications in the Reverse Genetics of Wild Type Negative Strand RNA Virus. BMC Microbiol 1: 34–44.

Birnbaum K, Shasha DE, Wang JY, Jung JW, Lambert GM, Galbraith DW, Benfey PN (2003) A gene expression map of the *Arabidopsis* root. Science 302: 1956–1960.

Brenner S, Johnson M, Bridgham J, Golda G, Lloyd DH, Johnson D, Luo S, McCurdy S, Foy M, Ewan M, Roth R, George D, Eletr S, Albrecht G, Vermaas E, Williams SR, Moon K, Burcham T, Pallas M, DuBridge RB, Kirchner J, Fearon K, Mao J, Corcoran K (2000) Gene expression analysis by massively parallel signature sequencing (MPSS) on microbead arrays. Nat Biotechnol 18: 630–634.

Brodersen P, Voinnet O (2006) The Diversity of RNA Silencing Pathways in Plants. Trends Genet 22: 268–280.

Burch-Smith TM, Anderson JC, Martin GB, Dinesh-Kumar SP (2004) Applications and advantages of virus-induced gene silencing for gene function studies in plants. Plant J 39: 734 –746.

Burch-Smith TM, Schiff M, Liu Y, Dinesh-Kumar SP (2006). Efficient virus-induced gene silencing in Arabidopsis. Plant Physiol 142: 21–27.

Chen L, Shiotani K, Togashi T, Miki D, Aoyama M, Wong HL, Kawasaki T, Shimamoto K (2010) Analysis of the Rac/Rop Small GTPase Family in Rice: Expression, Subcellular Localization and Role in Disease Resistance. Plant Cell Physiol 51: 585–595.

Cheung F, Town CD (2007) A BAC end view of the *Musa acuminata* genome. BMC Plant Biol 7: 29.

Chuang CF, Meyerowtiz EM (2000) Specific and Heritable Genetic Interference by Double-Stranded RNA in Arabidopsis thaliana. Proc Nat Acad Sci USA, 97: 985–4990.

Clendennen SK, May GD (1997) Differential gene expression in ripening banana fruit. Plant Physiol 115: 463–469.

Coemans B, Matsumura H, Terauchi R, Remy S, Swennen R, Sagi L (2005) Supersage combined with PCR walking allows global gene expression profiling of banana (*Musa acuminata*), a non-model organism. Theor Appl Genet 111: 1118–1126.

Colbert T, Till BJ, Tompa R, Reynolds S, Steine MN, Yeung AT, McCallum CM, Greene EA, Comai L, Henikoff S (2001) High-throughput screening for induced point mutations. Plant Physiol 126(2): 480–484.

Comai L, Young K, Till BJ, Reynolds SH, Greene EA (2004) Efficient discovery of DNA polymorphisms in natural populations by Ecotilling. Plant J 37: 778–786.

Constantin GD, Krath BN, MacFarlane SA, Nicolaisen M, Johansen IE, Lund OS (2004) Virus-induced gene silencing as a tool for functional genomics in a legume species. Plant J 40: 622–631.

Dalmay T, Hamilton AJ, Mueller E, Baulcombe DC (2000) Potato Virus X Amplicons in Arabidopsis Mediate Genetic and Epigenetic Gene Silencing. The Plant Cell 12: 369–380.

Davey MW, Graham NS, Vanholme B, Swennen R, May ST, Keulemans J (2009) Heterologous oligonucleotide microarrays for transcriptomics in a non-model species; a proof-of-concept study of drought stress in *Musa*. BMC Genom 10: 436.

Diatchenko L, Lau YFC, Campbell AP, Chenchik A, Moqadam F, Huang B, Lukyanov S, Lukyanov K, Gurskaya N, Sverdlov ED, Siebert PD (1996) Suppression subtractive hybridization: A method for generating differentially regulated or tissue-specific cDNA probes and libraries. Proc Natl Acad Sci USA, 93: 6025 6030.
Ding XS, Schneider WL, Chaluvadi SR, Mian MA, Nelson RS (2006) Characterization of a *Brome mosaic virus* strain and its use as a vector for gene silencing in monocotyledonous hosts. Mol Plant Microbe Interact 19: 1229–1239.
Drury R, Hortensteiner S, Donnison I, Bird CR, Seymour GB (1999) Chlorophyll catabolism and gene expression in the peel of ripening banana fruit. Physiol Plant 107: 32–38.
Dunoyer P, Himber C, Ruiz-Ferrer V, Alioua A, Voinnet O (2007) Intra- and intercellular RNA interference in *Arabidopsis thalian*a requires components of the microRNA and heterochromatic silencing pathways. Nat Genet 39: 848–856.
Escobar MA, Civerolo EL, Summerfelt KR, Dandekar AM (2001) RNAi-mediated oncogene silencing confers resistance to crown gall tumorigenesis. Proc Nat Acad Sci USA, 98: 13437–13442.
Fitzgerald A, Van Kha JA, Plummer KM (2004) Simultaneous silencing of multiple genes in the apple scab fungus *Venturia inaequalis,* by expression of RNA with chimeric inverted repeats. Fungal Genet Biol 41: 963–971.
Fizames C, Munos S, Cazettes C, Nacry P, Boucherez J, Gaymard F, Piquemal D, Delorme V, Commes T, Doumas P, Cooke R, Marti J, Sentenac H, Gojon A (2004) The Arabidopsis root transcriptome by serial analysis of gene expression. Gene identification using the genome sequence. Plant Physiol 134: 67–80.
Fofana IB, Sangare A, Collier R, Taylor C, Fauquet CM (2004) A geminivirus-induced gene silencing system for gene function validation in cassava. Plant Mol Biol 56: 613–624.
Gilchrist E, Haugh E (2005) TILLING without a plough: a new method with applications for reverse genetics. Curr Opin Plant Biol 8: 211–215.
Gilchrist EJ, Haughn GW, Ying CC, Otto SP, Zhuang J (2006) Use of Ecotilling as an efficient SNP discovery tool to survey genetic variation in wild populations of *Populus trichocarpa.* Mol Ecol 15: 1367–1378.
Gitlin L, Karelsky S, Andino R (2002) Short interference confers intracellular antiviral immunity in human cells. Nature 4: 418–430.
Godge MR, Purkayastha A, Dasgupta I, Kumar PP (2008) Virus-induced gene silencing for functional analysis of selected genes. Plant Cell Rep 27: 209–219.
Godoy AD, Cordenunsi BR, Lajolo FM, Nascimento JROD (2010) Differential display and suppression subtractive hybridization analysis of the pulp of ripening banana. Sci Hort 124: 51–56.
Goldoni M, Azzalin G, Macino G, Cogoni C (2004) Efficient Gene Silencing by Expression of Double Stranded RNA in *Neurospora crassa.* Fungal Genet Biol 41: 1016–1024.
Gossele VV, Fache II, Meulewaeter F, Cornelissen M, Metzlaff M (2002) SVISS—a novel transient gene silencing system for gene function discovery and validation in tobacco. Plant J 32: 859–866.
Hamada W, Spanu PD (1998) Co-Suppression of the Hydrophobin Gene *Hcf-1* is Correlated with Antisense RNA Biosynthesis in *Cladosporium fulvum.* Mol Gen Genet 259: 630–638.
Hammond TM, Keller NP (2005) RNA Silencing in *Aspergillus nidulans* is Independent of RNA-Dependent RNA polymerase. Genetics 169: 607–617.
Harrington JJ (2001) Creation of genome-wide protein expression libraries using random activation of gene expression. Nat Biotechnol 19: 440–445.
Hayes F, (2003) Transposon-based strategies for microbial functional genomics and proteomics. Annu Rev Genet 37: 3–29.
Henikoff S, Comai L (2003) Single-nucleotide mutations for plant functional genomics. Ann Rev Plant Biol 54: 375–401.
Henikoff S, Till JB, Comai L (2004) Tilling. Traditional mutagenesis meets functional genomics. Plant Physiol 135: 630–636.

Hiroyuki E, Izawa T, Kawahara M, Komatsu M, Koh S, Kyozuka J, Shimamoto K (1999) Ac as a tool for the functional genomics of rice. Plant J 19: 605–613.

Hřibová E, Neumann P, Matsumoto T, Roux N, Macas J, Doležel J (2010) Repetitive part of the banana (*Musa acuminata*) genome investigated by low-depth 454 sequencing. BMC Plant Biol 10:204 doi:10.1186/1471-2229-10-204

Iida S, Terada R (2004) A tale of two integrations, transgene and T-DNA: gene targeting by homologous recombination in rice. Curr Opin Biotechnol 15: 132–138.

Islam SMS, Miyazaki T, Tanno F, Itoh K (2005) Dissection of gene function by RNA silencing. Plant Biotechnol 22: 443–446.

Jacque JM, Triques K, Stevenson M (2002) Modulation of HIV-1 replication by RNA interference. Nature 418: 435–438.

Jain SM, Till B, Suprasanna P, Roux N (2011) Mutations and Cultivar Development of Banana In: M Pillay, A Tenkouano (eds) Banana Breeding: Progress and Challenges. CRC Press, Boca Raton, FL, USA, pp 203–217.

Jin ZQ, Xu BY, Liu JH, Su W, Zhang JB, Yang XL, Jia CH, Li M (2009) Identification of genes differentially expressed at the onset of the ethylene climacteric in banana. Postharvest Biol Technol 52: 307–309.

Kadotani N, Nakayashiki H, Tosa Y, Mayama S (2003) RNA Silencing in the Pathogenic Fungus *Magnaporthe oryzae.* Mol Plant-Microbe Interact 16: 769–776.

Kempin SA, Liljegren SJ, Block LM, Rounsley SD, Yanofsky MF, Lam E (1997) Inactivation of the *Arabidopsis* AGL5 MADS-box gene by homologous recombination. Nature 389: 802–803.

Kesaria R, Trivedia PK, Nath P (2007) Ethylene-induced ripening in banana evokes expression of defense and stress related genes in fruit tissue. Postharvest Biol Technol 46: 136–143.

Kim JA, Cho K, Singh R, Jung YH, Jeong SH, Kim SH, Lee JE, Cho YS, Agrawal GK, Rakwal R, Tamogami S, Kersten B, Jeon JS, An G, Jwa NS (2009) Rice OsACDR1 (*Oryza sativa* Accelerated Cell Death and Resistance 1) is a potential positive regulator of fungal disease resistance. Mol Cells 30: 431–439.

Lajolo FM (2004) Carbohydrate metabolism during fruit ripening: a functional genomics approach. www.pesquisaapoiada.fapesp.br/tematico accessed 5 March 2011.

Lawrence R, Pikaard SC (2003) Transgene-induced RNA interference: a strategy for overcoming gene redundancy in polyploids to generate loss-of-function mutations. Plant J 36: 114–121.

Lewin B (2004) Genes VII. Upper Saddle River, Pearson Prentice Hall NJ, USA.

Liu X, Shiomi S, Nakatsuka A, Kubo Y, Nakamura R, Inaba A (1999) Characterization of ethylene biosynthesis associated with ripening in banana fruit. Plant Physiol 121:1257–65.

Liu H, Cottrell TR, Pierini LM, Goldman WE, Doering TL (2002a) RNA interference in the pathogenic fungus *Cryptococcus neoformans*. Genetics 160: 463–470.

Liu P, Goh CJ, Loh CS, Pua EC (2002b) Differential expression and characterization of three metallothionein-like genes in Cavendish banana (*Musa acuminata*). Physiol Plant 114: 241–50.

Liu YL, Schiff M, Dinesh-Kumar SP (2002c) Virus induced gene silencing in tomato. Plant J 31: 777–786.

Lu R, Martin-Hernandez AM, Peart JR, Malcuit I, Baulcombe DC (2003) Virus-induced gene silencing in plants. Methods 30: 296–302.

Manrique-Trujillo SM, Ramırez-Lopez AC, Ibarra-Laclette E, Gomez-Lim MA (2007) Identification of genes differentially expressed during ripening of banana. J Plant Physiol 164: 1037–1050.

Martienssen RA (1998) Functional genomics:Probing plant gene function and expression with transposons. Proc Natl Acad Sci USA, 95: 2021–2026.

Marx J (2000) Interfering with gene expression. Science 288: 1370–1372.

Matsui A, Ishida J, Morosawa T, Mochizuki Y, Kaminuma E, Endo TA, Okamoto M, Nambara E, Nakajima M, Kawashima M, Satou M, Kim JM, Kobayashi N, Toyoda T, Shinozaki K,

Seki M (2008) Arabidopsis transcriptome analysis under drought, cold, high-salinity and ABA treatment conditions using a tiling array. Plant Cell Physiol 49: 1135–1149.

Matsumura H, Nirasawa S, Terauchi R (1999) Transcript profiling in rice (*Oryza sativa* L.) seedlings using serial analysis of gene expression (SAGE). Plant J 20: 719–726.

Matsumura H, Reich S, Ito A, Saitoh H, Kamoun S, Winter P, Kahl G, Reuter M, Kruger DH, Terauchi R (2003a) Gene expression analysis of plant host-pathogen interactions by SuperSAGE. Proc Natl Acad Sci USA, 100: 15718–15723.

Matsumura H, Nirasawa S, Kiba A, Urasaki N, Saitoh H, Ito M, Kawai-Yamada M, Uchimiya H, Terauchi R (2003b) Overexpression of Bax inhibitor suppresses the fungal elicitor-induced cell death in rice (*Oryza sativa* L.) cells. Plant J 33: 425–434.

Matzke MA, Matzke AJM (1995) How and why do plants inactivate homologues (trans)genes? Plant Physiol 107: 679–685.

Mbéguié-A-Mbéguié D, Hubert O, Fils-Lycaon B, Chillet M, Baurens FC (2008) EIN3-like gene expression during fruit ripening of Cavendish banana (*Musa acuminata* cv. Grande naine) Physiol Plant 133: 435–448.

McCallum CM, Comai L, Greene EA, Henikoff S (2000) Target Induced Local Lesions In Genomes (TILLING) for Plant Functional Genomics. Plant Physiol 123: 439–442.

Medina-Suarez R, Manning K, Fletcher J, Aked J, Bird CR, Seymour GB (1997) Gene expression in the pulp of ripening bananas. Plant Physiol 115: 453–61.

Meissner R, Chague V, Zhu Q, Elkind Y, Emmanuel E, Levy AA (2000) A high throughput system for transposon tagging and promoter trapping in tomato. Plant J 22: 265–274.

Milhavet O, Gary DS, Mattson MP (2003) RNA interference in biology and medicine. Pharmacol Rev 55: 629–648.

Miki D, Itoh R, Shimamoto K (2005) RNA silencing of single and multiple members in a gene family of rice. Plant Physiol 138: 1903–1913.

Mochida K, Shinozaki K (2010) Genomics and bioinformatics resources for crop improvement. Plant Cell Physiol 51: 497–523.

Nakayashiki H (2005) RNA Silencing in Fungi: Mechanisms and Applications. Fed Eur Biochem Soc Lett 579: 5950–5970.

Nieto C, Piron F, Dalmais M, Marco CF, Moriones E (2007) Ecotilling for the identification of allelic variants of melon eIF4E, a factor that controls virus susceptibility. BMC Plant Biol 7: 34.

Novina CD, Murray MF, Dykxhoorn DM, Beresford PJ, Riess J, Lee SK, Collman RG, Lieberman J, Shanker P, Sharp PA (2002) siRNA-directed inhibition of HIV-1 infection. Nat Mediter 8: 681–686.

Oliver S (1996) A network approach to the systematic analysis of yeast gene function. Trends Genet 12: 241–242.

Ortiz-Vázquez E, Kaemmer D, Zhang HB, Muth J, Rodríguez-Mendiola M, Arias-Castro C, James A (2005) Construction and characterization of a plant transformation-competent BIBAC library of the black Sigatoka-resistant banana *Musa acuminata* cv. Tuu Gia (AA). Theor Appl Genet 110(4): 706–13.

Piffanelli P, Vilarinhos AD, Safar J, Sabau X, Dolezel J (2008) Construction of bacterial artificial chromosome (BAC) libraries of banana (*Musa acuminata* and *Musa balbisiana*). Fruits 63: 375–379.

Pooggin M, Shivaprasad PV, Veluthambi K, Hohn T (2003) RNAi targetting of DNA viruses. Nat Biotechnol 21: 131–32.

Qu J, Ye J, Fang RX (2007) Artificial microRNA-mediated virus resistance in plants. J Virol 81: 6690–6699.

Radhamony RN, Prasad AM, Srinivasan R (2005) TDNA insertional mutagenesis in *Arabidopsis*: a tool for functional genomics. Elec J Biotechnol 8(1): 82–106.

Ratcliff F, Martin-Hernandez AM, Baulcombe DC (2001) Tobacco rattle virus as a vector for analysis of gene function by silencing. Plant J 25: 237–245.

Renner T, Bragga J, Driscoll HE, Cho J, Jackson AO, Specht CD (2009) Virus-induced gene silencing in the culinary ginger (*Zingiber officinale*): An effective mechanism for down-regulating gene expression in tropical monocots. Mol Plant 2: 1084–1094.

Roux NS (2004) Mutation induction in *Musa*—Review. In: SM Jain, R Swennen (eds) Banana Improvement: Cellular, Molecular Biology, and Induced Mutations. Science Publishers, Enfield, New Hampshire, USA, pp 23–32.

Ruiz MT, Voinnet O, Baulcombe DC (1998) Initiation and maintenance of virus-induced gene silencing. Plant Cell 10: 937–946.

Ryo A, Kondoh N, Wakatsuki T, Hada A, Yamamoto N, Yamamoto M, Yamamoto N (2000) A modified serial analysis of gene expression that generates longer sequence tags by nonpalindromic cohesive linker ligation. Analytical Biochemistry 277: 160–162.

Safar J, Noa-Carrazana JC, Vrana J, Bartos J, Alkhimova O, Sabau X, Simkova H, Lheureux F, Caruana ML, Dolezel J, Piffanelli P (2004) Creation of a BAC resource to study the structure and evolution of the banana (*Musa balbisiana*) genome. Genome 47: 1182–1191.

Sági L, Remy S, Coemans B, Thiry E, Santos E, Matsumura H, Terauchi R, Swennen R (2005) Functional analysis of the banana genome by gene tagging and SAGE. Plant & Animal Genomes XIII Conference, San Diego, CA, USA.

Saha S, Sparks AB, Rago C, Akmaev V, Wang CJ, Vogelstein B, Kinzler KW, and Velculescu VE (2002) Using the transcriptome to annotate the genome. Nat Biotechnol 20: 508–512.

Santos CMR, Martins NF, Horberg HM, de Almeida ERP, Coelho MCF, Togawa RC, da Silva FR, Caetano AR, Miller RN, Souza MT (2005) Analysis of expressed sequence tags from *Musa acuminata* ssp. *burmannicoides*, var. Calcutta 4 (AA) leaves submitted to temperature stresses. Theor Appl Genet 110: 1517–1522.

Santos E, Remy S, Thiry E, Windelinckx S, Swennen R, Sági L (2009) Characterization and isolation of a T-DNA tagged banana promoter active during *in vitro* culture and low temperature stress. BMC Plant Biol 9: 77.

Schweizer P, Pokorny J, Schulze-Lefert P, Dudler R (2002) Double Stranded RNA Interference with Gene Functions at the Single Cell in Cereals. Plant J 24: 895–903.

Shinozaki K, Yamaguchi-Shinozaki K (2007) Gene networks involved in drought stress response and tolerance. J Exp Bot 58(2): 221–227.

Stemple DL (2004) TILLING—a high-throughput harvest for functional genomics. Nat Rev Genet 5: 145–150.

Strosse H, Domergue R, Panis B, Escalant JV, Cote F (2003) Banana and plantain embryogenic cell suspensions. In: A Vezina and C Picq (eds). INIBAP Technical Guidelines & The International Network for the Improvement of Banana and Plantain, International Plant Genetic Resources Institute, Montpellier, France.

Sundaresan V, Springer P, Volpe T, Haward S, Jones JD, Dean C, Ma H, Martienssen R (1995) Patterns of gene action in plant development revealed by enhancer trap and gene trap transposable elements. Genes Dev 9: 1797–1810.

Tang W, Weidner DA, Hu BY, Newton RJ, Hu X (2006) Efficient Delivery of Small Interfering RNA to Plant Cells by a Nanosecond Pulsed Laser-Induced Wave for Post Transcriptional Gene Silencing. Plant Sci 171: 375–81.

Terada R, Urawa H, Inagaki Y, Tsugane K, Iida S (2002) Efficient gene targeting by homologous recombination in rice. Nat Biotechnol 20: 1030–1034.

Thomas DS, Turner D, Eamus D (1998) Independent effects of the environment on the leaf gas exchange of three banana (*Musa sp.*) Cultivars of different genomic constitution. Sci Hort 75(1-2): 41–57.

Tierney MB, Lamour KH (2005) An Introduction to Reverse Genetic Tools for Investigating Gene Function. The Plant Health Instructor DOI: 10.1094/PHI-A-2005-1025-01.

Till BJ, Reynolds SH, Greene EA, Codomo CA, Enns LC, Johnson JE, Burtner C, Odden AR, Young K, Taylor NE, Henikoff JG, Comai L, Henikoff S (2003) Large-Scale Discovery of Induced Point Mutations with High-Throughput TILLING. Genome Res 13: 524–530.

Till BJ, Zerr T, Bowers E, Greene EA, Comai L (2006a) High throughput discovery of rare human nucleotide polymorphisms by Ecotilling. Nucl Acids Res 34: e99.

Till BJ, Zerr T, Comai L, Henikoff S (2006b) A protocol for TILLING and Ecotilling in plants and animals. Nat Protoc 1: 2465–2477.

Till BJ, Jankowicz-Cieslak J, Nakitandwe J, Bado S, Mattijevic M, Afza R, Lagoda PGL, Mba C (2008) TILLING and EcoTILLING in Banana. Plant & Animal Genomes XVI Conference. San Diego, CA, USA.

Till BJ, Jankowicz-Cieslak J, Sagi L, Huynh OA, Utsushi H, Swennen R, Terauchi R, Mba C (2010) Discovery of nucleotide polymorphisms in the *Musa* gene pool by Ecotilling. Theor Appl Genet 121: 1381–1389.

Turnage MA, Muangsan N, Peele CG, Robertson D (2002) Geminivirus-based vectors for gene silencing in Arabidopsis. Plant J 30: 107–114.

Ullu E, Djikeng A, Shi H, Tschudi C (2002) RNA interference: Advances and questions. Phil Trans Roy Soc Lond Brit Biol Sci 29: 65–70.

Vanitharani R, Chellappan P, Fauquet CM (2003) Short interfering RNA-mediated interference of gene expression and viral DNA accumulation in cultured plant cells. Proc Natl Acad (USA) 100: 9632–9636.

Varshney RK, Tuberosa R (2007) Genomics-Assisted Crop Improvement: An Overview, Genomics Approaches and Platforms 1: 1–12.

Varshney RK, Graner A, Sorrells ME (2005) Genomics-assisted breeding for crop improvement. Trends Plant Sci 10 (12): 621–630.

Velculescu VE, Zhang L, Vogelstein B, Kinzler KW (1995) Serial analysis of gene expression. Science 270: 484–487.

Vilarinhos AD, Piffanelli P, Lagoda P, Thibivilliers S, Sabau X, Carreel F, D'Hont A (2003) Construction and characterization of a bacterial artificial chromosome library of banana (*Musa acuminata* Colla). Theor Appl Genet 106(6): 1102–1106.

Wang MB, Waterhouse PM (2002) Application of gene silencing in plants. Curr Opin Plant Biol 5: 146–150.

Wani SH, Sanghera GS (2010) Genetic engineering for viral disease management in plants. Notulae Sci Biol 2: 20–28.

Wani SH, Sanghera GS, Singh NB (2010) Biotechnology and plant disease control—Role of RNA interference. Amer J Plant Sci 1: 55–68.

Waterhouse PM, Graham MW, Wang MB (1998) Virus resistance and gene silencing in plants can be induced by simultaneous expression of sense and antisense RNA. Proc Nat Acad Sci USA, 95: 13959–13964.

Waterhouse PM, Wang MB, Lough T (2001) Gene silencing as an adaptive defense against viruses. Nature 411: 834–842.

Weil CF (2009) TILLING in grass species. Plant Physiol 149: 158–164.

Xu BY, Su W, Liu JH, Wang JB, Jin ZQ (2007) Differentially expressed cDNAs at the early stage of banana ripening identified by suppression subtractive hybridization and cDNA microarray. Planta 226(2): 529–39.

Yang Z, Tang L, Li M, Chen L, Xu J, Wu G, Li H (2010) Monitoring homologous recombination in rice (*Oryza sativa* L.). Mutation research Fund Mol Mech Mutagen 691: 55–63.

Yu H, Wang M, Tang D, Wang K, Chen F, Gong Z, Gu M, Cheng Z (2010) OsSPO11-1 is essential for both homologous chromosome pairing and crossover formation in rice. Chromosoma 119: 625–636.

10

Proteomics and Metabolomics of *Musa*

Hoang Lan Chi Dinh[1,a,*] and *Peer M. Schenk*[1,b]

ABSTRACT

Large amounts of data are currently produced via the "tooled up" foundation of "omic" technologies. Proteomics and metabolomics which are typical for post-genomic profiling techniques have been utilized in crop development. The current genomics projects of *Musa* spp. including banana and plantain are complemented with high-throughput metabolomics and proteomics approaches. While metabolic analyses globally focus on small molecules of plants under a specific condition, proteomics identify and quantify differentially expressed proteins. Data from metabolite and protein profiling are analyzed by bioinformatics and complement the more commonly used transcriptomics approaches. This chapter reviews the current developments in different areas of *Musa* metabolome and proteome studies. Experimental and analytical tools such as NMR, MS, 2D-gel electrophoresis, GC-MS, MALDI-TOF have been described as important requisites for the generation of *Musa* metabolome and proteome data. In general, systems biology then forms the expected integration of results from transcriptomics, metabolomics, proteomics and bioinformatics approaches in *Musa* genomics research.

Keywords: banana, "omics" technologies, plantain, systems biology

[1]School of Biological Sciences, The University of Queensland, St. Lucia, Queensland 4072, Australia.
[a]e-mail: *dinhhoanglanchi@gmail.com*
[b]e-mail: *p.schenk@uq.edu.au*
*Corresponding author

List of abbreviations after the text.

10.1 Introduction

Musa spp. including banana and plantain comprise the world's fourth most important crop. Global high-throughput approaches targeting the genome, transcriptome, proteome, metabolome and interactome have been developed as powerful tools to increase our understanding of cellular functions and plant systems biology (Fukushima et al. 2009). This chapter reviews the recent progress that has been made in *Musa* genomics research by making use of the omics technologies, proteomics and metabolomics.

Proteomics cover the global study of proteins produced in cells under various conditions (Primrose and Twyman 2006). The concept of "proteome" was defined by Wilkins and colleagues for the genome expression through protein isolation and identification (Wilkins et al. 1995). Research on *Musa* proteomics is based on large scale technologies for protein separation, isolation, detection and quantification. Metabolomics is the discovery and profiling of all metabolites within biological pathways of an organism under pleiotropic effects (Fiehn 2002). In agriculture, metabolite profiling for systems biology has been used to optimize trait development (Dixon et al. 2006). Metabolomics identifies and quantifies the global metabolites within a sample extracted from a particular organism aiming at constructing global biochemical networks (Primrose and Twyman 2006). Plant metabolomics refers to the unbiased identification and quantification of all the small molecules with molecular weights lower than 1000 Daltons. The metabolome comprises a set of biologically synthesized metabolites that are endproducts of gene expression and protein activity. This technology exploits typical metabolites, including primary metabolites as well as some secondary metabolite molecules involved in common characteristics such as pigments, nutrients, flavors and scents (Clark and Pazdernik 2009).

This chapter provides an overview of how recent advances in *Musa* proteomics and metabolomics integrate with other omic technologies to complement research in *Musa* genomics.

10.2 Tools Developed for Proteomics

Latest proteomics techniques support protein/peptide separation as well as the identification and characterization of these molecules, including two-dimensional gel electrophoresis (2-DE), sodium-dodecyl sulfate-polyacrylamide gel electrophoresis (SDS-PAGE), mass spectrometry (MS), tandem mass spectrometry (MS/MS), capillary electrophoresis (CE), surface-enhanced laser desorption/ionization (SELDI), fourier transform ion cyclotron resonance (FTICR), electrospray ionization mass spectrometry (ESI-MS/MS). The combination between 2-DE and MS (2-DE/MS) is the most important facility in plant proteomics research (Jorrín et al. 2007;

Thiellement et al. 2007). MS is currently used to identify large scale proteins isolated from 2-DE, which separates and visualizes proteins by iso-electric focusing (IEF) (O'Farrell 1975) before MS is used to identify and characterize the separated proteins (Fenn et al. 1989). The principle of the MS method is to match target proteins to gene sequences in databases via masses of fragments from MS instruments (Roberts 2002).

Regarding gel-based techniques, 2-DE has been used as a classical method of proteomics for protein separation (Fig. 10-1; Thongboonkerd 2007). The two steps for protein manipulation by 2-DE are protein separation and visualization. First, the crude protein is separated by pI and pH gradient polyacrylamide gels giving iso-electric points in the first dimension; then the isolated protein is separated by molecular mass-based SDS-PAGE. In the final step of protein isolation, a fluorescent dye is used to visualize the isolated proteins (Patton 2000). Data from gel-based applications are

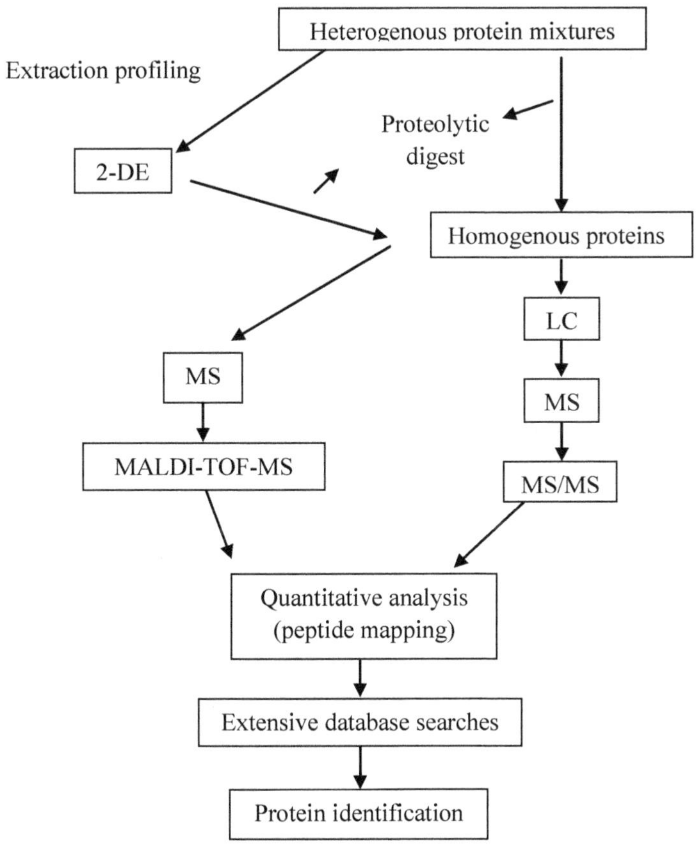

Figure 10-1 Experimental approach of plant proteomics.

subsequently analyzed by MS facilities (Weckwerth 2008). A gel-free method for protein identification includes the use of LC/MS as shotgun proteomics (Fig. 10-1; Thongboonkerd 2007; Weckwerth 2008). This technique not only identifies, but also separates proteins, thus is a promising tool for plant proteomics (Porubleva et al. 2001).

10.3 Proteomics for Marker Identification and Functional Analyses

Briefly, the common process of proteomic analysis includes protein extraction, separation by 2D-GE, and protein identification by MS before protein expression by bioinformatics. Proteomic analyses have been used as an efficient tool for the characterization of quantitative trait genetics by quantitative traits loci (QTL) mapping (de Vienne et al. 1999). The integration between 2-DE and *de novo* MS, currently, is the most effective tool for the generation of genetic markers (Samyn et al. 2007). Consequently, genetic mapping has become popular in *Musa* breeding research and many molecular markers have been identified in this manner and used over the last years. For example, the identification of 25 new microsatellite markers from *Musa balbisiana* has provided an up-to-date knowledge for linkage mapping in plantains and cooking banana (Buhariwalla et al. 2005). Another example is the use of 325 simple sequence repeat (SSR) markers within bacterial artificial chromosome (BAC) end-sequences of the *Musa acuminata* genome (Cheung and Town 2007). The linkage map of diploid bananas was drawn based on amplified fragment length polymorphism (AFLP), isozymes, RFLP (restriction fragment length polymorphism) and SSR data linked to a QTL map for Sigatoka resistance traits (Pillay and Tripathi 2007).

A variety of methods for proteomic profiling have already been used for *Musa* research. For example, NMR and MS were used by Carpentier and colleagues (2007) to study the *Musa* meristem proteome accumulated in response to osmotic stress. These techniques subsequently analyze amino acid sequence data and together with cross-species identification by sequence or structure homology searches have led to large-scale meristem protein identification and subsequent functional analyses. The importance of this *Musa* meristem study was to ideally produce a database for public access to all meristem proteome information. One of the most accessible *Musa* proteomics databases is pdata that is available at http://www.pdata.ua.ac.be/musa/

Another aspect of plant proteomics is the identification of derived peptides. Samyn et al. (2007) used integrated techniques, including 2-DE, MALDI-MS, *de novo* sequencing and homology-based search algorithms for the identification of 40 proteins and different iso-forms from banana. The authors also point towards the rising number of plant proteomes in databases, which results from the ongoing development of genome and EST sequencing. In addition to this research, the use of 2-DE for the release of different protein isoforms significantly explains some of the genetic differences in the *Musa* genome, including differences in gene loci, allele variations, subunit interactions and splice forms.

Various analyses of *Musa* subspecies and banana cultivars (Venkatachalam et al. 2008) show that the analysis of *Musa* transcriptomes, proteomes and metabolomes present appropriate strategies for crop improvement and breeding. Protein profiling in *Musa* aims to increase protein databases, functional knowledge and assists in marker searching. Sequence similarity has been used for research on non-sequenced genome organisms like banana in which databases are the decisive factor for comparing and screening of target proteomes.

10.4 Tools Developed for Metabolomics

In contrast to proteomics and transcriptomics in predicting gene activity, metabolomics is able to characterize individual plants at a phenotypic level based on morphology, physiology and disease susceptibility of plants (Allwood et al. 2008; Nikolau and Wurtele 2007). This non-target profiling technique is used to detect and quantify many metabolites, especially for identifying large intermediates of primary metabolism (Matthew et al. 2009). To ensure accuracy and reproducibility of metabolomic profiling, techniques such as nuclear magnetic resonance (NMR), gas chromatography (GC), liquid chromatography (LC) and the combination between mass spectrometry and gas chromatography (GC) or liquid chromatography (LC) are employed (Nikolau and Wurtele 2007). Other metabolomics technologies include matrix-assisted laser desorption/ionization (MALDI) (Li et al. 2008), fourier-transformed infrared spectroscopy (Hall et al. 2008), direct injection MS and flow injection MS, quadruple MS for the analysis of mass ions based on electric fields in a mass filter, tandem MS (MS-MS) for metabolite identity (Hanzlik et al. 2009), time of flight for the identification of analytes (TOF-MS), *de novo* methods, which includes the combination between LC to nuclear magnetic resonance (NMR) or MS for the identification of secondary plant metabolites (Samyn et al. 2007). Mathew et al. (2009) identified that much information from important metabolites in plants could be deciphered by the parallel pathways of NMR, GC/MS and thin layer chromatography (TLC). Technical procedures for plant metabolomics are mainly based on the integration between mass spectrometry and chromatography (Fig. 10-2; Moritz and Johansson 2008; Iijima et al. 2008).

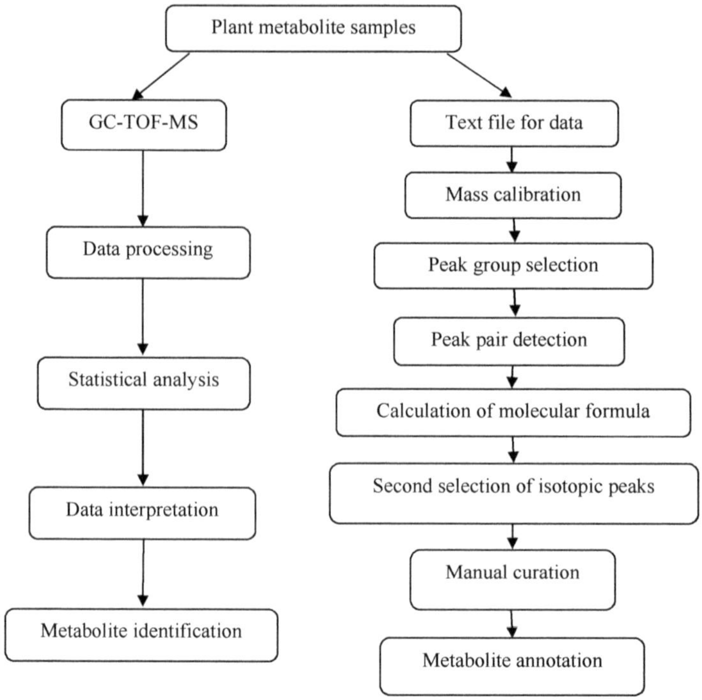

Figure 10-2 Experimental approach of plant metabolomics.

10.5 Applications of Metabolomics in Phenomics/Genomics-Assisted Breeding

Plant proteomics and metabolomics assist in QTL mapping (Subramaniam 2005). QTL mapping is used to clarify the number, phenotypic impact and interaction of loci by using particular markers or phenotypic variation (Nikolau and Wurtele 2007). QTL mapping typically uses DNA marker assays to evaluate how a specific locus affects the traits associated with the desired phenotype. This process can be significantly assisted by the use of expressed features of the plant including the transcriptome, proteome and metabolome. The stronger the integration of phenotype, genotype and heritability, the more reliable QTL mapping for a trait is identified (Fiehn 2002; Zeng et al. 2008).

Quantitative genetic mapping is currently still in its infancy for *Musa*. However, there are important considerations for several traits such as bunch weight, fruit quality parameters, fruit filling time and disease resistance (Ortiz 1995). The genetic mapping assisted-breeding program of *Musa* plants started in 1990 by crossing between wild diploid *M. acuminata* spp.

banksii x SF265. The first QTL map of 90 loci from 92 individuals of F_2 generation crossed between two *M. acuminata* accessions was established and published in 1992 (summarized by Heslop-Harrison and Schwarzacher 2007). Linkage maps and chromosome positions are currently integrated in banana genetic maps (Roux et al. 2008). A total of 120 molecular markers (AFLP, RFLP and SSR) were used to construct the genetic map of "Calcutta 4"x "Madang". Recently, the natural variances of metabolites were closely linked to phenotypic growth, so the utilization of quantitative genetics in metabolomics has created the capacity for genetic control (Keurentje et al. 2008). Breeders use QTL maps to create new *Musa* plants containing the improved polygenic traits such as high productivity as well as biotic and abiotic stress resistances. One example for QTL mapping is for R-genes within the *Musa* genome from diploid wild species such as *Musa acuminata* (A), *M. balbisiana* (B), *M. schizocarpa* (S), *M. textilis* (T), *M. velutina* and *M. ornate* (Azhar and Heslop-Harrison 2008). At the current time, microsatellite markers have been used widely in banana breeding programs (Creste et al. 2006).

Hundreds of samples can easily be produced for metabolite profiling and high-throughput automated analyses can be achieved by NMR (Allwood et al. 2008) and unknown compounds could be systematically identified by NMR (Fiehn 2002). Generally, proton nuclear magnetic resonance (H-NMR) is used to classify herbicides and their bioactive components (Schauer and Fernie 2006). NMR is also used to identify "steady-state concentrations of metabolites in different compartments *in vivo*" (Fiehn 2002; Roberts 2002). Regarding MS, the absorbance or redirection of electromagnetic (EM) radiation by chemical bonds-based technique provides a global metabolic fingerprint via rapid spectral acquisition and spatial measurement of metabolites (Allwood et al. 2008). The advantage of GC-MS is its ability to enhance metabolite volatilities by using chemical derivatization (Fiehn 2002). Indeed, MS detects metabolites that are volatile up to 2,500°C with polar non-volatile analysis (amino acids, sugars and organic acids; Allwood et al. 2008). One application of GC-MS and LC-MS is used to identify salt stress in plants (Kim et al. 2007). The integration of H-NMR, C-NMR, MS, HPLC was utilized to study the biosynthesis pathway of banana phytoalexins (Kamo et al. 2000). High performance liquid chromatography (HPLC) is capable of identifying differences between metabolites including their composition and quantity. For example, different phenolics (ferulic and sinapic acids) and phenolic polymers within banana roots were identified using HPLC by de Ascensao and Dubery (2003). The phenylpropanoid biosynthetic pathways reflect important metabolite activities of plants against pathogenic elicitors. Methods for extraction, identification of phenolics from banana roots to defense elicitors from *Fusarium oxysporum f.sp. cubense* (*Foc*) have been developed by de Ascensao and Dubery (2003).

Metabolomics continuously assist in metabolite discovery and profiling of different aspects of *Musa* research. The first approach is in crop/food analysis. Different technical facilities involving LC-MS, GC-MS and NMR are key techniques in identifying the quality and nutritional content of food, which is governed by the composition of various metabolites (Hall et al. 2008). Second, metabolomics is used for diagnostics purposes. For example, GC-MS is the most common technical strategy for screening the mode of action of herbicides (Schauer and Fernie 2006).

Musa metabolomics has great potential to assist in functional genomics and gene function annotations. However, for further applications of metabolomics, strategies to control and standardize technical processes and infrastructure of this science are necessary. Some standards for metabolite documenting are now established and described by The Metabolomics Standards Initiative (MSI) (Fiehn et al. 2008). The measurement parameters include study design, sample preparation, data acquisition, data processing, data analysis, and data integration (Fiehn et al. 2008).

Metabolomics plays an important role in designing better crops because molecular signatures associated with key traits identified by this technology help scientists to rearrange metabolic interactions as well as redesign/select crops with the desired phenotype and genotype (Hall et al. 2008). The study of entire phenotypes and genotypes refer to phenomics and genomics, respectively. Phenomics is a promising technique for "precision agriculture and predictive breeding", states Uli Schurr, Director of the Institute for Phytosphere Research, Germany (Finkel 2009). Metabolite profiling also supports genomics for future breeding programs in plants. For example, a metabolomic study on the phenylpropanoid pathway of banana and a comparative genomic study to Haemodoraceae plants reveal the close relationship between these two species (Kamo et al. 2000). Scientists can use information from the biosynthetic phenylpropanoid pathway retrieved from Haemodoraceae model plants for modifying the genetic basis of phytoalexins in banana fruits with implications to ripening traits.

Research on plant metabolites significantly expands our knowledge in both plant proteins and genetics; the possible changes in metabolite profiles explain related changes in proteins and genes that control the metabolic pathways (Roberts 2002). The breeding program from the International Institute of Tropical Agriculture (IITA) (Vuylsteke et al. 1997; Crouch et al. 1998) has been established for *Musa* species. NMR and MS can profile the status of several metabolites such as stress, nutrition factor, time and environmental perturbation (Dixon et al. 2006). Other techniques, such as microarray analyses support metabolic profiling, for example in measuring the intended and unintended adverse effects of genetically modified (GM) crops (Dixon et al. 2006).

10.6 Conclusion

Proteomics and metabolomics directly measure the expression of proteins and metabolites that are beneficial to the confirmation and construction of biological processes. The technologies aim at a better understanding of plant and cellular functions in *Musa*.

Various proteins have been identified from the *Musa* genome. For example, genes that encode ripening traits in banana, including the genotypes AAA (Cavendish), AAB (Rasthali and Poovan), AB (Kanthali), ABB (Monthan) (Choudhury et al. 2008). The advantages of using metabolomics and proteomics approaches include the option to shift to a semi-automatic manner with great capabilities for extension. For example, MS technology can be used in a high-throughput manner for proteomics, although it requires large investments for equipment. Its advantages over traditionally 2D-GE-based methods are less protein preparation and consumables, while reacting at a high rate and in very low quantities (Morot-Gaudry et al. 2007). The integrated databases allow deciphering of mass-protein profiles. The plant kingdom harbors vast chemical diversity and dynamic characteristics, thus metabolic profiling is still currently very limited (Tohge and Fernie 2009). MS-based metabolomics confronts the problem of dynamic range as described by Sumner (2003).

In comparison with proteomics, metabolomics presents higher throughput and more economical directions (Hollywood et al. 2006). The recent "omics" era is increasingly using microarrays, next generation sequencing and databases for metabolite and proteomic screening to assist plant breeding programs; thus this enhances the chance for the model crop plants such as *Musa* species for improvement of further generations. Further the emergences of proteomics and metabolomics are currently attracting and engaging young scientists who will upgrade and widen application areas of those sciences. The significance of studying proteomics and metabolomics is to further exploit pleiotropic effects on biochemical pathways of *Musa* species. The advanced "omics" facilities increase the chance for many proteins and metabolites within the *Musa* genome. In addition, procedures for plant metabolomics and proteomics are recently published as a basic resource for scientists to access (Nikolau and Wurtele 2007; Thiellement et al. 2007). The challenge for the *Musa* genome annotation is to choose effectively integral analysis tools for transcriptome, proteome and metabolome data sets. Lists of databases and programs used for systems biology research in *Musa* spp. or plant species are summarized in the *Musa* bioinformatics chapter of this book. The development of proteomics and metabolomics in plants is necessary for systems biology and plant improvements via genetic bases; however, technical management is required for both sectors. While architecture for metabolomics or ArMet (http://www.armet.org/publications/index.html) have been utilized as a data model for metabolite profiling (Jenkins et al. 2004), HUPO (Human Proteome Organization) Proteomics Standards Initiative or PSI (http://www.psidev.info/) is recently used as a standard for the evaluation of proteomics research. Many valuable resources of genomics sequences in different plant species help the areas of plant proteomics and metabolic progress with implications for *Musa* research.

"Omics" based biotechnology significantly improves the genetic exploitation of bananas and plantains due to its ability to identify powerful traits and dominant genes. The integration of transcriptomics, proteomics, metabolomics and bioinformatics provides comprehensive data for the *Musa* genome. Studies on proteins and metabolites significantly enhance the development of crop plants in tropical and subtropical regions around the world. Both proteins and metabolites, in addition to mRNA, improve the comprehensive and quantitative analysis of molecular phenotypes, including those of *Musa* species. Proteomics and metabolomics enhance the development of *Musa* at the molecular level which indirectly assists the breeding program in *Musa* plants; consequently enhancing the commercial value of this fruit crop.

Abbreviations

AFLP	:	amplified fragment length polymorphism
BAC	:	bacterial artificial chromosome
CE	:	Capillary Electrophoresis
ESI-MS/MS	:	Electrospray Ionization Mass Spectrometry
FTICR	:	Fourier Transform Ion Cyclotron Resonance
GC	:	gas chromatography
HPLC	:	High performance liquid chromatography
IEF	:	Iso-Electric Focussing
LC	:	liquid chromatography
MALDI	:	Matrix-Assisted Laser Desorption/Ionization
MS/MS	:	Tandem Mass Spectrometry
MS	:	Mass Spectrometry
NMR	:	nuclear magnetic resonance
QTL	:	quantitative trait loci
RFLP	:	restriction fragment length polymorphisms
SDS-PAGE	:	Sodium-Dodecyl Sulfate-Polyacrylamide Gel Electrophoresis
SELDI	:	Surface-Enhanced Laser Desorption/Ionization
SSR	:	simple sequence repeats
TOF	:	time of flight

Acknowledgements

We wish to thank Kevan Jones for useful comments and the Australian Research Council (DP1094749) for research funding.

References

Allwood JW, Ellis DI, Goodacre R (2008) Metabolomic technologies and their application to the study of plants and plant–host interactions. Physiol Plant 132: 117–135.

Azhar M, Heslop-Harrison JS (2008) Genomes, diversity and resistance gene analogues in *Musa* species. Cytogenet Genome Res 121: 59–66.

Buhariwalla HK, Jarret RL, Jayashree B, Crouch JH, Ortiz R (2005) Isolation and characterization of microsatellite markers from *Musa balbisiana*. Mol Ecol Notes 5: 327–330.

Carpentier SC, Witters E, Lauken K, Van Onckelen H, Swennen R, Panis B (2007) Banana (*Musa* spp.) as a model to study the meristem proteome: Acclimation to osmotic stress. Proteomics 7: 92–105.

Chen S, Harmon AC (2006) Advances in plant proteomics. Proteomics 6: 5504–55161.

Cheung F, Town CD (2007) A BAC end view of the *Musa acuminata* genome. BMC Plant Biol 7: 29.

Choudhury SR, Roy S, Saha PP, Singh SK., Sengupta DN (2008) Characterization of differential ripening pattern in association with ethylene biosynthesis in the fruits of five naturally occurring banana cultivars and detection of a GCC-box-specific DNA-binding protein. Plant Cell Rep 27: 1235–1249.

Clark DP, Pazdernik NJ (2009) Biotechnology: Applying the Genetic Revolution. Academic Press/Elsevier, Amsterdam, The Netherlands.

Creste S, Benatti TR, Orsi MR, Risterucci AM, Figueira A (2006) Isolation and characterization of microsatellite loci from a commercial cultivar of *Musa acuminata.* Mol Ecol Notes 6: 303–306.

Crouch JH, Vuylsteke D, Ortiz R (1998) Perspectives on the application of biotechnology to assist the genetic enhancement of plantain and banana (*Musa* spp.). Electronic J Biotechnol 1: 11–22.

De Ascensao AR, Dubery IA (2003) Soluble and wall-bound phenolics and phenolic polymers in *Musa acuminata* roots exposed to elicitors from *Fusarium oxysporum f.sp. cubense.* Phytochemistry 63: 679–686.

De Vienne D, Leonardi A, Damerval C, Zivy M (1999) Genetics of proteome variation for QTL characterization: application to drought-stress responses in maize. J Exp Bot 50: 303–309.

Dixon RA, Gang DR, Charlton AJ, Fiehn O, Kuiper HA, Reynolds TL, Tjeerdema RS, Jeffery EH, German JB, Ridley WP, Seiber JN (2006) Applications of metabolomics in agriculture. J Agric Food Chem 54: 8984–8994.

Fenn JB, Mann M, Meng CK, Wong SF, Whitehouse CM (1989) Electrospray ionization for mass spectrometry of large biomolecules. Science 246: 64–71.

Fiehn O (2002) Metabolomics—the link between genotypes and phenotypes. Plant Mol Biol 48: 155–171.

Fiehn O, Wohlgemuth G, Scholz M, Kind T, Lee DY, Lu Y, Basil SM (2008) Quality control for plant metabolomics: reporting MSI-compliant studies. Techniques for molecular analysis. Plant J 53: 691–704.

Finkel E (2009) With 'Phenomics,' plant scientists hope to shift breeding into overdrive. Science 325: 380.

Fukushima A, Kusano M, Redestig H, Arita M, Saito K (2009) Integrated omics approaches in plant systems biology. Curr Opin Chem Biol 13: 532–538.

Hall RD, Brouwerand ID, Fitzgerald MA (2008) Plant metabolomics and its potential application for human nutrition. Physiol Plant 132: 162–175.
Hanzlik RP, Fang J, Koen YM (2009) Filling and mining the reactive metabolite target protein database. Chem-Biol Interact 179: 38–44.
Heslop-Harrison JS, Schwarzacher T (2007) Domestication, Genomics and the Future for Banana. Ann Bot 100: 1073–1084.
Hollywood K, Brison DR, Goodacre R (2006) Metabolomics: Current technologies and future trends. Proteomics 6: 4716–4723.
Iijima Y, Nakamura Y, Ogatal Y, Tanaka K, Sakurai N, Suda K, Suzuki T, Suzuki H, Okazaki K, Kitayama M, Kanaya S, Aoki K, Shibata D (2008) Metabolite annotations based on the integration of mass spectral information. Plant J 54: 949–962.
Jenkins H, Hardy N, Beckmann M, Draper J, Smith AR, Taylor J, Fiehn O, Goodacre R, Bino RJ, Hall R, Kopka J, Lane GA, Lange BM, Liu JR, Mendes P, Nikolau BJ, Oliver SG, Paton NW, Rhee S, Roessner-Tunali U, Saito K, Smedsgaard J, Sumner LW, Wang T, Walsh S, Wurtele ES, Kell DB (2004) A proposed framework for the description of plant metabolomics experiments and their results. Nat Biotechnol 22: 1601–1606.
Jorrín JV, Maldonado AM, Castillejo MA (2007) Plant proteome analysis: A 2006 update. Proteomics 7: 2947–2962.
Kamo T, Hirai N, Tsuda M, Fujioka D, Ohigashi H (2000) Changes in the content and biosynthesis of Phytoalexins in banana fruits. Biosci Biotechnol Biochem 64: 2089–2098.
Keurentjes JJB, Koornneef M, Vreugdenhil D (2008) Quantitative genetics in the age of omics. Curr Opin Plant Biol 11: 123–128.
Kim JK, Bamba T, Harada K, Fukusaki E, Kobayashi A (2007) Time course metabolic profiling in Arabidopsis thaliana cell cultures after salt stress. J Exp Bot 58: 415–424.
Lakshmanan V, Venkataramareddy SR, Neelwarne B (2008) The use of genetic markers for detecting DNA polymorphism, genotype identification and phylogenetic relationships among banana cultivars. Mol Phylogenet Evol 47: 974–985.
Li Y, Shrestha B, Vertes A (2008) Atmospheric pressure infrared MALDI imaging mass spectrometry for plant metabolomics. Anal Chem 80: 407–420.
Matthew T, Zhou W, Rupprecht J, Lim L, Thomas-Hall SR, Doebbe A, Kruse O, Hankamer B, Marx UC, Smith SM, Schenk PM (2009). The metabolome of *Chlamydomonas reinhardtii* following Induction of Anaerobic H2 Production by Sulfur Depletion. J Biol Chem 284: 23415–23425.
Moritz T, Johansson AI (2008) Plant Metabolomics. In: WJ Griffiths (ed) Metabolomics, Metabonomics and Metabolite Profiling. The Royal Society of Chemistry. Cambridge, UK, pp 254–272.
Morot-Gaudry JF, Lea P, Briat JF (2007) Functional Plant Genomics. Science Publishers. Enfield, NH, USA.
Nikolau BJ, Wurtele ES (2007) Concepts in Plant Metabolomics. Springer. The Netherlands.
O' Farrell PH (1975) High resolution two dimensional electrophoresis of protein. J Biol Chem 250: 4007–4021.
Ortiz R (1995) *Musa* genetics. In: S Gowen (ed) Banana and Plantains. Chapman and Hall, London, UK, pp 84–109.
Patton WF (2000) A thousand points of light: the application of fluorescence detection technologies to two-dimensional gel electrophoresis and proteomics. Electrophoresis 21: 1123–1144.
Pillay M, Tripathi L (2007) Banana breeding. In: MS Kang, PM Priyadarshan (eds) Breeding Major Food Staples. Blackwell Publishing, Boston, MA, USA, pp 393–428.
Porubleva L, Velden V, Kothari KS, Livier DJ, Chitnis PR (2001) The proteome of maize: use of gene sequence and expressed sequence tag data for identification of proteins with mass fingerprints. Electrophoresis 22: 1724–1738.
Primrose SB, Twyman RM (2006) Principles of Gene Manipulation and Genomics. 7th edn. Blackwell, Oxford, UK.

Roberts JKM (2002) Proteomics and a future generation of plant molecular biologists. Plant Mol Biol 48: 143–154.
Roux N, Baurens FC, Doleze J, Hribova E, Heslop-Harrison P, Town C, Sasaki T, Matsumoto T, Aert R, Remy S, Souza M, Lagoda P (2008) Genomics of Banana and Plantain (*Musa* spp.), major stable crops in the tropics. In: PH Moore, R Ming, DP Delmer (eds) Plant Genetics and Genomics of Tropical Crop Plants, vol 1. Springer, New York, USA, pp 83–111.
Samyn B, Sergeant K, Carpentier S, Debyser G, Panis B, Swennen R, Beeumen JV (2007) Functional proteome analysis of the banana plant (*Musa* spp.) using *de novo* sequence analysis of derivatized peptides. J Proteome Res 6: 70–80.
Schauer N, Fernie AR (2006) Plant metabolomics: towards biological function and mechanism. Trends Plant Sci 11: 508–516.
Subramaniam S (2005) Bioinformatics. In: M J Dunn, LB Jorde, PFR Little, S Subramaniam (eds) Encyclopedia of Genetics, Genomics, Proteomics and Bioinformatics Encyclopedia of Genetics, Genomics, Proteomics and Bioinformatics 7 Wiley, New York, USA, pp 3117–3126.
Sumner LW, Mendes P, Dixon RA (2003) Plant metabolomics: large-scale phytochemistry in the functional genomics era. Phytochemistry 62: 817–836.
Thiellement H, Zivy M, Damerval C, Mechin V (2007) Plant proteomics: Methods and Protocols. Methods in Molecular Biology, vol. 355. Human Press, Inc., Totowa, NJ, USA.
Tohge T, Fernie AR (2009) Web-based resources for mass-spectrometry-based metabolomics: A user's guide. Phytochemistry 70: 450–456.
Thongboonkerd V (2007) Practical points in urinary proteomics. J Proteome R 6: 3881–3890.
Venkatachalam L, Sreedhar RV, Bhagyalakshmi N (2008) The use of genetic markers for detecting DNA polymorphism, genotype identification and phylogenetic relationships among banana cultivars. Molecular Phylogenetics and Evolution 47: 974–985.
Vilarinhos A (2004) Genetic and cytogenetic mapping in bananas: characterization of translocations. InfoMusa 13: 1.
Vuylsteke D, Ortiz R, Ferris RSB, Crouch JH (1997) Plantain improvement. Plant Breed Rev 14: 267–320.
Weckwerth W (2008) Integration of metabolomics and proteomics in molecular plant physiology—coping with the complexity by data-dimensionality reduction. Physiol Plant 132: 176–189.
Wilkins MR, Sanchez JC, Gooley AA, Appel RD, Humphery-Smith I, Hochstrasser DF, Williams KL (1995) Progress with proteome projects: why all proteins expressed by a genome should be identified and how to do it. Biotechnol Genet Eng Rev 13: 19–50.
Zeng Y, Li J, Wang C, Chang M, Yang R, Wu RL (2008) Genetic mapping of quantitative trait loci. In: C Kole, AG Abbott (eds) Principles and Practices of Plant Genomics. I. Genome Mapping, Science Publishers, Northants, UK, pp 175–204.

11

Role of Bioinformatics as a Tool

Mathieu Rouard,[1,a,*] *Sebastien Christian Carpentier,*[2] *Stephanie Bocs,*[3,a] *Gaëtan Droc,*[3,b] *Xavier Argout,*[3,c] *Nicolas Roux*[1,b] and *Manuel Ruiz*[3,d]

ABSTRACT

Bioinformatics plays an essential role in today's plant science mainly due to the exponential growth of genomic sequences generated by high-throughput sequencing technologies. The success of genomics has also fostered the emergence of complementary "omics" research areas and has led to the diversification of data. In this context, various approaches, software and databases have been developed to transform biological data into meaningful information, and some of them are used on a daily basis by scientists. Compared to model plant species, *Musa* is still at its early stage but useful tools have been established and are ready for much larger datasets that are expected in the near future. In this chapter, we introduce the resources and tools available to support *Musa* research, and bioinformatics topics such as gene annotation, transcriptomics, proteomics, and data integration are addressed.

Keywords: Bioinformatics, databases, genome annotation, proteomics, molecular markers

[1]Bioversity International, Commodities for Livelihood Program, 34397 Montpellier, France.
[a]e-mail: *m.rouard@cgiar.org*
[b]e-mail: *n.roux@cgiar.org*
[2]Department of Biosystems, Division of Crop Biotechnics, Katholic University of Leuven, Leuven, Belgium; e-mail: *sebastien.carpentier@biw.kuleuven.be*
[3]CIRAD, Biological Systems Department, UMR AGAP-TA40/03, 34398, Montpellier, France.
[a]e-mail: *stephanie.sidibe-bocs@cirad.fr*
[b]e-mail: *gaetan.droc@cirad.fr*
[c]e-mail: *xavier.argout@cirad.fr*
[d]e-mail: *manuel.ruiz@cirad.fr*
*Corresponding author

List of abbreviations after the text.

11.1 Introduction

This chapter on bioinformatics of *Musa* covers the latest information on activities developed by the Global *Musa* Genomics Consortium (GMGC). The following sections describe how genetic, genomic, transcriptomic and proteomic data are exploited in various information systems. These developments are intended to facilitate the understanding of the genetic basis of important agronomic traits which, among other outcomes, will help breeding efforts.

11.2 Gene and Genome Databases

Unlike genomics of model plant species, such as *Arabidopsis thaliana* or *Oryza sativa*, *Musa* genomics is still in its early stage, although large-scale projects (e.g., whole genome sequencing) will soon be generating significant amounts of data for this species. Until now, most of the information available in the database maintained by the GMGC (http://www.musagenomics.org/) has been produced by research projects focusing on biotic and abiotic stresses or comparative genomics. These sequences were obtained by screening bacterial artificial chromosome (BAC) libraries for genes of interests. The BAC libraries used are of the diploid wild species at the origin of most cultivars—*Musa acuminata*, which donated the A genome, and *Musa balbisiana*, which donated the B genome—as well as of the triploid cultivar "Grande Naine", which is of *acuminata* origin (Vilarinhos et al. 2003; Safar et al. 2004; Ortiz-Vazquez et al. 2005).

11.2.1 Automatic Annotation

Many analyses have been performed on portions of the *Musa* genome to predict genes and transposable elements (TEs), and to infer gene function. The prediction programs or methods used were combined in pipelines to facilitate sequence analysis. A non-exhaustive list of the software used in annotation pipelines is provided in Table 11-1. Gene structure prediction was performed using the EuGène software (Foissac et al. 2008) using rice-specific parameters This software combines *ab initio* gene finders and extrinsic similarity search programs. Despite the lack of experimental evidence from *Musa*, genome annotation has improved due to the continuously increasing availability of expressed sequence tags (ESTs). Computational analyses have also helped to identify and annotate transposable elements (TEs) that are known to have contributed to the structure and evolution of the genome.

Functional information on these gene products has been predicted by searching protein databases like UniProtKB (Schneider et al. 2005) and Interpro (Hunter et al. 2009). The UniProtKB/Swiss-Prot knowledge base

Table 11-1 List of software used in sequence annotation pipelines. Genes were predicted using several gene finders and weighted by cDNA alignments, if any. A consensus was sought using the Eugène combiner. Repetitive elements were also detected by different tools such as the pipeline Repet. Sequences were searched against the TrEMBL and SwissProt protein databases.

Program	URL
Eugène	http://www.inra.fr/mia/T/EuGene/
Fgenesh	http://linux1.softberry.com/
GeneMark	http://exon.gatech.edu/GeneMark/
SpliceMachine	http://bioinformatics.psb.ugent.be/webtools/splicemachine/
Sim4	http://globin.cse.psu.edu/dist/sim4/
RepeatMasker	http://www.repeatmasker.org/
Repseek	http://wwwabi.snv.jussieu.fr/public/RepSeek/
Repet	http://www.inra.fr (not available on line)
BlastX	http://blast.ncbi.nlm.nih.gov/
Exonerate	http://www.ebi.ac.uk/~guy/exonerate/
InterproScan	http://www.ebi.ac.uk/Tools/InterProScan/

provides reviewed entries and therefore ensures a high quality of annotation while its complementary section, UniProtKB/TrEMBL provides insights into sequences not yet curated.

Based on the analysis of 64 BACs (~7 Mbp), 1,508 genes have been predicted. Basic statistics computed on this dataset have refined the previous insights on *Musa* genome organization (Aert et al. 2004; Cheung and Town 2007; Lescot et al. 2008). Mean G+C content is about 40% while gene density is estimated at one per 4.7 kbp. Twenty seven percent of the genomic sequences analyzed were composed of transposable elements. Using the Repet pipeline (Abad et al. 2008), 3,140 TEs were identified. They were mainly from class I (LTR—long terminal repeat, LARD—Large retrotransposon derivative, LINE - long interspersed nuclear element, SINE —short interspersed nuclear element) (Wicker et al. 2007). Of these TEs, 39% were not well characterized. These results highlight the importance of manual annotation to fully harness the content of the database.

11.2.2 Data Storage and Viewers

The previously described automated annotations were stored in a PostgreSQL database under the Chado schema provided by the Generic Model Organism System Database Project (GMOD). We then set up the Web-based genome viewer, GBrowse (Stein et al. 2002), to display the genomic features along a reference sequence generated by the automatic analysis (Fig. 11-1). The genome browser makes it possible for users to navigate the genomic regions and to display different tracks of information, including automated prediction and manual curation of genes and repetitive elements. Users can search for a specific gene product and then download

Figure 11-1 Screenshot of the GNPAnnot genome browser showing a *Musa* BAC sequence. Genome features are dispatched in different tracks according the kind of analyses performed (e.g., Eugene, Repet, BlastX). By clicking one of the features, a pop-up window provides links to a detailed report. Links to the Apollo and Artemis software are available on the track dedicated to manual curation.

Color image of this figure appears in the color plate section at the end of the book.

the sequence. Expressed sequences (*Musa* ESTs) and protein sequences (UniprotKB/Swiss-prot and UniprotKB/TrEMBL) with significant similarities have also been compiled. The GMGC website runs the latest version of GBrowse, which includes new user-friendly features. For instance, drag and drop features can be used to move tracks and organize them in a convenient way. GBrowse offers many options to customize the view and to facilitate browsing genomes. The gene database is accessible via the GMGC website.

11.2.3 Manual Curation

The accuracy of a predicted gene structure and protein function improves with manual checking. The most reliable evidence to validate the structure of a protein-coding gene is when a locus sequence aligns with its cognate full-length cDNA (Aubourg et al. 2006) whereas for its function, it is homologous relationships. For instance, GreenPhyl (Conte et al. 2008a, b) proposes orthologous gene predictions based on phylogenomic methods that allow transferring annotation from well studied genes to uncharacterized genes. Having such procedures in place is important for *Musa* genome analysis because genome sequencing is underway and several projects have already generated 454/Roche and Solexa/Illumina data. Thus, the gene database has been also connected to the mainstream tools (Fig. 11-2) for sequence annotation, Artemis (Carver et al. 2008) and Apollo (Lewis et al. 2002). Both of them are available because usually once a community of users is acquainted to one, it is difficult to convince them to adopt another. Until now, annotators loaded flat files stored on their own computers and rarely shared and synchronized data. From the genome browser, annotators can

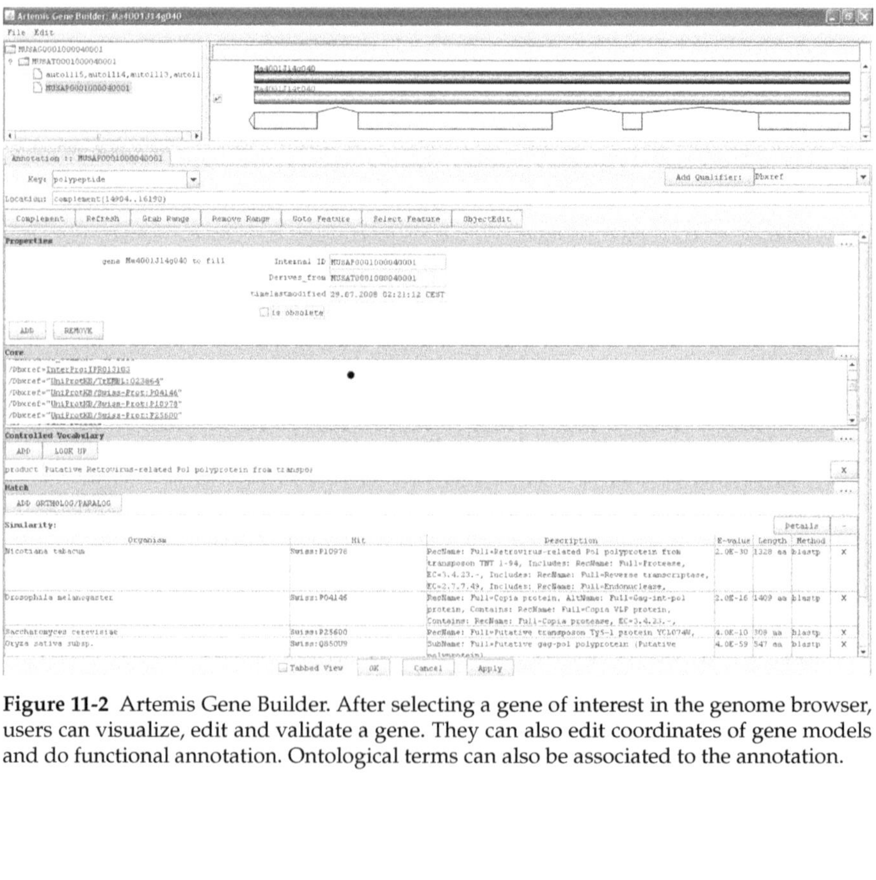

Figure 11-2 Artemis Gene Builder. After selecting a gene of interest in the genome browser, users can visualize, edit and validate a gene. They can also edit coordinates of gene models and do functional annotation. Ontological terms can also be associated to the annotation.

search genome features of interest and launch annotation editors by clicking on a link in a popup. These sequence annotation tools can be used to modify, for instance, boundaries genes and transposable elements. Upstream and downstream untranslated regions (e.g., promoters, regulatory elements) may be annotated as well. Annotations are directly read and saved on a remote server. Such infrastructure should greatly enhance collaborative initiatives in *Musa* genomics.

11.3 Gene Expression Databases

A database of ESTs is the starting point for transcriptomic studies because a large EST collection, once clustered, represents a potentially significant proportion of the gene-coding components of the genome. In addition, they have a strong value to identify gene regions and to refine gene predictions. Consequently, as done for many other organisms, developing these resources has been one of the strategies of the GMGC to characterize the *Musa* genome (Roux et al. 2008). Public ESTs are available in the TIGR Plant Transcript Assemblies (Childs et al. 2007) and in ESTtik, for expressed sequence tag treatment and investigation kit (Argout et al. 2008) accessible from the GMGC website. The latter is particularly used within the consortium to provide different access levels to ESTs, depending on whether they are in the public domain, accessible to GMGC members only or to members who have signed a Material Transfer Agreement. GMGC members, however, are encouraged to release their data in the public domain as much as possible.

11.3.1 Expressed Sequence Tags (EST) Database

ESTtik is an information system composed of a pipeline for data processing and a website to explore the EST collection (http://esttik.cirad.fr). Considering the high rate of sequence errors and redundancy in ESTs, sequences are processed through a pipeline that addresses these issues. The different steps including base calling, sequence cleaning, assembling and clustering are illustrated in Fig. 2-1. The obtained contigs were annotated using BLAST (Altschul et al. 1990), and Gene Ontology terms (Ashburner et al. 2000) were assigned automatically to the transcripts. Many groups have contributed to the still modest number of sequences that come from a variety of accessions and from both the A and B genomes. Currently, the *Musa* EST database contains approximately 100,000 sequences. ESTs are significantly shorter than the length of the average plant gene (Rudd 2003), on average 609 bp. The datasets were divided among several projects to facilitate the identification of genes with different expression levels (e.g., leaf without and under stress). ESTs have been mined for SSR markers and more than 2,000 SSRs have been uncovered.

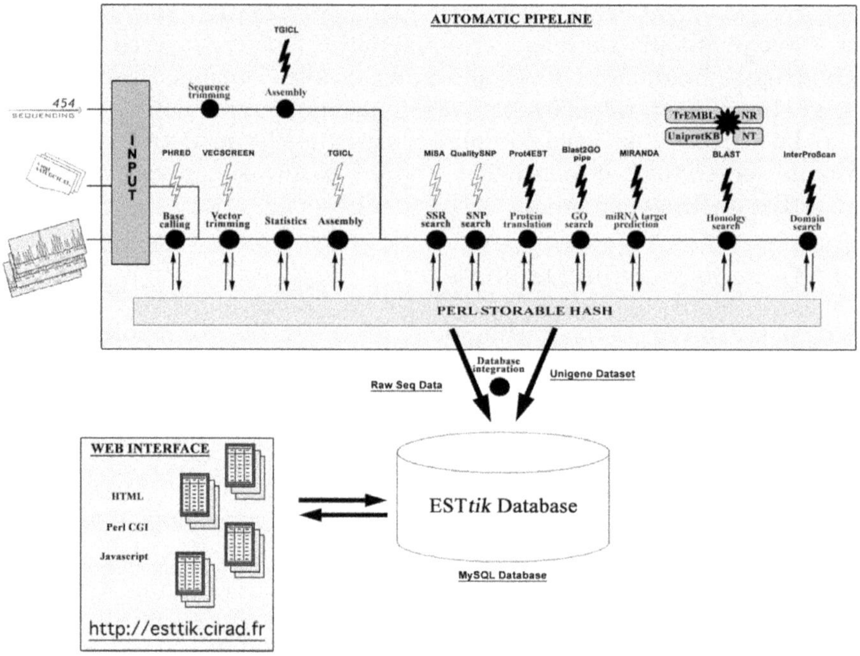

Figure 11-3 The ESTtik pipeline executes a series of programs to assess quality and nucleotides from chromatograms, then edits, and assembles the input cDNA sequence into a non-redundant data set. This set of Unigene is then mined for SSR and SNP and Gene Ontology terms and Blast annotation are assigned to the clusters.

The ESTtik database accessible via the GMGC website provides methods to search for individual sequences on the basis of their respective cDNA library and their preliminary functional annotation as the database stores the 10 best Blast hits from the non-redundant peptide sequence public database (NR) maintained by NCBI. Alignments of translated EST against protein sequence (BlastX) were performed because protein comparison is known to be better at identifying more distantly related protein coding genes. Highly conserved motifs are usually shared across species as a result of common ancestry. Conservation of structure probably implies conservation of function.

For instance, by typing the keyword "ethylene" in ESTtik, a hormone regulating fruit ripening, more than 30 sequences annotated as putative ethylene response factor or transcription factor will be obtained. Results can be filtered on sequences having SSR. Clustered data or raw data can be then downloaded for complementary experiments that could lead to the validation of the gene function. However, since large public databases contain huge amounts of sequences that are all not well characterized, it is recommended to carefully look at similar EST sequences. A BLAST server was included in the website to allow users to search the database on the basis of similarity to their own sequences.

A tentative assembly of all the sequences (Town et al. unpubl.) resulted in 12,385 contigs and 14,533 singletons for a total of 26,918 unigenes. It is expected that the whole collection will soon be in the public domain and as such should become a valuable resource for discovery of new genes.

11.4 Molecular Marker and Genetic Map Databases

More than 1,200 *Musa* cultivars, representing much of the diversity of the crop, are maintained at Bioversity's International Transit Centre (ITC) in Belgium. Most of the accessions conserved in the genebank are in the public domain and freely available for research and breeding purposes. Genetic studies using molecular markers are therefore very helpful to characterize conserved materials and check its integrity. Moreover, assessing the diversity of the collection is a prerequisite contributing to its rationalization.

11.4.1 Genetic Diversity

Since the 1980s, wild and cultivated accessions of banana have been studied using molecular markers to evaluate genetic diversity. Various kind of markers have been produced including RAPD, Isozymes, RFLP, AFLP (e.g., Horry and Jay 1988; Jarret et al. 1992; Lebot et al. 1993; Pillay et al. 2000; Wong et al. 2001; Carreel et al. 2002; Ude et al. 2002; Nwakanma et al. 2003), and more recently SSR and DArT (Hippolyte et al. 2010). All these studies clearly underscored the genetic differentiation between different sections of *Musa* species (Roux et al. 2008).

TropGENE-DB (Ruiz et al. 2004) has been conceived to manage genetic information about tropical crops and includes a module on banana (http://tropgenedb.cirad.fr/en/banana.html). Web user interfaces have been created to allow quick and complex queries and user-friendly results representation.

Currently, the TropGENE-DB banana module contains around 1,100 accession entries. They are linked to detailed phenotypic information: geographic origin, collection description, agro-morphological data, disease, pest and abiotic reactions, parentage, breeding data, etc. Associated molecular and genetic data comprise the genotypes at various markers (e.g., RFLP, AFLP, microsatellites, isozymes), genetic maps, information on the markers themselves linked to probes, primers, sequences data, etc. Pictures of gel profiles, photos of fruits or plant disease reactions are also available.

11.4.2 Genetic Mapping

The first attempts at producing genetic maps of *Musa* (Fauré et al. 1993; Noyer et al. 1997) have been stored in TropGENE-DB. The comparative map viewer CMap, developed as part of the GMOD project (http://gmod.org/wiki/Cmap), has been incorporated into TropGENE-DB to support comparative analyses of these linkage map data (Fig. 11-4). CMap allows users to choose any map as a reference map in a given comparison. Comparative maps can

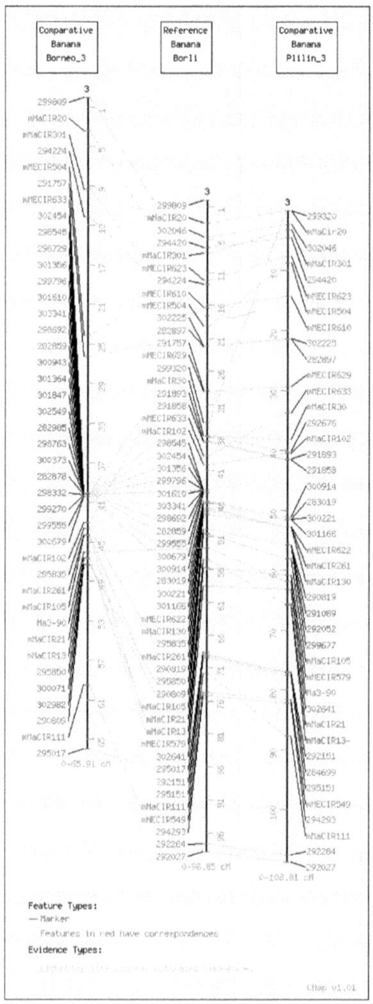

Figure 11-4 Graphical representation in CMap of correspondences between two banana linkage groups.

Color image of this figure appears in the color plate section at the end of the book.

then be added to the left or right of the reference map from a list of all maps in the database having a minimal set of correspondences to the reference map. Lines indicating relationships between features are drawn between the corresponding features on maps in a comparison. Indeed CMap can highlight non-homoeologous and homoeologous duplications, syntenic duplications, conflicts due to inversions, and evidence for translocations.

Table 11-2 Number of markers by type of markers in the banana module of TropGENE-DB. *(Note: The database contains mostly generated at CIRAD but a Memorandum of understanding (MoU) was signed with Bioversity and any partners willing to share data can be inserted in the database.)*

Type of markers	Number of markers
AFLP	358
Isozyme	8
SSR	126
RAPD	8
RFLP	151

11.5 Proteomics Tools and Protein Databases

Gene expression profiling and understanding gene function can be approached via several techniques. RNA-based system biology approaches have largely been applied to the classical model organisms. These so-called transcriptomics approaches are extremely powerful and highly automated, allowing massive screening of hundreds of genes simultaneously. However, the success of those approaches depends greatly on the genomic progress. Successful techniques like cDNA microarrays, cDNA amplified fragment length polymorphism (AFLP) and serial analysis of gene expression (SAGE) are in practice restricted to model organisms or species that are already characterized to a great extent. The power of those transcript based techniques is lost in non-model organisms like banana due to the lack of genomic information or due to the sequence divergence from a related model organism (Carpentier et al. 2008b). Gene sequences are rarely identical from one species to another and orthologous genes are usually riddled with nucleotide substitutions. An alternative for examining gene expression is studying its end products, the proteins. Protein sequences are more conserved making the high throughput identification of non-model gene products by comparison to well known orthologous proteins quite efficient. "Proteomics is the endeavour to understand gene function and to characterize the molecular processes of the living cell through the large-scale study of proteins found in a specific biological/physiological context" (Liska and Shevchenko 2003a). As its name suggests, proteomics is intricately linked to allied technologies such as genomics, transcriptomics and metabolomics. The proteome is the complete set of proteins present in a cell, tissue or organism at a specific time point under specific conditions.

In order to start proteomics on banana, we had to focus on: the development of wet lab techniques (Carpentier et al. 2005) and the implementation of different software tools for statistical analysis (Carpentier et al. 2008a) and protein identification. This identification may be achieved using two different approaches:

- high throughput **database dependent** searching (cross species and EST-based) (Carpentier et al. 2007)
- **database independent** *de novo* sequencing combined with error tolerant searching (Samyn et al. 2007).

Within the scope of this chapter, we will focus on the implementation of tools for protein identification.

At the end of the 80s, a shift in methods of protein identification occurred. Till then, Edman sequencing was the principal method of protein identification (Edman and Begg 1967). In order to determine the sequence of a peptide, the N-terminal amino acid is labeled and cleaved from the peptide in a cyclic manner till a full peptide sequence is obtained. Although Edman *de novo* sequencing was and still is used with considerable success, the method remains relatively slow and insensitive. The introduction of the soft ionization techniques matrix assisted laser desorption ionization (MALDI) and electrospray ionization (ESI) mass spectrometry (MS) subsequently provided faster and more sensitive tools. For the first time, a technology was available that allowed to ionize big molecules (such as peptides) that can subsequently be analyzed by mass spectrometry (MS).

11.6 Database Dependent Searching

The banana genome is still relatively poorly characterized. To illustrate, a query against the UniprotKB (Swissprot and TrEMBL) database (March 2009) using the sequence retrieval system with the restriction: organism = *Musa** returns only 1,302 entries. This exercise was repeated by way of comparison with the restriction: organism = Arabidopsis* and returned 88,312 entries. Cross species identification is the only option for protein identification whenever a genome is poorly characterized (Wilkins and Williams 1997; Lester and Hubbard 2002; Mathesius et al. 2002; Liska and Shevchenko 2003a; Liska and Shevchenko 2003b; Witters et al. 2003; Carpentier et al. 2008c). In this approach, proteins are identified by comparing the proteins of interest to orthologous proteins of species that are well characterized. Cross-species identification is successful when the mass information of multiple peptides is used to compare orthologous proteins of closely related species.

Peptide mass fingerprinting (PMF) allows for protein identification (Wilkins and Williams 1997). The concept behind peptide mass fingerprint analysis is quite simple and was independently implemented by several groups at approximately the same time (Henzel et al. 1993; James et al. 1993; Mann et al. 1993; Pappin et al. 1993; Yates et al. 1993. Via PMF, individual proteins are digested into smaller peptides (e.g., by trypsin) and the masses of these proteolytic peptides are subsequently measured. In the database search, each protein sequence is *in silico* digested according to the specificity of the enzyme or substrate. The masses of the resulting *in silico* peptides of each individual database entry are compared with the measured mass spectrum. Only a subset of all peptides from a protein digest needs to match to get a significant hit. The peptides that have amino acid substitutions or mass shifts due to post-translational modifications (PTM) that do not fall within the chosen tolerance window or correspond to the selected PTM's are not recognized and do not contribute to the identification. Several software tools are available for PMF (Fenyo 2000). The simplest and most obvious scoring methods for peptide mapping is to count the number of measured peptide masses that correspond to the theoretical peptide masses of each protein in the database. However, those methods have a bias toward bigger proteins. More sophisticated methods take into account this same basic principle but they also attempt to make better use of the mass spectrometric information compensating for example effects of protein size. The MOWSE (molecular weight search) score (Pappin et al. 1993) takes into account the relative abundance of the peptides in the database when calculating the score. As such, non-redundant peptides will contribute to a higher degree to the score.

However, when organisms are more phylogenetically distant, orthologous genes retain a lower percentage of identity. Hence, the chance of finding significant and conserved peptides decreases and PMF fails or results in false positive hits (Mathesius et al. 2002). Therefore additional information is essential and masses of derived fragment ions need to be measured in MS/MS mode. Several sophisticated algorithms, such as Mascot (Perkins et al. 1999) and SEQUEST (Eng et al. 1994), have been developed for identifying proteins from peptide MS/MS data. Peptides are identified by correlating the uninterpreted MS/MS spectra with theoretical (predicted) product ion spectra derived from peptides of the same mass contained in the available databases. More specific MS/MS analysis provides confident cross-species identifications for proteins with a lower sequence identity compared to available sequences. Only a few orthologous peptides might be sufficient for a reliable identification.

Mascot supports PMF and MS/MS ions search (Perkins et al. 1999). The acquisition of peptide mass fingerprints and MS/MS spectra on the same instrument and the combined interpretation enhances the identification rates significantly. The main advantage of the Mascot probability-based score is that one can estimate the chance of a false positive hit. The probability (P) is calculated in such a way that the observed match between an experimental data set and each sequence database entry is a chance event. This probability is then reported as a probability based MOWSE score: $-10\log_{10}(P)$. However, the issue of multiple testing when reporting large data sets may not be underestimated and the validity of the probabilities calculated by Mascot should be tested by repeating the searches against a randomized decoy database. We use PMF and sequence information in the same Mascot search in an automated way to correlate the un-interpreted spectra with the theoretical product ion spectra from different databases. Many different protein sequence databases are available and can offer additional information that yield better matches. The querying of multiple databases results in a good validation of the identification and can in some cases make the difference between a false negative and a significant positive hit. However, the reduction of false positive assignments remains a major issue in automated protein identification. Although the banana genome is still poorly characterized, several EST database have become available. Mascot translates nucleic acid databases in all six reading frames and predicts the theoretical fragment ions for all the possibilities without looking for a start codon and restarts after a stop codon. The EST data base is extremely useful to confirm the positive cross species results and can emerge in a positive identification in cases where the search against orthologous proteins is unsuccessful.

11.7 Database Independent Searching

When orthologous genes retain a low percentage of identity, the chance of finding significant and conserved peptides decreases and a PMF-MS/MS approach might fail. Within a fixed peptide mass and peptide fragment mass tolerance window, all algorithms require a non-error tolerant matching of the analyzed peptides to the database entries. Setting these tolerance windows too narrow, may result in false negative results and a too loose setting may result in false positive hits. A method that can extract protein-identifying information directly from spectra without being dependent on existing databases would be extremely valuable. In our case, *de novo* sequences of each individual banana protein are obtained that can afterwards be used in homology based search algorithms for identification. The *de novo* derived (partial) sequence information from each spot is combined in one search query and analyzed using search algorithms as: MS-BLAST (Shevchenko and Shevchenko 2001), FASTS (Mackey et al. 2002) and MS-Homology

(Clauser et al. 1999). FASTS searches databases using peptide sequences of unknown order, thereby evaluating all possible peptide arrangements. Because the true order of the query peptides used by FASTS is not known, FASTS only requires that the peptides do not overlap. The algorithm is based on the heuristic FASTA comparison strategy and uses alignment probability, rather than a similarity score, as the criterion for alignment (Mackey et al. 2002). MS-BLAST utilizes redundant, degenerate, and partially inaccurate peptide sequence data. MS-BLAST does not allow gaps within individual peptides, while gaps between peptides are not penalized and can be of arbitrary length. Therefore, all peptide sequences obtained by the interpretation of acquired MS/MS are assembled into a single search string in an arbitrary order (Shevchenko and Shevchenko 2001). MS-Homology (Protein Prospector 4.0.5) is a program that allows the comparison of a number of *de novo* derived peptide sequences against a selected database. The peptide sequences are entered followed by an indication of the maximum number of amino acid substitutions allowed for each sequence. It is also possible to enter a part of the sequence as a mass, along with a tolerance factor (Clauser et al. 1999). A database search will look for proteins containing peptides identical or homologous to the listed sequences. The results from all search algorithms depends on the number of peptides sequenced and the accuracy of the sequence information entered, as well as on database completeness and species to species sequence variability for the peptides entered. When trying to identify proteins by sequence similarity searches, the number of peptides recognized from a digested protein determines the success of the identification. It has been calculated that as more peptides are analyzed and matched, proteins of less similarity to database sequences can be identified with the limit being around 50% identity (Mackey et al. 2002).

The optimized workflow for a non-model organism comprises (i) the combination of different complementary protein fractionation, separation and quantification techniques to maximize the resolution and to cover the proteome as good as possible and (ii) the usage of different complementary MS techniques and error tolerant database searches. Cross-species identification by combining database dependent and independent methods are expected to improve the exploitation of the Bioversity's International Musa Germplasm Collection, which is currently stored at the Laboratory of Tropical Crop Improvement (K.U. Leuven, Belgium), under the auspices of the Food and Agriculture Organization (FAO). To facilitate future proteome analysis of banana tissue and to make the data accessible, an on-line database (http://www.pdata.ua.ac.be/*Musa*/) was constructed using two-dimensional gel electrophoresis (2-DE).

1. *Integration of Different Data*

A key problem for scientists is the integration of diverse and dispersed biological data and their analysis using diverse analytical tools (Stein 2003). The Generation Challenge Program (GCP) bioinformatics platform seeks to alleviate this problem by providing an informatics platform, which allows data integration via an agreed domain model and a workbench of interoperable applications (Bruskiewich et al. 2006; Bruskiewich et al. 2008; Wanchana et al. 2008). In this context, a visualization and analysis workbench for genetic diversity, GenDiversity, has been developed. We will restrict the content of this section to the description of its features.

2. GenDiversity: A Workbench for Analysis of Genetic Diversity

Very large datasets involving genotypes of thousands of accessions are now available. The present goal of geneticists is indepth analyses of these data to answer a broad range of questions about domestication, genetic organization of species, and comparison between different crop models. Furthermore, the integration of multi-level data like genotypes, passports and phenotypes will help to identify novel and diverse variants of genes involved in complex stress tolerance traits.

GenDiversity, is a query and analysis web application for

(a) querying decentralized genotyping data from different databases
(b) combining two or more different genotyping studies
(c) converting outputs to different file formats for diversity analysis: DARwin (Perrier 2006), Convert (Glaubitz 2004), Structure (Falush et al. 2007), Haploview (Barrett et al. 2005)
(d) filtering by passport data
(e) computation of allelic frequencies
(f) computation of linkage disequilibrium
(g) checking for duplicates. SSR and SNP genotyping studies are currently available.

These data comes from the TropGENE-DB database (Ruiz et al. 2004), and the GCP central repository, which stores all the data produced by the GCP projects, http://gcpcr.grinfo.net/. GenDiversity is accessible by internet (http://gendiversity.cirad.fr/Home).

Filtering Based on Name of Germplasm/Markers

For each genotyping study, users have access to the list of names of germplasm and markers analyzed. Users can select in the list the group of (Germplasm x Marker) they want to extract. They can also enter a list of germplasm and/or a list of markers separated by a semicolon (Fig. 4-1). Users can define the ploidy of the species studied. The ploidy is set to diploid by default. If GenDiversity detects inconsistencies between the real genotyping data and the ploidy, warnings are raised. Users can define a threshold for the rate of missing data allowed. All germplasm having a rate of missing data higher than the rate indicated by users will be pre-checked in the results matrix. The prechecked data can be easily removed. By default, GenDiversity will pre-check all the germplasm that have at least one missing data.

Filtering Based on Passport Data

Users can filter genotyping data by names of germplasm, names of markers and/or by passport information: GermplasmID, Collections, Cultivars, Pedigree, Collecting Locations, Descriptors an so on (Fig. 11-5).

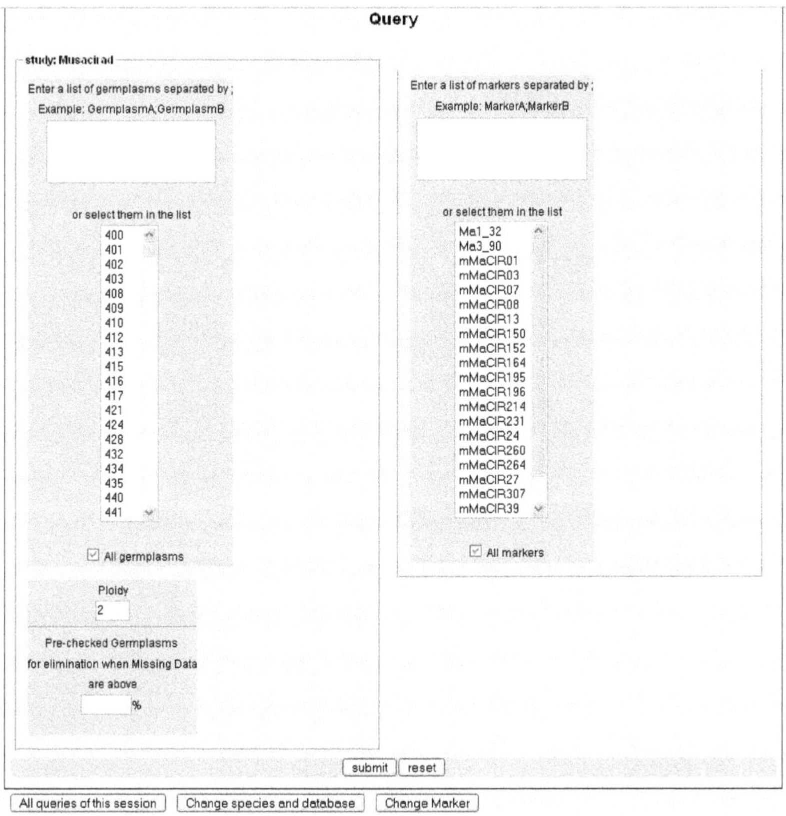

Figure 11-5 List of germplasm and markers names for a *Musa* genotyping study.

After clicking on "Filter on passport and phenotyping data", GenDiversity indicates which passport studies are available. Only one study can be selected. The filtering based on passport data is available with only one study. After selecting the passport study, a list of passport attributes is displayed. Users can select attributes to be used for filtering data. Only attributes having values can be selected with checkboxes. For each selected attribute, the list of different values is displayed, with the number of hits between parentheses. Then users can filter by selecting the values (Fig.11-6).

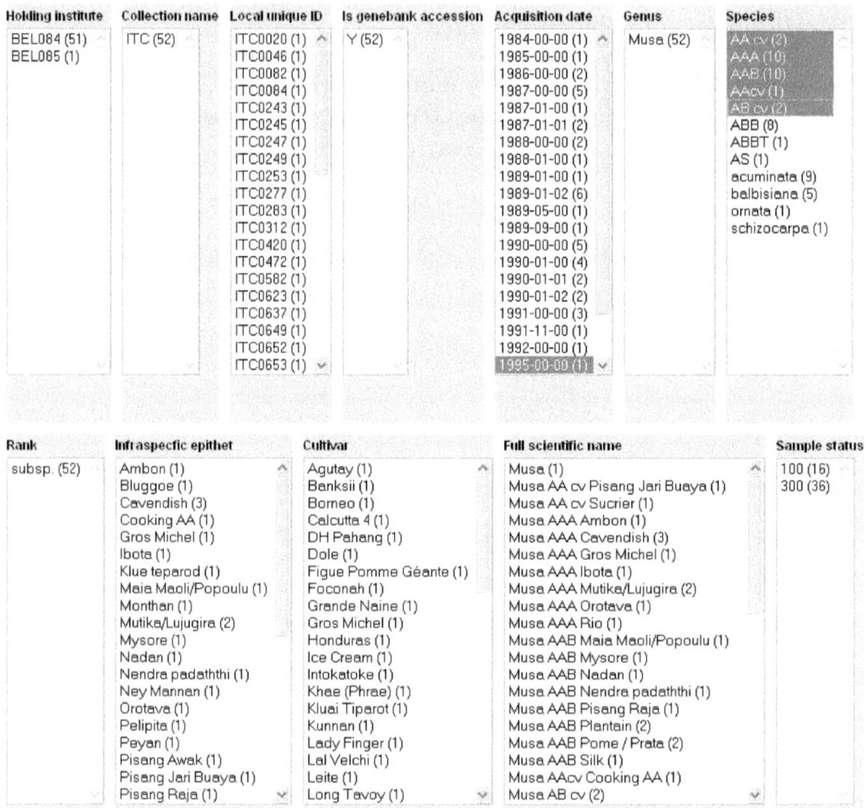

Figure 11-6 Users can select passports attributes to be used for filtering data. For each attribute, the list of different values is displayed, with the number of hits between parentheses.

Results Display

In the results display, users have link to the matrix, to their statistics, to allelic frequencies, to linkage disequilibrium values, and to indication of monomorphic locus (Fig. 11-7). Missing data are highlighted. Several features are available: rate of missing data by markers, rate of missing data by germplasm, checkbox selected by default to delete the germplasm with rate of missing data, checkbox to select or deselect all markers, checkbox to select or deselect all germplasm, and rate of overall missing data.

Users can delete selected germplasm and markers with a click on the button "delete checked germplasm and markers". The results will be a new matrix of results removed from the selected markers and germplasm. If users are satisfied with the new matrix, they can export it in several formats: DarWin, Structure, Convert, and Haploview (for SNP).

Germplasm	mMaCIR01		mMaCIR03		mMaCIR07		mMaCIR08		mMaCIR13		mMaCIR150	
400	-99	-99	122	124	-99	-99	261	267	286	286	257	261
401	254	264	122	124	-99	-99	261	267	286	296	257	257
402	258	266	122	127	158	160	261	261	286	286	257	257
403	250	254	120	124	170	170	261	264	286	298	257	259
408	254	258	122	124	158	170	261	267	286	286	257	261
409	254	258	122	124	158	170	261	267	286	286	257	261
410	258	266	122	127	158	160	261	261	286	286	257	257
412	-99	-99	122	122	-99	-99	261	261	286	286	257	257
413	256	298	122	127	158	170	261	261	270	270	257	264
415	258	298	120	122	158	172	261	261	286	286	257	257
416	292	304	121	127	160	166	270	270	292	292	261	261
Missing Data	18.2%	18.2%	0.0%	0.0%	27.3%	27.3%	0.0%	0.0%	0.0%	0.0%	0.0%	0.0%
Monomorphic	No		No		No		No		No		No	
☐	☐		☐		☐		☐		☐		☐	
	mMaCIR01		mMaCIR03		mMaCIR07		mMaCIR08		mMaCIR13		mMaCIR150	

Alleles frequencies are here

Linkage disequilibrium are here

Figure 11-7 Matrix of a genotyping study with names of germplasm per names of markers, and allele values into the cells. Missing data are in red.

Fusion of Different Results Matrix

If there are several matrices from different genotyping studies on the same page, users can merge them. Four options are available:

1. The complete fusion that merges results without reflecting their common points. It will generate missing data where there is no

information at the intersections (for example for a germplasm only present in a study and a marker only present in another study).

2. The fusion using only common germplasm. There will be a loss of information for any germplasm not present in both studies.
3. The fusion using only common markers.
4. The fusion using common germplasm and common markers. This is important for checking data consistency between different studies.

The result of these fusions may highlight inconsistencies between different studies, for a same couple of germplasm/marker.

Abbreviations

AFLP	:	Amplified Fragment Length Polymorphism
BAC	:	Bacterial Artificial Chromosomes
BLAST	:	Basic Local Alignment Search Tool
EST	:	Expressed Sequence Tags
GCP	:	Generation Challenge Program
GMGC	:	Global *Musa* Genomics Consortium
GMOD	:	Generic Model Organism System Database
LARD	:	LArge Retrotransposon Derivative
LINE	:	long interspersed nuclear element
LTR	:	Long Terminal Repeat
MOWSE	:	MOlecular Weight SEarch
MS	:	Mass Spectrometry
PMF	:	Peptide Mass Fingerprinting
PTM	:	Post-Translational Modifications
RAPD	:	Random Amplification of Polymorphic DNA
RFLP	:	Restriction Fragment Length Polymorphism
SAGE	:	Serial Analysis of Gene Expression
SINE	:	Short Interspersed Nuclear Element
SNP	:	Single Nucleotide Position
SSR	:	Simple Sequence Repeat
TE	:	Transposable Elements

Acknowledgements

The platform for gene annotation is part of the GNPAnnot project (http://www.gnpannot.org) supported by the French National Research Agency (ANR) and involving the following organizations: CIRAD, INRA and Bioversity International.

GenDiversity is supported by the Generation Challenge Program (GCP). We also thank Anne Vezina for her editorial suggestions on the manuscript.

References

Abad P, Gouzy J, Aury J-M, Castagnone-Sereno P, Danchin EGJ, Deleury E, Perfus-Barbeoch L, Anthouard V, Artiguenave F, Blok VC, Caillaud M-C, Coutinho PM, Dasilva C, De Luca F, Deau F, Esquibet M, Flutre T, Goldstone JV, Hamamouch N, Hewezi T, Jaillon O, Jubin C, Leonetti P, Magliano M, Maier TR, Markov GV, McVeigh P, Pesole G, Poulain J, Robinson-Rechavi M, Sallet E, Segurens B, Steinbach D, Tytgat T, Ugarte E, van Ghelder C, Veronico P, Baum TJ, Blaxter M, Bleve-Zacheo T, Davis EL, Ewbank JJ, Favery B, Grenier E, Henrissat B, Jones JT, Laudet V, Maule AG, Quesneville H, Rosso M-N, Schiex T, Smant G, Weissenbach J, Wincker P (2008) Genome sequence of the metazoan plant-parasitic nematode Meloidogyne incognita. Nat Biotech 26: 909.

Aert R, Sági L, Volckaert G (2004) Gene content and density in banana (*Musa acuminata*) as revealed by genomic sequencing of BAC clones. Theoretical and Applied Genetics 109: 129–139.

Altschul SF, Gish W, Miller W, Myers EW, Lipman DJ (1990) Basic local alignment search tool. Journal of Molecular Biology 215: 403–410.

Argout X, Fouet O, Wincker P, Gramacho K, Legavre T, Sabau X, Risterucci A, Da Silva C, Cascardo J, Allegre M, Kuhn D, Verica J, Courtois B, Loor G, Babin R, Sounigo O, Ducamp M, Guiltinan M, Ruiz M, Alemanno L, Machado R, Phillips W, Schnell R, Gilmour M, Rosenquist E, Butler D, Maximova S, Lanaud C (2008) Towards the understanding of the cocoa transcriptome: Production and analysis of an exhaustive dataset of ESTs of Theobroma cacao L. generated from various tissues and under various conditions. BMC Genomics 9: 512.

Ashburner M, Ball CA, Blake JA, Botstein D, Butler H, Cherry JM, Davis AP, Dolinski K, Dwight SS, Eppig JT, Harris MA, Hill DP, Issel-Tarver L, Kasarskis A, Lewis S, Matese JC, Richardson JE, Ringwald M, Rubin GM, Sherlock G (2000) Gene ontology: tool for the unification of biology. The Gene Ontology Consortium. Nature Genetics 25: 25–29.

Aubourg S, Delseny M, Lecharny A (2006) L'organisation des génomes végétaux révélée par leur annotation. Biofuture 265: 33–37.

Barrett JC, Fry B, Maller J, Daly MJ (2005) Haploview: analysis and visualization of LD and haplotype maps. Bioinformatics 21: 263–265.

Bruskiewich R, Davenport G, Hazekamp T, Metz T, Ruiz M, Simon R, Takeya M, Lee J, Senger M, McLaren G, Van Hintum T (2006) Generation Challenge Programme (GCP): standards for crop data. Omics: A Journal of Integrative Biology 10: 215–219.

Bruskiewich R, Senger M, Davenport G, Ruiz M, Rouard M, Hazekamp T, Takeya M, Doi K, Satoh K, Costa M, Simon R, Balaji J, Akintunde A, Mauleon R, Wanchana S, Shah T, Anacleto M, Portugal A, Ulat VJ, Thongjuea S, Braak K, Ritter S, Dereeper A, Skofic M, Rojas E, Martins N, Pappas G, Alamban R, Almodiel R, Barboza LH, Detras J, Manansala K, Mendoza MJ, Morales J, Peralta B, Valerio R, Zhang Y, Gregorio S, Hermocilla J, Echavez M, Yap JM, Farmer A, Schiltz G, Lee J, Casstevens T, Jaiswal P, Meintjes A, Wilkinson M, Good B, Wagner J, Morris J, Marshall D, Collins A, Kikuchi S, Metz T, McLaren G, van Hintum T (2008) The Generation Challenge Programme Platform: Semantic Standards and Workbench for Crop Science. International Journal of Plant Genomics.

Carreel F, Gonzalez de León D, Lagoda P, Lanaud C, Jenny C, Horry JP, Tezenas du Montcel H (2002) Ascertaining maternal and paternal lineage within *Musa* by chloroplast and mitochondrial DNA RFLP analysis. Genome 45: 679–692.

Carpentier SC, Witters E, Laukens K, Deckers P, Swennen R, Panis B (2005) Preparation of protein extracts from recalcitrant plant tissues: an evaluation of different methods for two-dimensional gel electrophoresis analysis. Proteomics 5: 2497–2507.

Carpentier SC, Witters E, Laukens K, Van Onckelen H, Swennen R, Panis B (2007) Banana (*Musa* spp.) as a model to study the meristem proteome: acclimation to osmotic stress. Proteomics 7: 92–105.

Carpentier SC, Coemans B, Podevin N, Laukens K, Witters E, Matsumura H, Terauchi R, Swennen R, Panis B (2008a) Functional genomics in a non-model crop: transcriptomics or proteomics? Physiologia Plantarum 133: 117–130.
Carpentier SC, Panis B, Swennen R, Lammertyn J (2008b) Finding the significant markers: statistical analysis of proteomic data. Methods in Molecular Biology (Clifton, NJ) 428: 327–347.
Carpentier SC, Panis B, Vertommen A, Swennen R, Sergeant K, Renaut J, Laukens K, Witters E, Samyn B, Devreese B (2008c) Proteome analysis of non-model plants: a challenging but powerful approach. Mass Spectrometry Reviews 27: 354–377.
Carver T, Berriman M, Tivey A, Patel C, Bohme U, Barrell BG, Parkhill J, Rajandream M-Al (2008) Artemis and ACT: viewing, annotating and comparing sequences stored in a relational database. Bioinformatics. Oxford, England 24: 2672–2676.
Cheung F, Town CD (2007) A BAC end view of the *Musa acuminata* genome. BMC Plant Biology 7: 29.
Childs KL, Hamilton JP, Zhu W, Ly E, Cheung F, Wu H, Rabinowicz PD, Town CD, Buell CR, Chan AP (2007) The TIGR Plant Transcript Assemblies database. Nucleic Acids Research 35: D846–851.
Clauser KR, Baker P, Burlingame AL (1999) Role of accurate mass measurement (+/– 10 ppm) in protein identification strategies employing MS or MS/MS and database searching. Analytical Chemistry 71: 2871–2882.
Conte MG, Gaillard S, Droc G, Perin C (2008a) Phylogenomics of plant genomes: a methodology for genome-wide searches for orthologs in plants. BMC Genomics 9: 183.
Conte MG, Gaillard S, Lanau N, Rouard M, Périn C (2008b) GreenPhylDB: a database for plant comparative genomics. Nucleic Acids Research 36: D991–998.
Edman P, Begg G (1967) A protein sequenator. European Journal of Biochemistry/FEBS 1: 80–91.
Eng JK, McCormack AL, Yates Iii JR (1994) An approach to correlate tandem mass spectral data of peptides with amino acid sequences in a protein database. Journal of the American Society for Mass Spectrometry 5: 976–989.
Falush D, Stephens M, Pritchard JK (2003) Inference of Population Structure Using Multilocus Genotype Data: Linked Loci and Correlated Allele Frequencies. Genetics 164: 1567–1587.
Falush D, Stephens M, Pritchard JK (2007) Inference of population structure using multilocus genotype data: dominant markers and null alleles. Molecular Ecology Notes 7: 574–578.
Fauré S, Noyer JL, Horry JP, Bakry F, Lanaud C, Gonzalez de León D (1993) A molecular marker-based linkage map of diploid bananas (*Musa acuminata*). Theoretical and Applied Genetics 87: 517–526.
Fenyo D (2000) Identifying the proteome: software tools. Current Opinion in Biotechnology 11: 391–395.
Foissac S, Gouzy J, Rombauts S, Mathe C, Amselem J, Sterck L, Van de Peer Y, RouzÃ© P, Schiex T (2008) Department of Plant B, Genetics Genome annotation in plants and fungi: EuGene as a model platform.
Glaubitz JC (2004) convert: A user-friendly program to reformat diploid genotypic data for commonly used population genetic software packages. Mol Ecol Notes 4: 309–310.
Henzel WJ, Billeci TM, Stults JT, Wong SC, Grimley C, Watanabe C (1993) Identifying proteins from two-dimensional gels by molecular mass searching of peptide fragments in protein sequence databases. Proc Nat Acad Sci USA, 90: 5011–5015.
Hippolyte I, Bakry F, Seguin M, Gardes L, Rivallan R, Risterucci A-M, Jenny C, Perrier X, Carreel Fo, Argout X, Piffanelli P, Khan IA, Miller RNG, Pappas GJ, MbÃ©guiÃ©-A-MbÃ©guiÃ© D, Matsumoto T, De Bernardinis V, Huttner E, Kilian A, Baurens F-C, D'Hont Al, Cote Fo, Courtois B, Glaszmann J-C (2010) A saturated SSR/DArT linkage map of Musa acuminata addressing genome rearrangements among bananas. BMC Plant Biology 10: 65.

Horry J, Jay M (1988) Distribution of anthocyanins in wild and cultivated banana varieties. Phytochemistry 27: 2667–2672.

Hunter S, Apweiler R, Attwood TK, Bairoch A, Bateman A, Binns D, Bork P, Das U, Daugherty L, Duquenne L, Finn RD, Gough J, Haft D, Hulo N, Kahn D, Kelly E, Laugraud Al, Letunic I, Lonsdale D, Lopez R, Madera M, Maslen J, McAnulla C, McDowall J, Mistry J, Mitchell A, Mulder N, Natale D, Orengo C, Quinn AF, Selengut JD, Sigrist CJA, Thimma M, Thomas PD, Valentin F, Wilson D, Wu CH, Yeats C (2009) InterPro: the integrative protein signature database. Nucleic Acids Research 37: D211–215.

James P, Quadroni M, Carafoli E, Gonnet G (1993) Protein identification by mass profile fingerprinting. Biochemical and Biophysical Research Communications 195: 58–64.

Jarret R, Gawel N, Whittemore A, Sharrock S (1992) RFLP-based phylogeny of *Musa* species in Papua New Guinea. Theor Appl Genet 84: 579–584.

Lebot V, Aradhya KM, Manshardt R, Meilleur B (1993) Genetic relationships among cultivated bananas and plantains from Asia and the Pacific. Euphytica 67: 163–175.

Lescot M, Piffanelli P, Ciampi A, Ruiz M, Blanc G, Leebens-Mack J, da Silva F, Santos C, D'Hont A, Garsmeur O, Vilarinhos A, Kanamori H, Matsumoto T, Ronning C, Cheung F, Haas B, Althoff R, Arbogast T, Hine E, Pappas G, Sasaki T, Souza M, Miller R, Glaszmann J-C, Town C (2008) Insights into the *Musa* genome: Syntenic relationships to rice and between *Musa* species. BMC Genomics 9: 58.

Lester PJ, Hubbard SJ (2002) Comparative bioinformatic analysis of complete proteomes and protein parameters for cross-species identification in proteomics. Proteomics 2: 1392–1405.

Lewis SE, Searle SMJ, Harris N, Gibson M, Lyer V, Richter J, Wiel C, Bayraktaroglir L, Birney E, Crosby MA, Kaminker JS, Matthews BB, Prochnik SE, Smithy CD, Tupy JL, Rubin GM, Misra S, Mungall CJ, Clamp ME (2002) Apollo: a sequence annotation editor. Genome Biology 3: RESEARCH0082.

Liska AJ, Shevchenko A (2003a) Combining mass spectrometry with database interrogation strategies in proteomics. Trends in Analytical Chemistry 22: 291–298.

Liska AJ, Shevchenko A (2003b) Expanding the organismal scope of proteomics: Cross-species protein identification by mass spectrometry and its implications. PROTEOMICS 3: 19–28.

Mackey AJ, Haystead TAJ, Pearson WR (2002) Getting more from less: algorithms for rapid protein identification with multiple short peptide sequences. Molecular & Cell Proteomics 1: 139–147.

Mann M, Hojrup P, Roepstorff P (1993) Use of mass spectrometric molecular weight information to identify proteins in sequence databases. Biological Mass Spectrometry 22: 338–345.

Mathesius U, Imin N, Chen H, Djordjevic MA, Weinman JJ, Natera SHA, Morris AC, Kerim T, Paul S, Menzel C, Weiller GF, Rolfe BG (2002) Evaluation of proteome reference maps for cross-species identification of proteins by peptide mass fingerprinting. Proteomics 2: 1288–1303.

Noyer JL, Dambier D, Lanaud C, Lagoda P (1997) The saturated map of diploid banana *Musa acuminata*. Plant and Animal Genomes V Conference, January 12–16, San Diego, CA, USA.

Nwakanma DC, Pillay M, Okoli BE, Tenkouano A (2003) Sectional relationships in the genus *Musa* L. inferred from the PCR-RFLP of organelle DNA sequences. Theor Appl Genet 107: 850–856.

Ortiz-Vázquez E, Kaemmer D, Zhang HB, Muth J, Rodríguez-Mendiola M, Arias-Castro C, James A (2005) Construction and characterization of a plant transformation-competent BIBAC library of the black Sigatoka-resistant banana *Musa acuminata* cv. Tuu Gia (AA). Theor Appl Genet 110: 706–713.

Pappin DJ, Hojrup P, Bleasby AJ (1993) Rapid identification of proteins by peptide-mass fingerprinting. Current Biology 3: 327–332.

Perkins DN, Pappin DJ, Creasy DM, Cottrell JS (1999) Probability-based protein identification by searching sequence databases using mass spectrometry data. Electrophoresis 20: 3551–3567.

Perrier X, Jacquemoud-Collet JP (2006) DARwin software.

Pillay M, Nwakanma DC, Tenkouano A (2000) Identification of RAPD markers linked to A and B genome sequences in *Musa* L. Genome 43: 763–767.

Roux N, Baurens F-C, Doležel J, Hribová E, Heslop-Harrison P, Town C, Sasaki T, Matsumoto T, Aert R, Remy S, Souza M, Lagoda P (2008) Genomics of Banana and Plantain (*Musa* spp.), Major Staple Crops in the Tropics. in Moore PH, Ming R, Genomics of Tropical Crop Plants. New York, NY: Springer New York, p 83–111.

Rudd S (2003) Expressed sequence tags: alternative or complement to whole genome sequences? Trends in Plant Science 8: 321–329.

Ruiz M, Rouard M, Raboin LM, Lartaud M, Lagoda P, Courtois B (2004) TropGENE-DB, a multi-tropical crop information system. Nucleic Acids Research 32: D364–367.

Šafář J, Noa-Carrazana JC, Vrána J, Bartoš J, Alkhimova O, Sabau X, Šimková H, Lheureux F, Caruana M-L, Dolezel J, Piffanelli P (2004) Creation of a BAC resource to study the structure and evolution of the banana (*Musa balbisiana*) genome. Genome 47: 1182–1191.

Sàgi L, Volckaert G, Aert R (2004) Gene content and density in banana (*Musa acuminata*) as revealed by genomic sequencing of BAC clones. Theor Appl Genet 109: 129–139.

Samyn B, Sergeant K, Carpentier S, Debyser G, Panis B, Swennen R, Van Beeumen J (2007) Functional proteome analysis of the banana plant (*Musa* spp.) using de novo sequence analysis of derivatized peptides. J Proteome Research 6: 70–80.

Schneider M, Bairoch A, Wu CH, Apweiler R (2005) Plant Protein Annotation in the UniProt Knowledgebase. Plant Physiology 138: 59–66.

Shevchenko A, Shevchenko A (2001) Evaluation of the efficiency of in-gel digestion of proteins by peptide isotopic labeling and MALDI mass spectrometry. Analytical Biochemistry 296: 279–283.

Stein LD (2003) Integrating biological databases. Nat Rev Genet 4: 337–345.

Stein LD, Mungall C, Shu S, Caudy M, Mangone M, Day A, Nickerson E, Stajich JE, Harris TW, Arva A, Lewis S (2002) The generic genome browser: a building block for a model organism system database. Genome Research 12: 1599–1610.

Ude G, Pillay M, Nwakanma D and Tenkouano A (2002) Analysis of genetic diversity and sectional relationships in *Musa* using AFLP markers. Theoretical and Applied Genetics 104: 1239–1245.

Vilarinhos AD, Piffanelli P, Lagoda P, Thibivilliers S, Sabau X, Carreel F, D'Hont A (2003) Construction and characterization of a bacterial artificial chromosome library of banana (*Musa acuminata* Colla). Theor Appl Genets 106: 1102–1106.

Wanchana S, Thongjuea S, Ulat VJ, Anacleto M, Mauleon R, Conte M, Rouard M, Ruiz M, Krishnamurthy N, Sjolander K, van Hintum T, Bruskiewich RM (2008) The Generation Challenge Programme Comparative Plant Stress-responsive Gene Catalogue. Nucleic Acids Research 36: D943–946.

Wicker T, Sabot F, Hua-Van A, Bennetzen JL, Capy P, Chalhoub B, Flavell A, Leroy P, Morgante M, Panaud O, Paux E, SanMiguel P, Schulman AH (2007) A unified classification system for eukaryotic transposable elements. Nat Rev Genet 8: 973–982.

Wilkins MR, Williams KL (1997) Cross-species protein identification using amino acid composition, peptide mass fingerprinting, isoelectric point and molecular mass: a theoretical evaluation. J Theor Biol 186: 7–15.

Witters E, Laukens K, Deckers P, Van Dongen W, Esmans E, Van Onckelen H (2003) Fast liquid chromatography coupled to electrospray tandem mass spectrometry peptide sequencing for cross-species protein identification. Rapid Communications in Mass Spectrometry: RCM 17: 2188–2194.

Wong C, Kiew R, Loh JP, Gan LH, Set O, Lee SK, Lum S, Gan YY (2001) Genetic Diversity of the Wild Banana *Musa acuminata* Colla in Malaysia as Evidenced by AFLP. Ann Bot 88: 1017–1025.

Yates JR, 3rd, Speicher S, Griffin PR, Hunkapiller T (1993) Peptide mass maps: a highly informative approach to protein identification. Analytical Biochemistry 214: 397–408.

12

Genomics of *Mycosphaerella fijiensis*—The Causal Agent of Black Sigatoka in *Musa* spp.

Cláudia Fortes Ferreira,[1,a,*] *Hermínio Souza Rocha,*[1,b] *Ricardo Franco Cunha Moreira*[2] and *Lahyre Izaete Silveira Gomes*[3]

ABSTRACT

Black Sigatoka, also known as black leaf streak disease (BLSD), is caused by *Mycospherella fijiensis.* This airborne fungal leaf spot disease is considered one of the most devastating diseases in bananas. It is considered extremely aggressive. Losses can range from 20 to 90% and it is globally replacing the existing *Mycospherella musicola,* pathogen that causes yellow Sigatoka. The genome sequencing of this pathogen, initiated by a solid partnership, has opened new windows as far as worldwide disease strategy control of black Sigatoka is concerned. Although the first draft of the genome sequence of *Mycosphaerella fijiensis* was released only four years ago, much work has been initiated and much more is expected in the coming years. In this chapter, we highlight the main topics of genomics and other tools being used to hasten and improve *M. fijiensis* studies; especially considering the advancements in biotechnology and the progress in WGS (Whole Genome Sequencing) in the last years.

Keywords: Black Sigatoka, genetics, genomics, disease control strategies

[1]Molecular Biology Department, Rua Embrapa, C.P. 007, 44380-000, Cruz das Almas, Bahia, Brazil.
[a]e-mail: *claudiaf@cnpmf.embrapa.br.*
[b]e-mail: *herminio@cnpmf.embrapa.br*
[2]Federal University of Bahia Recôncavo, Center for Agricultural Sciences, Environmental and Biological, Rui Barbosa, 710-University Campus, CEP 44380-000, Cruz das Almas, Bahia, Brazil; e-mail: *ricardofcm@ufrb.edu.br*
[3]Federal University of Viçosa–UFV, Department of Phytopathology, Rua PH Rolfs 36570-000 Viçosa–MG, Brazil; e-mail: *lahyreizaete@yahoo.com.br*
*Corresponding author

List of abbreviations after the text.

12.1 Introduction

The banana Sigatoka complex involves three fungi found in bananas: Mycosphaerella fijiensis, *M. musicola* and *M. eumusae* (Jones 2003). Black Sigatoka (Fig. 12-1), also known as black leaf streak disease (BLSD), is caused by *M. fijiensis* (anamorph *Pseudocercospora fijiensis*) Morelet; a haploid, sexual, hemibiotrophic filamentous Ascomycete, with heterothallic mating system, and is considered one of the most devastating diseases in bananas (Stover and Simmonds 1987; Mourichon and Fullerton 1990; Mobambo et al. 1993; Cordeiro et al. 2004; Conde-Ferraez et al. 2007). This airborne fungal leaf spot disease decreases leaf area photosynthesis, reducing fruit size and inducing early maturation in susceptible banana cultivars. *Mycospherella fijiensis* is a relatively new pathogen and still actively colonizes new areas where bananas are grown. It is extremely aggressive leading to losses ranging from 20 to 90% (Stover and Simmonds 1987; Mobambo et al. 1993) and is globally replacing the existing *M. musicola* pathogen that causes yellow Sigatoka disease. The main control method of BLSD still relies on excessive application of fungicides (up to 70 times per year in some countries in Central America), contributing nearly 27% of the total production cost (Mobambo et al. 1993), leading to critical health and environmental issues and therefore this disease is considered one of the most costly banana diseases worldwide (Carlier et al. 2000).

Figure 12-1 Black leaf streak symptoms in the variety BRS Maravilha (picture by Edson Perito Amorim).

Color image of this figure appears in the color plate section at the end of the book.

In this chapter, we highlight on some of genomics and other tools used to better understand the pathogen, *M. fijiensis*. These include the advanced biotechnological approaches and the whole genome sequencing (WGS) strategy to study the pathogen.

12.2 Development of Molecular Markers and other Biotechnological Tools for Studies on *M. fijiensis*

In the last two decades, DNA marker technologies have revolutionized the strategies for plant pathogen genomic analyses and have been extensively employed in many fields of molecular plant pathology. Molecular markers offer the possibility of faster and accurate identification and early detection of plant pathogens and have been widely used in pathogenic fungal studies (Michelmore and Hulbert 1987; Meng et al. 1999; Armengol et al. 2010; Bhat and Browne 2010; Daval et al. 2010; Dubey et al. 2010; Gladieux et al. 2010; Njambere et al. 2010). The use of molecular techniques for detection, identification and study of genetic diversity and other studies of *M. fijiensis* are presented below.

The fungi that cause the banana Sigatoka disease complex are morphologically very similar, making morphological evaluation alone very laborious, ambiguous and sometimes unreliable; hindering preventive management strategies for adequate control of *M. fijiensis* (Arzanlou et al. 2008).

Molecular studies concerning the detection and identification of *M. fijiensis* using molecular markers began in the early 90s. Johanson and Jeger (1993) used polymerase chain reaction (PCR) methods to rapidly and accurately distinguish between *M. musicola* and *M. fijiensis* using primers from a variable region identified in the nucleotide sequence of the internally transcribed spacer (ITS)1 region of ribosomal DNA (rDNA) from both the species. The PCR-based technique of random amplification of polymorphic DNA (RAPD) was used with the same purpose (Johanson et al. 1994).

Arzanlou et al. (2007) developed rapid and reliable species-specific molecular-based diagnostic tools for the detection and quantification of the three species of *Mycosphaerella* using PCR primers developed on the basis of the actin gene; being able to detect as little as 100 pg/µL of *M. fijiensis* DNA. In addition, a TaqMan real-time PCR assay was developed based on the beta-tubulin gene, which can detect as low as 1 pg/ul DNA of each *Mycosphaerella* sp. These assays certainly contribute to strengthen the detection of these pathogens in banana infested areas, and can be promptly and easily used for quarantine strategies of control. The intricacies of the Sigatoka disease complex in bananas requires new assays such as these, so that more laboratories worldwide can have access to quick diagnosis, especially by using conventional PCR methods.

Interesting new results regarding mating type genes have also been presented (Conde-Ferràez 2007; Arzanlou et al. 2010; Conde-Ferràez et al. 2010). Conde-Ferràez et al. (2007) isolated and characterized the mating type loci (*mat1-1* and *mat1-2*) from *M. fijiensis*. This type of characterization is required to understand the reproduction and recombination process, and the epidemiology and the interaction of this pathogen with other species. Furthermore, the results confirmed high identity in inversion regions within both idiomorphs suggesting unique inversion events that may have played a key role in the evolution of *M. fijiensis* (Conde-Ferràez et al. 2007).

Recently, Conde-Ferràez et al. (2010) developed a multiplex-PCR method to identify the distribution and frequency of the mating types of *M. fijiensis* in Mexico and different size fragments for *mat1-1* and *mat1-2* genes were reported. These same primers were used to amplify fragments of the idiomorphs of *M. musicola*, which were sequenced and showed to be highly similar to those from *M. fijiensis*, implying that the methodology could be applied to both species.

Further analysis of the mating type loci in *Mycosphaerella* spp. were investigated by PCR and chromosome walking approaches (Arzanlou et al. 2010). The sequences were compared to previously published mating type loci of *M. fijiensis* and other *Mycosphaerella* spp. Two new *Mycosphaerella*-specific genes (*MATORF1* and *MATORF2*) were reported. The proteins encoded by these genes showed higher inter- than intra-species homology (Arzanlou et al. 2010). These findings are highly significant to allow a better understanding of the dynamics and mechanisms underlying the evolutionary path and mating processes of these pathogens and can be very useful in population structure analyses, co-evolution and pathogen differentiation studies.

Knowledge of the pathogenic cycle and virulence factors (Pariaud et al. 2009) are crucial for developing effective crop protection strategies, including the development of resistant plant genotypes through classical plant breeding, genetic engineering or the use of biological control strategies (González-Fernández et al. 2010).

It is well known that environmental variables such as air temperature, atmospheric pressure, light and humidity, influence the aerobiology of fungi and disease epidemiology. Understanding disease progression, dispersal patterns and analysis of population structure of pathogens, are the key for reaching adequate disease control strategies (reviewed in churchill 2010), especially when considering countries located in the tropics. Furthermore, given the heterothallic nature of *M. fijiensis*, recombination events increases the chances of the development of more adaptive isolates to changing environmental conditions and even as to fungicide applications, making population structure and genetic diversity studies crucial.

On the other hand, with the advances of molecular biology techniques many complex questions could be answered such as the sources of inoculum, changes in their population structures and the population dynamics of the disease. Methodologies for accessing *M. fijiensis* virulence have also been studied (Donzelli and Churchhill 2007). Important applications of molecular biology include analyses of population genetic diversity, genetic variation characterization, taxonomy and phylogeny of plant pathogen taxa, gene mapping and tagging, genome fingerprinting and genome evolution.

Many genetic studies of *M. fijiensis* have been carried out in various parts of the world using different molecular markers. Restriction fragment length polymorphism (RFLP) was used in global population studies of *M. fijiensis* isolates from South-East Asia, Africa, Latin America and the Pacific Islands (Carlier et al. 1996). This study showed that random sexual reproduction occurred in these populations and that South-East Asia is the center of origin of this fungus and from there spread to other parts of the world relatively recently.

Zapater et al. (2004) developed PCR-RFLP markers for studying the population structure in *M. fijiensis*. These markers were used in isolates from different geographical origins and proved to be easily transferred to laboratories in banana-growing countries especially because of the use of 2.0% agarose gels which demand only basic laboratory equipment. It is important to emphasize the importance of the development of conventional PCR methodologies that can be easily transferred to laboratories in developing countries where the diseases are destroying small scale banana plantations.

Microsatellite molecular markers have been increasingly preferred in recent years because of their high level of polymorphism, and locus specificity. Many microsatellite markers have already been developed for *M. fijiensis* and have been used for population studies (Yang and Zhong 2008; Zapater et al. 2008).

The genetic variability of Nigerian isolates of *M. fijiensis* was investigated using di- and tri- nucleotide microsatellite markers (Muller et al. 1997). These markers showed genetic variation within a lesion, between lesions on the same plant, between plants and cultivars, and between geographical locations.

In 1999 Neu et al. characterized Nigerian and Mexican isolates with 11 microsatellite markers and confirmed the results found by Carlier et al. (1996), showing the low within-region diversity and strong geographical differentiation between isolates of two places.

Rivas et al. (2004) inferred that gene flow and dissemination patterns of *M. fijiensis* on a continental scale from analyzing population structures in regions recently invaded by the pathogen using PCR-RFLP and microsatellite markers. By analyzing isolates collected from banana plantations in 13 countries in Latin America, the Carribean and Africa, the authors reported that founder effects accompanied the recent invasion of *M. fijiensis* in the Latin-America-Carribean regions, suggesting stochastic spread of the disease at the continental level that could either be caused by limited dispersal of ascospores or by translocation of infected plant material. This information is very useful and these techniques should be used in epidemiological investigations and disease monitoring programs carried out worldwide.

Among the new markers developed for *M. fijiensis* are those based on retrotransposons IRAP (Inter-Retrotransposon Amplified Polymorphism) and REMAP (Retrotransposon-Microsatellite Amplified Polymorphism), developed for the first time for the characterization of a population of *M. fijiensis* in Brazil. These markers besides being highly polymorphic, are also very cheap. The M. fijiensis population in Brazil has been characterized with different molecular markers and data revealed the presence of three structured populations despite recent introduction of this pathogen in the Country (Silva, GF, personal communication).

Recently using PCR-RFLP and microsatellite markers, Fahleson et al. (2009) characterized a collection of isolates from Uganda, and showed that the *M. fijiensis* population in the Lake Victoria Basin constitutes a homogenous population that probably resulted from a small population of a recent founder event.

The population structure of *M. fijiensis* in two recently colonized areas in Costa Rica and Cameroon was analyzed using 19 microsatellite markers. Low-to-moderate genetic differentiation among populations in both countries and isolation by distance in Cameroon was reported, suggesting continuous range expansion at the scale of banana-production area through gradual dispersal of spores (Halkett et al. 2010). This report throws a new light that contributes to enrich data as to colonization pathways and dispersal processes.

Single nucleotide polymorphisms (SNPs) was used in rDNA to determine the genetic structure of geographically distinct populations of *Mycosphaerella* spp. collected from Nigeria, identifying 14 SNP haplotypes of *M. fijiensis*. This work also suggested that *M. musicola* had been replaced predominantly by *M. fijiensis*, with a smaller representation of *M. eumusae* in the population (Zandjanakou-Tachin et al. 2009).

In addition to molecular markers, tools such as BAC (Bacterial artificial chromosome) libraries are considered as true genetic information reservoirs for physical mapping, map-based cloning and genome sequencing, whereas large insert BAC libraries are considered as essential tools in modern genome research (Canto-Canché et al. 2007). These libraries are the key for isolation and characterization of gene clusters associated with metabolic pathways, for synteny analysis and gene copy number. A BAC library of

M. fijiensis was constructed and consisted of 1,920 clones representing 4.9x of genome equivalents. The strategy made use of a non-spaeroplasting technique eliminating some limitations of the previously used spaeroplastic techniques, which can be laborious, time consuming and somewhat empirical. According to the authors, this is the first BAC library constructed using a non-spaeroplasting technique, providing a simple and cost-efficient method for genomic studies such as map-based cloning, among others (Canto-Canchè et al. 2007).

Three cDNA libraries from two *M. fijiensis* isolates (MF5-Hilo and MF11-Hilo) under different culture conditions were constructed. A total of 3,771 ESTs, assembled into a set of 1,945 unique sequences, was generated. Fifty putative genes involved in pathogenicity were identified (Cho et al. 2008) and ESTs were compared to gene databanks from sequenced genomes from 18 other fungi. Gene discovery studies are lacking for the *M. fijiensis* and this work will certainly pave the way for the elucidation of the genes involved in pathogenicity, fungicide resistance and life cycle of this pathogen.

Genetic linkage maps are important tools to be used in evolutionary studies and positional cloning of genes of interest, such as those responsible for plant resistance (Drader and Kleinhofs 2010; Kang and Rouf Mian 2010; Ubayasena et al. 2010). Genetic linkage maps have been constructed using molecular markers for many species of pathogenic fungi (Kema et al. 2002; Zhong et al. 2002; Kuhn et al. 2006, Xu et al. 2009).

The first genetic linkage map of *M. fijiensis* was reported in 2008. A cross between the CIRAD86 (Cameroon) and CIRAD139A (Colombia) isolates was analyzed using 298 AFLPs, 16 SSRs and the MAT locus. The map spans over 23 linkage groups covering 1,878 cM (Manzo-Sánchez et al. 2008). The estimated total haploid genome size was calculated using the genetic mapping data at 4,298.2 cM.

Other approaches and markers have been used to study many aspects of *M. fijiensis*, such as electrophoretic karyotype profiles (Rodriguez-Garcia et al. 2006), multi-gene nucleotide sequence data set and morphology (Arzanlou et al. 2008).

12.3 *Mycosphaerella fijiensis* Genome Sequencing

Recently, the advances in sequencing technologies have generated a large amount of information regarding pathogenic fungal genes that are important or crucial to plant infection and colonization. A better understanding of fungus-plant interactions is the key to the development of efficient control strategies. Since knowledge of the molecular mechanisms underlying fungal pathogenesis is still somewhat limited, comparative and functional genomic studies offer a new perspective for the understanding of host-pathogen interactions (Xu et al. 2006).

Established in 2003, the International *Mycosphaerella* Genomics Consortium (IMGC; http://imgc.inibap.org/) made an effort to take *M. graminicola* (Goodwin et al. 2004) as the model to generate more genetic and genomic research information for *M. fijiensis*. This consortium led to the initiation of an international project, which featured to sequence the genome of the *M. fijiensis* virulent strain, CIRAD86, and 40,000 ESTs (Kema 2009).

The first draft of the *M. fijiensis* genome was made public by the Department of Energy Joint Genome Institute (JGI) in August 2007 and the second in May 2010.

The 1.0 version of the whole genome shotgun reads generated a coverage of ~ 7.11× with 602,555 reads assembled into 395 genome scaffolds totaling 73.4 Mb. A total of 13,903 genes were structurally and functionally annotated after the release of the second version with the genome at 74.1 Mb (http://genome.jgi-psf.org/Mycfi2/Mycfi2.info.html). For a more thorough review regarding the entire concept for the *M. fijiensis* sequencing project, refer to Kema (2009). For more information regarding the strong participation of Brazilian researchers in the MyGene project, which covers comparative studies of the genome of *M. graminicola* and *M. fijiensis*, and also annotation of pathogenic genes in *M. fijensis*, please refer to Martins et al. (2008).

The *M. fijiensis* genome sequence information generated by the JGI pipeline has led to remarkable results and since its 2.0 version was released in May 2010, more information is expected to be available soon.

The first draft of the *M. fijiensis* genome was used to develop 14 highly polymorphic microsatellite primers with average of 6.43 alleles per locus (Yang and Zhong 2008).

Recently, VNTR primers from the database provided by the *M. fijiensis* genome were developed. Five VNTR markers showing multiple alleles were optimized in a reference set of isolates from different parts of the world (Indonesia, Tanzania, Burundi, the Philippines, Gabon, Colombia, Cameroon, and New Caledonia) and a population from a banana plantation in Costa Rica. The PIC values ranged from 0.6414 to 0.7544 and 0.0400 to 0.7373, for the reference set and the population set, respectively, with 80% of the PIC values above 0.60 (Garcia et al. 2010). These values indicate that the markers are highly informative and can be used for fungicide resistance management and to develop breeding strategies to control black Sigatoka in bananas. This is the first report of VNTR—minisatellites developed directly from the *M. fijiensis* genome sequence pipeline (v1.0; Garcia et al. 2010).

A genetic linkage map of individuals from the cross between the CIRAD86 and CIRAD139A isolates, consisting of 19 linkage groups (LOD >10) using 87 microsatellite markers, 3 VNTR markers, the mating type (Mat) locus and 235 DArT markers covering 1,417 cM, was constructed and is now in its final stages of was constructed and is now in its final stages of conclusion and should soon be published (Ferreira et al. 2009).

The study of plant-pathogen interactions and understanding the molecular mechanisms of fungicide resistance (Ma and Michailides 2005) are crucial for breeding strategies and other means of control. ESTs derived from a *M. fijiensis* x banana interaction were analyzed by suppressive subtractive hybridization (SSH) cDNA library, which identified transcripts induced at the late stages of infection in the host and pathogen (Portal et al. 2011). Portal et al. (2011) identified specific genes involved in plant-fungal interactions and quantitative-PCR experiments revealed that antifungal genes encoding pathogenesis related (PR) proteins and GDSL-like lipase are only transiently induced 30 days post-inoculation (dpi), indicating that the fungus may be actively repressing plant defense. These data also suggest that new information may be fed into genome banks increasing data on host/pathogen interactions.

12.4 Future Perspectives

Studies of fungal pathogens and their interactions with plants have been performed using several approaches, from classical genetics, cell biology, and biochemistry, to the modern, holistic, and high-throughput omics techniques accompanied by proper bioinformatic tools (González-Fernández et al. 2010).

In recent years, the study of fungal plant pathogens has been greatly promoted by the availability of their genomic sequences and resources for functional genomics analysis, especially concerning the complex mechanisms involving pathogenesis and host avoidance (Ma et al. 2005; Stergiopoulos et al. 2010; Portal et al. 2011). Also by studies including transcriptomics, proteomics, and metabolomics, which, in combination with targeted mutagenesis or transgenic studies, are unraveling molecular host-pathogen cross-talk.

As shown above, significant progress has been made in unveiling the *M. fijiensis* genome in recent years, especially due to the rapid advancements of biotechnological tools. A transformation system for *M. fijiensis* to study host-pathogen interactions was developed and the growth of transformants expressing green fluorescent protein (GFP) throughout multiple stages of plant infection was characterized (Balint-Kurti et al. 2001). This report is extremely important for understanding the mechanisms underlying host-pathogen modes of infection.

As it is well known, phytopathogenic fungi are exposed to reactive oxygen species (ROS) secreted either by host plants or fungicides. Beltran -Garcia et al. (2009) analyzed the oxidative stress response of *M. fijiensis* to hydrogen peroxide (H_2O_2) and paraquat. The oxidative stress responses to H_2O_2, paraquat and hydroquinone were characterized to catalase and superoxidase dismutase (SOD) activities. Results suggested that catalase is important for rapid decomposition of H_2O_2 and has a direct bearing on cell viability and also that the superoxide anion formed through the redox or paraquat and hydroquinone has greater effect than H_2O_2 on the cellular viability of *M. fijiensis* (Gàrcia-Beltràn et al. 2009). Studies of this nature are very few in the literature and are important for understanding the defense mechanisms involved during pathogen exposure to fungicides and by the secretion of their own toxins used to facilitate their penetration and successful colonization of plant host tissues (Gàrcia-Beltràn et al. 2009). The defense mechanisms against m. fijiensis in banana plantas was investigated. cavalcante et al. (2011) evaluated the role of the peroxidase enzyme and H_2O_2 and its association with a hypersensitivy-like reaction in the resistant genotype calcutta 4.

Recently, aqueous infusions from healthy bananas (cv. Grande Naine) leaves with antifungal activity against *M. fijiensis* were detected suggesting the production of phytoprotectants by the plant (Cruz-Cruz et al. 2010).

Undoubtebly, much needs to be discovered in the search for answers of plant-pathogen interactions, methylation processes involved in *Musa* tolerance to black Sigatoka (Gimenèz et al. 2006), secretome analysis (Chuc-Uc et al. 2011) and the development of basic biology tools to help elucidate other aspects of the three main species making up the banana black Sigatoka disease complex (Diaz-Trujillo et al. 2009).

With the recent expansion of phytopathogen genome sequence data banks, locus-to-phenotype or gene-to-phenotype reverse genetics tools, such as knock-out, RNAi, ATMT, REMI, and TILLING, have become increasingly attractive methods to elucidate the molecular basis of host-pathogen interactions (compatible or incompatible), phytopathogen development, and virulence and pathogenicity (Bhadauria et al. 2009).

These reverse genetics tools can efficiently decode genome information into biological information (Weld et al. 2006; Nakayashiki and Nguyen 2008). In the post-genomics era, gene targeting (knock-out) by homologous recombination has become the most influential reverse genetics tool to identify gene function.

However, it is clear that much progress has been made for the *M. fijiensis* pathogen and most of the advances are in part due to the combined effort of successful partnerships (Kema 2009). With the release of the second version of the sequenced genome of *M. fijiensis* more can be expected in the coming years as far as breeding and control strategies of this devastating disease are concerned. The genome sequence is the first step to this understanding and the new information being generated from this data has only scratched the surface of the intricacies of the Sigatoka disease complex in bananas.

The information provided by the genome sequencing pipeline will set the standards for widening studies regarding genes involved in pathogen colonization and upgrade control strategies, as there are many challenges for controlling black Sigatoka (Churchill 2010). The next step will certainly involve, as it has occurred in other species that had their genome sequenced, proteomic studies will definitely refine all the information being generated and provide more information to be used by breeders enhancing banana breeding programs worldwide.

Abbreviations

ATMT	:	*Agrobacterium tumefaciens* mediated transformation
BAC	:	bacterial Artificial Chromosome
BLSD	:	black leaf streak disease
DArT	:	diversity array technology
ESTs	:	expressed Sequence Tags
GFP	:	green fluorescent protein
PCR	:	polymerase Chain Reaction
RAPD	:	random amplification of polymorphic DNA
REMI	:	restriction enzyme-mediated integration
RFLP	:	Restriction fragment length polymorphism
RNAi	:	RNA interference
ROS	:	reactive oxygen species
SNPs	:	single nucleotide polymorphisms
SSH	:	suppressive Subtractive Hybridization
TILLING	:	targeting induced local lesions in Genomes
VNTR	:	variable number of tandem repeat
WGS	:	whole Genome Sequencing

References

Armengol J, Vicent A, León M, Berbegal M, Abad-Campos P, García-Jiménez J (2010) Analysis of population structure of *Rosellinia necatrix* on *Cyperus esculentus* by mycelial compatibility and inter-simple sequence repeats (ISSR). Plant Pathol 59: 179–185.

Arzanlou M, Abeln ECA, Kema GHJ, Waalwijk C, Carlier J, de Vries I, Guzman M, Crous PW (2007) Molecular diagnostics for the Sigatoka diease complex of banana. Phytopathol 97: 1112–1118.

Arzanlou M, Groenewald JZ, Fullerton RA, Abeln ECA, Carlier J, Zapater MF, Budenhagen IW, Viljoen A, Crous PW (2008) Multiple genealogies and phenotypic characters differentiate several novel species of Mycosphaerella and related anamorphs on banana. Persoonia 20: 19–37.

Arzanlou M, Crous PW, Zwiers LH (2010) Evolutionary dynamics of mating-type loci of *Mycosphaerella* spp. Occurring on banana. Eukar Cell 9(1): 164–172.

Balint-Kurti PJ, May GD, Churchill ACL (2001) Development of a transformation system for *Mycosphaerella* pathogens of banana: a tool for the study of host/pathogen interactions. FEMS Microbiol Lett 195: 9–15.

Bhadauria V, Banniza S, Wei Y, Peng YL (2009) Reverse Genetics for Functional Genomics of Phytopathogenic Fungi and Oomycetes. Comp Funct Genom Article ID 380719, 11 pages.

Bhat RG, Browne GT (2010) Specific detection of Phythophthors cactorum in diseased strawberry plants using nested polymerase chain reaction. Plant Pathol 59: 121–129.

Beltran-Garcia MJ, Manzo-Sanches G, Guzmán-Gonzalez S, Arias-Castro C. Rodrigues-Mendiola M, Avila-Miranda M, Ogura T (2009) Oxidative stress response of Mycosphaerella fijiensis, the causal agent of black leaf streak disease in banana plants, to hydrogen peroxide and paraquat. Can J Microb 55: 887–894.

Carlier J, Lebrun MH, Zapater MF, Dubois C, Mourichon X (1996) Genetic structure of the global population of banana black leaf streak fungus, *Mycosphaerella fijiensis*. Mol Ecol 5: 499–510.

Carlier JE, Fouré F, Gauhl DR, Jones R, Lepoivre P, Mourichon X, Pasberg-Gauhl C, Romero RA (2000) Black leaf streak. In: DR Jones (ed) Diseases of Banana, Abacá and Enset, CABI Publishing, Wallingford, UK, pp 37–79.

Canto-Canché B, Guillén-Maldonado DK, Peraza-Echeverría L, Conde-Ferráez L, James-Kay A (2007) Construction and characterization of a bacterial artificial chromosome library of the causal agent of black Sigatoka fungal leaf spot disease of banana and plantain, *Mycosphaerella fijiensis*. Mol Biotechnol 36: 64–70.

Cavalcante, MJB, Escoute J, Madeira JP, Romero RE, Nicole MR, Oliveira LC, Hamelin C, Lartaud M, Verdeil JL (2011) Reactive oxygen species and cellular interactions between Mycosphaerella fijiensis and banana. Tropical Plant Biology 4: 134–143.

Cho Y, Hou S, Zhong S (2008) Analysis of expressed sequence tags from the fungal banana pathogen *Mycosphaerella fijiensis*. The Open Mycol J 2: 61–73.

Chuc-Uc J, Brito-Argáez L, Canto-Canché B, Tzec-Simá M, Garcia-Rodriguez C, Peraza-Echeverría L, Peraza-Echeverría S, James-Kay A, Cruz-Cruz A, Pería-Rodriguez M, Isias-Flores I (2011) The *in vitro* secretome of *Mycosphaerella fijiensis* induces cell death in banana leaves. Plant Physiol Biochem DOI: 10.1016/j.plaphy. 2011.02.006.

Churchill ACL (2010) Mycosphaerella fijiensis, the black leaf streak pathogen of banana: progress towards understanding pathogen biology and detection, disease development, and the challenges of control. Mol Plant Pathol 1: 22.

Conde-Ferràez L, Waalwijk C, Canto-Canchè BB, Kema GHL, Crous PW, James AC, Abeln ECA (2007) Isolation and characterization of the mating type locus of *Mycosphaerella fijiensis*, the causal agent of black leaf streak disease of banana. Mol Plant Pathol 8: 111–120.

Conde-Ferráez L, Grijalva-Arango R, Canto-Canché BB, Manzo-Sanchéz G, Canul-Salazar MI, James AC (2010) The development of mating type-specific primers for *Mycosphaerella fijiensis*, the causal agent of black Sigatoka of banana, and analysis of the frequency of idiomorph types in Mexican populations. Aust Plant Pathol 39: 217–225.

Cordeiro ZJMC, Matos APM, Silva SO, Gasparotto L, Cavalcante MJB (2004) Impact and management of Black Sigatoka in Brasil. XVI Reunião Internacional ACORBAT 63–69.

Cruz-Cruz C, Ramirez-Tec G, Garcia-Sosa K, Escalante-Erosa F, Hill L, Osbourn AE, Pena-Rodriguez LM (2010) Phytoanticipins from banana (*Musa acuminata* cv. Grande Naine) plants with antifungal activity against *Mycosphaerella fijiensis*, the causal agent of black-Sigatoka. Eur J Plant Pathol 126: 459–463.

Daval S, Lebreton L, Gazengel K, Guillerm-Erckelboudt AY, Sarniguet A (2010) Genetic evidence for differentiation of Gaeumannomyces graminis var. tritici into two major groups. Plant Pathol 59: 165–178.

Diaz-Trujillo C, Kobayashi A, Zwiers LH, Souza Jr. MT (2009) Developing tools for *Mycosphaerella fijiensis* studies. Gewasbescherming Jaargang 40: 3.

Donzelli, BGG and Chuchill, ACL (2007). A quantitative assay using mycelial fragments to assess virulence of *Mycosphaerella fijiensis*. Phytopathology 97 (8): 916–929.

Drader T, Kleinhofs A (2010) A synteny map and disease resistance gene comparison between barley and the model monocot *Brachypodium distachyon*. Genome 53: 406–417.

Dubey SC, Tripathi A, Singh SR (2010) ITS-RFLP fingerprinting and molecular marker for detection of *Fusarium oxysporum* f. sp. *ciceris*. Folia Microbiol 55(6): 629–634.

Fahleson J, Nakyanzi M, Okori M, Seal L, Kenyon L, Dixelius C (2009) Genetic analysis of *Mycosphaerella fijiensis* in the Ugandan Lake Victoria region. Plant Pathol 58: 888–897.

Ferreira CF, Van der Lee TAJ, Zapater MF, Carlier J, Goodwin B, Souza Jr. MT, Kema GHJ (2009) A genetic linkage map of *Mycosphaerella fijiensis* using SSR and DArT markers. Fungal Genet Rep 565 Abstract. N° 513.

Garcia SAL,Van der Lee TAJ, Ferreira CF, Te Hekkert Lintel B, Zapater MF, Goodwin SB, Guzmán M, Kema GHJ, Souza Jr. MT (2010) Variable number of tandem repeat markers in the genome sequence of *Mycosphaerella fijiensis*, the causal agent of black leaf streak disease of banana (*Musa* spp). Genet Mol Res 9(4): 2207–2212.

Giménez C, Palacios G, Colmenares M (2006) *Musa* methylated DNA sequences associated with tolrance to Mycosphaerella fijiensis toxins. Plant Mol Biol Repor 24: 33–43.

Gladieux P, Zhang XG, Róldan-Ruiz I, Caffier V, Leroy T, Devaux M, Van Glabeke S, Coart E, Le Cam B (2010) Evolution of the population structure of *Venturia inaequalis*, the apple scab fungus, associated with the domestication of its host. Mol Ecol 19: 658–674.

Goodwin SB, Waalwijk C, Kema GHJ (2004) Genetics and genomics of *Mycosphaerella graminicola*: a model for the Dothideales. In: DK Arora, GG Khachatourians (eds) Applied Mycology and Biotechnology. Elsevier, New York, USA, pp 315–330.

González-Fernández R, Prats E, Jorrín-Novo JJ (2010) Proteomics of Plant Pathogenic Fungi. J Biomed Biotechnol Article ID 932527.

Halkett F, Coste D, Rivas Platero GF, Zapater MF, Abadie C, Carlier J (2010) Genetic discontinuities and disequilibria in recently established populations of the plant pathogenic fungus *Mycosphaerella fijiensis*. Mol Ecol 19: 3909–3923.

Johanson A, Jeger MJ (1993) Use of PCR for detection of *Mycosphaerella fijiensis* and *M. musicola*, the causal agents of Sigatoka leaf spots in banana and plantain. Mycol Res 97: 670–674.

Johanson A, Crowhurst RN, Rikkerink EHA, Fullerton RA, Templeton MD (1994) The use of species-specific DNA probes for the identification of *Mycosphaerella fijiensis* and *M. musicola*, the causal agents of Sigatoka disease of banana. Plant Pathol 43: 701–707.

Jones DR (2003) The distribution and importance of the *Mycosphaerella* leaf spot diseases of banana. *In:* Jacome L, Lepoivre P, Marin D, Ortiz R, Romero R, Escalant JV (eds) Mycosphaerella leaf spot diseases of bananas: present status and outlook. Proceedings of the Second International Workshop on Mycosphaerella leaf spot diseases of bananas, San José, Costa Rica: 25–42. INIBAP, France.

Kang ST, Rouf Mian MA (2010) Genetic map of the powdery mildew resistance gene in soybean PI 243540. Genome 53: 400–405.

Kema GHJ (2009) Sequencing the major *Mycosphaerella* pathogens of wheat and banana. Acta Hort 828: 147–152.

Kema GHJ, Goodwin SG, Hamza S, Verstappen ECP, Cavaletto JR, Van der Lee TAJ, Weerdt M, Bonants JM, Waalwijk C (2002) A combined amplified fragment length polymorphism and randomly amplified polymorphim DNA genetic linkage map of *Mycosphaerella graminicola*, the *Septoria Tritii* leach blotch pathogen of wheat. Genetics 161: 1497–1505.

Kuhn ML, Gout L, Howlett BJ, Melayah D, Meyer M, Balesdent Mh, Rouxel T (2006) Genetic linkage map and genomic organization in *Lettospheria maculans*. Eur J Plant Pathol 114: 17–31.

Ma A, Michailides TJ (2005) Advances in understanding molecular mechanisms of fungicide resistance and molecular detection of resistant genotypes in phytopathogenic fungi. Crop Protec 24: 853–863.

Martins NF, Togawa Rc, Souza Jr. MT, Brammer S, Bonato AL, Nhani A (2008) Identificação e caracterização functional de genes em estudos comparatives dos genomas de *Mycosphaerella graminicola* e *Mycosphaerella fijiensis*. Bol Pesq Desenvol 237.

Manzo-Sánchez G, Zapater MF, Luna-Martínez F, Conde-Ferráez L, Carlier J, James-Kay A, Simpson J (2008) Construction of a genetic linkage map of the fungal pathogen of banana *Mycosphaerella fijiensis*, causal agent of black leaf streak disease.Curr Genet 53: 299–311.

Meng XQ, Shoemaker RC, Yang XB (1999) Analysis of pathogenicity and genetic variation among *Phytophthora sojae* isolates using RAPD. Mycol Res 103(2): 173–176.

Michlemore RW, Hulbert SH (1987) Molecular markers for genetic analysis of pathogenic fungi. Annu Rev Phytopathol 25: 383–404.

Mobambo KN, Gauhl F, Vuylsteke D, Ortiz R, Pasberg-Gauhl C, Swennen R (1993) Yield loss in plantain from black Sigatoka leaf spot and field performance of resistant hybrids. Field Crop Res 35: 35–42.

Mourichon X, Fullerton RA (1990) Geographical distribution of two types of *Mycosphaerella musicola* Leach (*Cercospora musae*) and *Micosphaerella fijiensis* (*Cercospora fijiensis*), respectively, agents of Sigatoka disease and Black Leaf Streak Disease in bananas and plantains. Fruits 445: 213–218.
Muller R, Pasberg-Gauhl C, Gauhl F, Ramser J, Kahl G (1997) Oligonucleotide fingerprinting detects genetic variability at different levels in Nigerian *Mycosphaerella fijiensis*. J Phytopathol 154: 25–30.
Nakayashiki H, Nguyen QB (2008) RNA interference: roles in fungal biology. Curr Opin Microbiol (11)6: 1–9.
Neu C, Kaemmer D, Kahl G, Fischer D, Weising K (1999) Polymorphic microsatellite markers for the banana pathogen *Mycosphaerella fijiensis*. Mol Ecol 8: 523–525.
Njambere EN, Vandemark G, Chen W (2010) Development and characterization of microsatellite markers of the fungal plant pathogen *Sclerotinia trifoliorum*. Genome 53: 494–500.
Rivas GG, Zapater MF, Abadie C, Carlier J (2004) Founder effects and stochastic dispersal at the continental scale of the fungal pathogen of bananas *Mycosphaerella fijiensis*. Mol Ecol 13: 471–482.
Rodrigues-Garcia CM, Raigosa-Flores N, Conde-Ferraz L, Peraza-Echeverria L, Canto-Canchê B, James-Kay A (2006) Variation in eletrophoretic karyotype among Mexican isolates of *Mycosphaerella fijiensis* Can J Plant Pathol 28: 236–241.
Pariaud B, Ravigné V, Halkett F, Goyeau H, Carlier J, Lannou C (2009) Aggressiveness and its role in the adaptation of plant pathogens. Plant Pathology (58) 3: 409–424.
Portal O, Yovanny I, de Vleesschauwer D, Sánchez-Rodriguez A, Mendoza-Rodrigues M, Acosta-Suárez M, Ocaría B, Jiménez E, Höfte M (2011) Analysis of expressed sequence tags derived from a compatible *Mycosphaerella fijiensis*-banana interaction. Plant Cell Rep DOI: 10.1007/s00299-011-1008-z.
Stergiopoulos I, van den Burg HA, Ökmen B, Beenen HG, van Liere S, Kema GHJ, De Wit PJGM (2010) Tomato Cf resistance proteins mediate recognition of cognate homologous effectors from fungi pathogenic on dicots and moncots. Proc Natl Acad Sci USA, 107: 7610–7615.
Stover RH, Simmonds NW (1987) Bananas. 3rd edn. Longman Scientific and Technical Essex, England, UK.
Ubayasena L, Bett K, Tar'an B, Vijayan P, Warkentin T (2010) Genetic control and QTL analysis of cotyledon bleaching resistance in green field pea (*Pisum sativum* L.). Genome 53: 346–359.
Weld RJ, Plummer KM, Carpenter MA, Ridgway HJ (2006) Approaches to functional genomics in filamentous fungi. Cell Res (16)1: 31–44.
Xu JR, Peng YL, Dickman MB, Sharon A (2006) The dawn of fungal pathogen genomics. Annu Review Phytopathol 44: 337–366.
Xu X, Roberts T, Barbara D, Harvey NG, Gao L, Sargent DJ (2009) A genetic linkage map of Venturia inaequalis, the causal agent of apple scab. BMC Res Notes 2: 163.
Yang BJ, Zhong B (2008) Fourteen polymorphic microsatellite markers for the fungal banana pathogen *Mycospharella fijiensis*. Mol Ecol Res 8: 910–912.
Zandjanakou-Tachin M, Vroh-Bi I, Ojiambo PS, Tenkouano A, Gumedzoe YM, Bandyopadhyay R (2009) Identification and genetic diversity of *Mycosphaerella* species on banana and plantain in Nigeria. Plant Pathol 58: 536–46.
Zapater MF, Rakotonantoandro S, Cohen S, Carlier L (2004) Polymerase chain reaction—restriction fragment length polymorphism markers for the fungal banana pathogen *Mycospharella fijiensis*. Mol Ecol Notes 4: 80–82.
Zapater MF, Duchemin M, Dussart JF, Coste D, Brottier P, Carlier J (2008) Microsatellite markers for the fungal banana pathogens *Mycosphaerella fijiensis*, *Mycosphaerella musicola* and *Mycosphaerella eumusae*. Mol Ecol Resour 8: 1121–1125.
Zhong S, Steffenson BJ, Martinez JP, Cluffetti LM (2002) A molecular map and eletrophoretic karyotype of the plant pathogenic fungus Cochliobolus sativus. Mol Plant-Microbe Interact 15(5): 481–92.

13

Genomics of *Fusarium oxysporum* f. sp. *cubense* Causing Wilt Disease in Banana (*Musa* spp.)

Raman Thangavelu,[1,a,*] *Ganga Devi Perumal,*[1,b] *Mohammed Mustaffa,*[1,c] *Subramaniam Sreeramanan*[2] and *Xavier Rathinam*[3]

ABSTRACT

Panama wilt caused by *Fusarium oxysporum* f.sp. *cubense* (*Foc*) is the most devastating disease of banana causing severe yield losses, especially in commercial Cavendish group of bananas throughout the world. The infection by the pathogen results in wilting and death of the plants. This fungus survives for long time in the soil. Based on the infection in the differential host, the *FOC* has been grouped in to four races. Among these, the race 4 (VCG 01213/01216) is considered as an important and severe strain as it affects the Cavendish group of bananas which are important in the export industry. Although several methods are available, the *FOC* has been characterized widely by VCG and DNA fingerprinting methods. So far, 21 VCGs have been identified worldwide and in Asia alone, 15 VCGs have been reported. The molecular characterization of *FOC* collected from different parts

[1]National Research Centre for Banana, Thiruchirapalli 620102, Tamil Nadu, India.
[a]e-mail: *bananathanga@rediffmail.com*, *rtbanana@gmail.com*
[b]e-mail: *ganga_gns@yahoo.com*
[c]e-mail: *directornrcb@gmail.com*
[2]School of Biological Sciences, Universiti Sains Malaysia (USM), Georgetown, 11800, Penang, Malaysia; e-mail: *sreeramanan@usm.my*
[3]Department of Biotechnology, AIMST University (AIMST), Semelling, 08000, Kedah, Malaysia; e-mail: *rxavier77@yahoo.com*
*Corresponding author

List of abbreviations after the text.

of the world grouped all the VCG's into 2 groups. Group 1 contained the VCG's of race 4 and group 2 contained the VCG's of race 1 and race 2. The isolates within each VCG produced a very similar banding pattern and they were closely related. The close relationship between the isolates within each group suggests that each has a common progenitor. There was a cross reaction among different VCG isolates belonging to the same race. Similarly cross reaction between VCG's of race 1 and race 2 has also been reported under field conditions. To identify and diagnoze the *FOC* pathogen at its early stage of infection in the plant and in the soil, SCAR as well as rDNA- ITS/IGS region targeted molecular markers have been developed which will be useful for quarantine purposes and also for implementation of management practices effectively. Biotechnology offers another scientific approach to create fungus-resistant banana cultivars by using antifungal genes from beyond the species barrier. Various agronomically important genes can be introduced in to elite banana germplasm. Tremendous scientific progress made in genetic engineering of banana for enhanced resistance to fungal pathogens especially in *FOC* is an indication of the high level of interest in the banana specific community.

13.1 Introduction

Panama wilt of banana caused by *Fusarium oxysporum* f.sp. *cubense* (Foc) Synder and Hanson is one of the most devastating diseases of banana in the world (Ploetz and Pegg 1997). The disease destroyed the main export cultivar "Gros Michel" (AAA genome) in the Central American/Caribbean region in the 1950s (Stover 1962). "Gros Michel" was then replaced with Cavendish cultivars (AAA genome), which are resistant to *FOC* race 1. Unfortunately, this cultivar has also succumbed to wilt attack because of the emergence of a virulent form of *FOC* (VCG 01213/16), which is now posing a serious threat to the multibillion-dollar banana export industry and also to the livelihood of small-scale banana growers (Ploetz 2005).

The fungus infects the roots of banana plants, colonizing the vascular system of the rhizome and pseudostem inducing characteristic wilting symptoms usually after 5–6 months of planting (Blomme et al. 2011). The symptoms are expressed both externally (yellowing and wilting of leaves, longitudinal splitting in the pseudostem, etc.) and internally (reddish-brown discoloration of vascular tissue) (Wardlaw 1961; Stover 1962). Infected plants generally do not produce a bunch and under severely diseased condition, the entire plant dies.

The fungus survives as chlamydospores in the soil and on plant debris and invades the host through the root hairs, root tips and natural wounds along the lateral root base. It also enters through the wounds caused by farm implements, insect pests and parasitic nematodes. The micro-conidia and also the hyphae are capable of infecting the host. After entry, the fungus colonizes the cortex and enters the xylem in the vascular system (Rishbeth 1955). Once the fungus enters inside the vascular system, *FOC* may move from the root system into the corm and into the pseudostem (Rishbeth 1955). After colonization, the pathogen blocks the plant's vascular system, that leads to wilting and finally plant mortality (Ploetz and Pegg 2000).

Spread of the pathogen is affected by the transfer of infected corms or suckers, soil attached to infested planting material or implements/vehicles (Ploetz 1994) surface water and also by roots (Moore et al. 1995). Wind is also involved in spreading the *FOC* through dust and trash.

13.2 Race Classification in *FOC*

Isolates of *FOC* have been traditionally grouped into four physiological races based on pathogenicity to a small number of differential banana host cultivars in the field (Moore et al. 1995). This racial classification is an artificial grouping and does not reflect a genetically defined relationship, as the genetics of virulence and resistance are poorly understood in this pathosystem. However, it provides some useful information. The characteristics of *FOC* races are:

Race 1: occurs throughout the world and nearly destroyed the world banana export industry in the tropics, which was dominated by cv. "Gros Michel". Fusarium wilt caused by race 1 was the most serious problem for more than 60 years in tropical America. The disease was under control only after the replacement of the susceptible "Gros Michel" with the resistant Cavendish banana cultivar in the 1960s (Stover 1962). It also attacks cultivars such as "Silk" (AAB), "Pome" (AAB) "abacá", "Maqueño"(AAB), "Pisang Awak" (ABB) and I.C.2' (AAAA) (Ploetz et al. 1990, Bentley et al. 1995). Recently, it was found at the National Research Center for Banana, (NRCB), India that some of the VCG's of race 1 attacks the race 2 suspects and vice versa. In addition, VCG 0124 of race-1 was found to attack the race-4 differential cultivar "Grand Naine" in India (Thangavelu and Mustaffa 2010).

Race 2: is also widely distributed in all the banana growing regions. It is pathogenic to "Bluggoe" (ABB), "Monthan" (ABB) and other closely related cooking bananas (ABB). It also affects enset (*Ensete ventricosum*) and some bred tetraploids such as "Bodles Altafort".

Race 3: Waite (1963) first identified the wilt disease in several species of *Heliconia* in Central and South America in the mid-1900s and named the causal strain as race 3 of *F. oxysporum* f. sp. *cubense*. An outbreak of Fusarium wilt on *H. chartacea* in Australia was also reported (Bentley et al. 1998; Moore et al. 2001). The wilt disease in *Heliconia* spp. was also recorded in

Honduras and Costa Rica. This *Heloconia* strain of *FOC* is nonpathogenic or weakly pathogenic on *Musa* spp. (Ploetz 1990).

Race 4: occurs in most of the banana growing regions like Canary Islands, Vietnam, Taiwan, Indonesia, China, Australia, the Philippines, South Africa, Cambodia, Malaysia, Brazil etc. (Ploetz 1990). It is also found in the islands of Sumatra, Java, Sulawesi, Halhamera, and in the province of Papua near the border with Papua New Guinea (Shivas and Philemon 1996; Davis et al. 2000). However, race 4 has not been reported in India. This race is most destructive since it affects race 1 and race 2 susceptible clones as well as the Cavendish cultivars and "Pisang Mas" (AA).

Race 4 is separated into two groups, viz. sub-tropical race 4 (SR4) (VCG's 0120, 0121, 0129, 01211 found in Australia, 0120 in the Canaries and South Africa and 0122 in the Philippines) and tropical race 4 (TR4) (VCG 01213-01216 complex). SR4 affects Cavendish plants that have been predisposed to disease by cold temperatures in the subtropics. Whereas, TR4 attacks Cavendish more aggressively under tropical conditions in the absence of any predisposing factors. TR4 is distinguished from sub-tropical race 4, because it is genetically distinct and damages Cavendish in the tropics (Ploetz 2004). Although the host range is similar, the clone Pisang lilin (AA) is affected only by TR 4. The strain TR4 (VCG 01213) was first designated in Taiwan on the basis of wilt in Cavendish (Su et al. 1986) and was considered to have originated in the Malay peninsula and in Sumatra (Buddenhagen 2009). Though the presence of the TR4 strain is found only in Indonesia (Halmahera, Irian Jaya, Java, Sulawesi and Sumatra), Peninsular Malaysia, South China, Taiwan (Simmonds and Shepherd 1955; Shivas and Philemon 1996; Ploetz and Pegg 2000) and northern Australia, it is of grave concern to the Cavendish dependent export trade. It has caused severe damage to the Cavendish banana industry in Taiwan since 1967, destroyed new plantations of Cavendish banana in Indonesia and Malaysia since early 1990s and caused field epidemics in the northern territory of Australia between 1997 and 1999. In 2004, severe *FOC* infection in banana plantations in Guangdong province, South China, heightened the TR4 threat. Recently, this race (TR4) has destroyed more than 20,000 ha Cavendish banana in China and is threatening the huge export industry of the Philippines (Daniells 2009).

In the case of SR4, the strain VCG 0120–01215 caused severe damage to Cavendish cultivars in sub-tropical Australia (New South Wales and Queensland), the Canary and Madeira Islands and South Africa (Natal and Transvaal). There are no reports of occurrence of *FOC* TR4 in the major banana growing areas in India (Thangavelu et al. 2001). Although TR4 is not found in Latin America, it is a major threat to the export banana industry including Latin America, which grows only Cavendish banana.

13.3 Pathogen Diversity

The analysis of genetic variation within *FOC* has led to the advancement of research toward long-term disease management strategies, such as host plant resistance, bio-control, and activators of systemically acquired resistance. Additionally, the inference of evolutionary relationships between VCG's of *FOC* may help to gauge the potential for the development of novel pathotypes. Moreover, phylogenetic studies will facilitate the development of accurate diagnostic tools, which are vital for the maintenance of quarantine management and for the rapid identification of introduced exotic strains of *FOC*.

Generally, *F. oxysporum* has no known sexual stage. In the absence of meiotic recombination, variation in the fungus is thought to result through mutation (Kistler and Miao 1992). *FOC* is a highly variable pathogen (Ploetz 1990), and hence, complete understanding of pathogen diversity within *FOC* is necessary for the selection or breeding of banana cultivars with durable resistance to *Fusarium* wilt and also for the implementation of meaningful international and local quarantine restrictions to limit the spread of the pathogen (Bentley et al. 1995). Moreover, such studies would be useful to identify the origins of the pathogen and study how population of the fungus varies within and between different geographic areas and races. This will enable an understanding of how pathogenicity might change and new races evolve.

Several techniques have been used to characterize *FOC*, to assess the genetic variation within *FOC* and for the determination of phylogenetic relationships between isolates of *FOC*. Vegetative compatibility grouping (VCG) analysis (Ploetz and Correll 1988; Brake et al. 1990, Ploetz 1990; Moore et al. 1993, Moore 1994), production of volatile organic compounds (Stover 1962; Moore et al. 1991); electrophoretic karyotype (Boehm et al. 1994; Miao 1990); random amplified polymorphic DNA (RAPD) analysis (Bentley et al. 1995); restriction fragment length polymorphism (RFLP) analysis (Koenig et al. 1997); DNA amplification fingerprinting (DAF) analysis (Bentley and Bassam 1996; Bentley et al. 1998) and DNA sequencing of nuclear and mitochondrial genes (O'Donnell et al. 1998) have been used to characterize Foc isolates. These methods have been useful for pathotype determination (Pegg et al. 1995) for which no small-plant pathogenicity test exists. Although several of the above methods have been developed, the methods such as volatile production in culture, vegetative compatibility group (VCG) analysis, and molecular characterization techniques are being followed widely to determine the diversity of *FOC*.

13.3.1 Volatile Production

The production of volatile compounds by *FOC* pathogens is a consistent and repeatable trait used to differentiate strains of *FOC* pathogen (Stover 1962). When isolates of *FOC* are cultured on a starch substrate of sterilized, steamed rice, they produce odorous compounds, "aldehydes", in the headspace above the cultures and this can be analyzed using gas chromatography. Those isolates that produced the aldehydes were referred to as cultivar "Odoratum", whereas those that did not were referred as cultivar "Inodoratum". It was found that isolates from "Bluggoe", which is affected by race 2, did not produce an odor; whereas isolates in race 1 belong to either the "odoratum" or "inodoratum" type (Stover 1962). Pegg et al. (1996) described that VCG's 0120, 0121, 0122, 0126, 0129, 01210, 01211, 01213, 01215, 01216 and 01219 belong to the "odoratum" group and VCG's 0123, 0124, 0125, 0128, 01212, 01214, 01217, 01218 and 01220 belong to inodoratum group. Moore et al. (1991) also analyzed the production of volatile compounds with high-pressure liquid chromatography using Australian *FOC* isolates and found that VCG isolates 0120, 0129, and 01211, 01213/16 which belong to race 4, produced characteristic volatile profiles, whereas VCG isolates 0123, 0124, 0124/0125, 0125, 01220 of race 1 and VCG 0128 of race 2 did not. This technique has also been used to characterize Asian isolates of *FOC* (Moore et al. 1991; Pegg et al. 1993). There was an absolute correlation between the production of volatile substances, VCG, and pathogenicity in the Australian isolates. This method is a useful and reliable means for characterizing isolates within a population of *FOC* and provides a more rapid and inexpensive method for initial characterization of field isolates (Moore et al. 1991).

13.3.2 Vegetative Compatibility Grouping

Vegetative compatibility has been used to study the diversity, genetics, ecology and population biology of several plant pathogenic fungi including *FOC*. Individual isolates of *FOC* within races can fuse asexually to form a stable heterokaryon (Puhalla 1985) are said to belong to the same vegetative compatibility group (VCG) (Ploetz and Correll 1988). This trait is genetically controlled. Conventionally, this is determined by the ability of nitrate non-utilizing auxotropic mutants (*nit*) complement to one another for nitrate utilization (Fig. 13-1). If two different strains of a fungus are unable to form a heterokaryon when they are paired, they are said to be vegetatively incompatible (Puhalla 1985). Vegetatively compatible strains are much more likely to be genetically similar than vegetatively incompatible strains. For example, strains of a fungus that are vegetatively compatible are quite

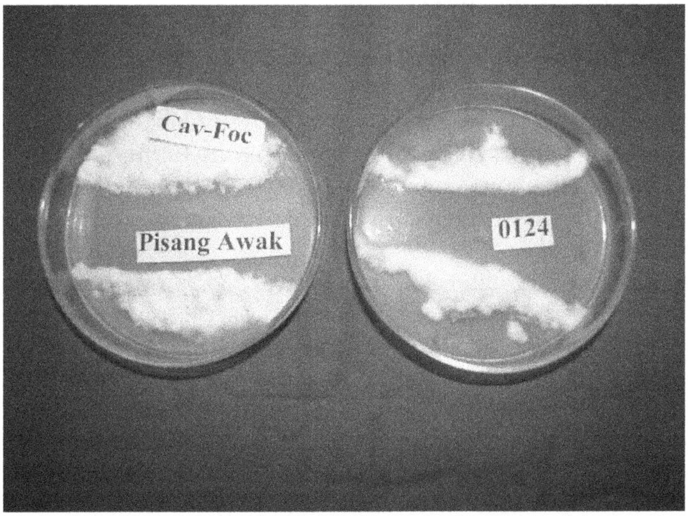

Figure 13-1 Heterokaryon formation between nit-M testers of a) Pisang Awak & b) VCG-0124 and *nit*-1 of *Foc* from cv. Grand Naine (AAA-Cavendish group) in Vegetative compatibility grouping test.

similar to each other with respect to some traits such as colony size (Croft and Jinks 1977; Correll et al. 1986) antibiotic production (Croft and Jinks 1977), virulence (Correll et al. 1985, 1986; Gorden et al. 1986; Bosland and Williams 1987) and isozyme patterns (Bosland and Williams 1987) compared to vegetatively incompatible strains (Moore et al. 1995).

VCG's for *FOC* were first reported by Puhalla in 1985 and to date 21 VCG's have been identified worldwide (Correll et al. 1987; Ploetz and Correll 1988; Moore 1994; Pegg et al. 1995) (Table 13-1). In Asia alone, 15 VCG's have been identified (Ploetz 1990; Moore 1994; Pegg et al. 1995) and the greatest number of VCG's were recorded in the Indo-Malaysian region where *Musa acuminate* occurs in large numbers (Pegg et al. 1994, 1996). Only a few VCG's are found in Africa and the Americas (Ploetz 1993). In Australia, Brake et al. (1990) have used VCG analysis to delineate six VCG's among 245 Australian isolates of *FOC*. Race 1 strains were present in VCG's 0124 and 0125; race 2 strains in VCG 0128 and race 4 strains in VCG's 0120, 0129 and 01211. Correlation between VCG and pathogenicity has also been demonstrated for Australian isolates of *FOC* in field reactions and glasshouse pathogenicity tests. Based on the VCG data, Pegg et al. (1993) suggested that the ancestral VCG's 0120 and 0124–0125 were evolved independently in the centers of diversities for *M. acuminate* and *M. balbisiana*, respectively, and less widely distributed VCG's probably might have arose from the pregenitor VCG's as per the mutations hypothesis proposed by Ploetz (1990).

13.3.3 Molecular Characterization

Although VCGs provide a useful means of sub-dividing *FOC* into genetically isolated groups, it does not give any indication of genetic relatedness among isolates belonging to different VCGs, or even of isolates within the same VCG. Isolates that are vegetatively compatible are thought to have identical alleles at each of the "*vic*" loci (Correll 1991). A mutation at a single "*vic*" locus, however, could result in closely related isolates becoming vegetatively incompatible (Bentley et al. 1995, 1998). Also, some VCGs produce VCG complexes by forming heterokaryons between separate VCGs (Ploetz 1990; Bentley et al. 1998). Under these circumstances, molecular characterization techniques can be used to determine the genetic similarity between isolates within each VCG and the genetic relatedness among the VCGs (Pegg et al. 1996). Bentley et al. (1995) reported that RAPD-PCR banding patterns were generally VCG-specific and it was possible to determine the genetic relatedness between the different VCGs and of isolates within each VCG. By comparing the RAPD-PCR banding patterns of worldwide collection of isolates of *F. oxysporum* f. sp. *cubense* representing different VCGs and races both visually and by phonetic analysis, the *Foc* isolates has been sub-divided into two major groups. Group-1 contains isolates belonging to VCGs 0120, 0121, 0122, 0126, 0129, 01210, 01211, 01213, 01215, 01216 and 01219 and Group 2 contains isolates belonging to VCGs 0123, 0124, 0125, 0128, 01212, 01214, 01217, 01218 and 01220. Isolates within each VCG produced similar banding pattern and were closely related (Bentley et al. 1995). The two groups differentiated by RAPD-PCR analysis, correspond with previous classifications based on volatile production and electrophoretic karyotyping. Among Australian isolates of *Foc*, there was a direct correlation between DNA fingerprint group and race. Similarly, the characterization of Indian *Foc* isolates by RAPD analysis using the primer OPB-07 indicated the presence of wide variation among the *Foc* isolates and grouped into 23 different groups. However, there was no correlation based on host source/ race/ geographical origin (Fig. 13-2), Sorensen et al. (1993) analyzed the genetic variation of *Foc* which represented the 3 races and 7 VCGs present in Australia by RAPD-PCR and found two distinct clusters: one cluster contained race 4 (VCGs 0120, 0129 and 01211) and another cluster contained both race 1 and 2 (VCGs 0128 of race 2 and 0124, 01245 and 0125 of race 1). For Asian isolates, the RAPD-PCR banding pattern of all race 4 isolates were similar to that of race 4 isolates from Australia, the Canary Islands, South Africa, and to putative race 4 isolates from Honduras (Pegg et al. 1994).

Bentley et al. (1998) examined 350 isolates of *Foc* using modified DNA fingerprinting analysis and further divided into lineages. The genetic similarity between the two major groups of *Foc* was 60%. The average genetic similarity between VCGs within the lineage was 80%. These lineages

Table 13-1 Vegetative compatibility groups, races and origins of strains of *Fusarium oxysporum* f. sp. *cubense* (*FOC*) and the varieties from which it was identified (Jones 2000; Ploetz 2005; Thangavelu 2008).

VCG	VCG complex	Race	Cultivars with genomic groups	Origins
0120	0120-01215	1, 4	AA: SH-3142, SH-3362; AAA: Gros Michel, Highgate, Pisang Ambon Putih, Pisang Ambon, Dwarf Cavendish, Williams, Mons Mari, Grand Nail, Lacatan; AAB: Prata, Lady Finger, Pacovan, Hua Moa, Silk	South Africa, Canada, Islands, Australia, Brazil, Costa Rica, Honduras, Indonesia, Jamaica, Malaysia, Nigeria, Portugal, Spain, Taiwan, Florida (USA), France
0121	None	4	AAA: Gros Michel, Cavendish	Indonesia, Taiwan
0122	None	4?	AAA: Cavendish; ABB: Saba	The Philippines
0123	None	1	AAA: Gros Michel, Grand Nail; AAB: Silk, Latundan, Pisang Keling; ABB: Pisang Awak, Kluai Namwa	Malaysia, the Philippines, Taiwan, Thailand
0124	0124-0125 0128-01220	1,2	AAA: Williams, Grand Naine; AAB: Silk AB: Ney Poovan ABB: Pisang Awak, Monthan	Australia, Brazil, Burundi, China, Cuba, Democratic Republic of Congo, Haiti, Honduras, India, Jamaica, Malawi, Malaysia, Mexico, Nicaragua, Rwanda, Thailand, Tanzania, Uganda, Florida (USA)
0125	None	1	AAAA: Jamaica 1242 AAB: Silk	Australia, Brazil, Burundi, China, Cuba, Democratic Republic of Congo, Haiti, Honduras, India, Jamaica, Malawi, Malaysia, Mexico, Nicaragua, Rwanda, Thailand, Tanzania, Uganda, Florida (USA)
0126	None	1	AA: Pisang Berlin, AAA: Highgate; AAB: Maqueno, Pisang Manrung	Honduras, Indonesia, Papua New Guinea, the Philippines

Table 13-1 contd....

Table 13-1 contd....

VCG	VCG complex	Race	Cultivars with genomic groups	Origins
0128	0124-0125-0128-01220	1, 2	AAB: Lady Finger, Maca, Manzano, Maqueno, Silk ABB: Pisang Awak, Monthan	Australia, Brazil, Burundi, China, Cuba, Democratic Republic of Congo, Haiti, Honduras, India, Jamaica, Malawi, Malaysia, Mexico, Nicaragua, Rwanda, Tanzania, Thailand, Australia, Uganda, Florida (USA)
0129	None	4	AAA: Mons Mari AAB: Lady Finger	Australia
01210	None	1	AAA: Gros Michel AAB: Manzano	Cayman Islands, Cuba, Florida (USA)
01211	None	4	AA: SH-3142 ABB: Monthan, Pisang Awak	Australia, India
01212	None	?4	AB: Ney Poovan: AAB: Silk, Kisubi ABB: Pisang Awak, Bluggoe	Tanzania, India
01213	01213-01216	T4	AA : Pisang Lilin, Pisang Mas	Australia, Indonesia, Malaysia, Taiwan
01214	None	2	ABB: Harare, Mbufu	Malawi
01215	0120-01215	1, 4	AA : SH-3142,SH-3362 AAA: Gros Michel, Highgate, Pisang Ambon Putih, Pisang Ambon, Dwarf Cavendish, Williams, Mons Mari, Grand Naine, Lacatan AAB: Prata, Lady Finger, Pacovan, Hua Moa, Silk	Canary Island, Australia, Brazil, Costa Rica, Honduras, Indonesia, Jamaica, Malaysia, Nigeria, Portugal, Spain, Taiwan, Florida (USA)

01216	01213-01216	T4	AAA: Pisang Ambon, Valery, Williams, Grand Naine, Novaria, Red, Pisang Udang, Pisang Susu, Pisang Nangka, Pisang Barangan AAB : Pisang Raja Serah, Pisang Rastali, Pisang Rajah, Relong ABB : Pisang Awak Legor, Saba, Pisang Kepok, Pisang Caputu, Pisang Kosta, Unknown: Pisang Batan	Australia Indonesia, Taiwan, Malaysia
01217	None	?	ABB : Harare, Mbufu	Malaysia
01218	None	?	AAB : Pisang Raja Serah; Silk ABB: Pisang Awak, Kluai Namwa, Pisang Kepok, Pisang Siam	Indonesia, Malaysia, Thailand, India
01219	None	?	AAA : Pisang Ambon, Pisang Ambon Putih Unknown: PIsang Raja Garing	Indonesia,
01220	0124-01216 0128-01220	4?	ABB : Pisang Awak, Ducasses, Kayinga, Zambia, Kluai Namwa, Bluggoe, Harare, Kholobowa, Dwarf Bluggoe, Mbufu, Burrro Criolla, Pelipita, Ice cream AAB: Silk	Australia, Brazil, Burundi, China, Cuba, Democratic Republic of Congo, Haiti, Honduras, India, Jamaica, Malawi, Malaysia, Mexico, Nicaragua, Rwanda, Thailand, Tanzania, Uganda, Florida(USA)
01221	None	?	ABB: Kluai Namwa	Thailand

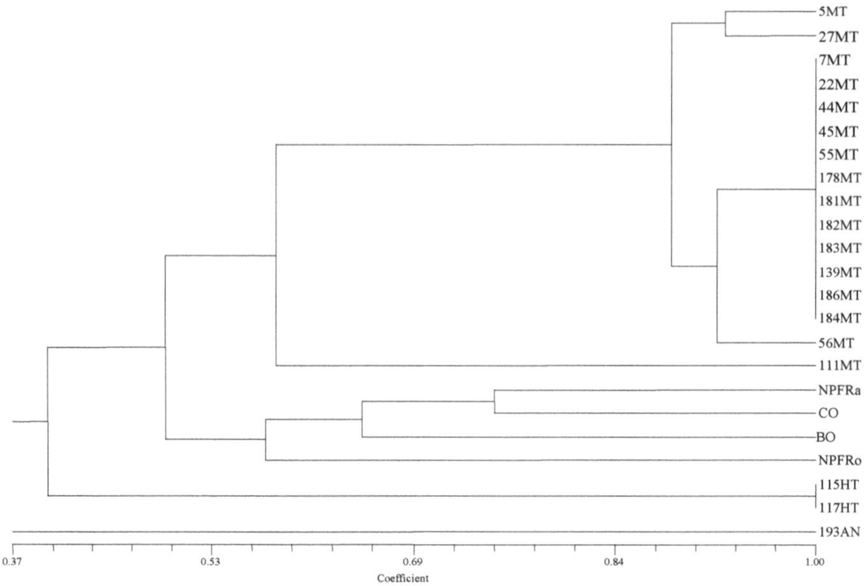

Figure 13-2 Dendrogram showing variation among the *FOC* isolates obtained from cv. Monthan-ABB.

of *Foc* are therefore different from each other as they are from other formae specialis of *F. oxysporum*, which may indicate independent origins for the individual lineages of *Foc*. Koenig et al. (1997) studied the phylogenies of 165 strains in 13 VCGs/VCG complexes with restriction fragment length polymorphisms (RFLPs) of genomic DNA and 19 anonymous, single-copy DNA probes and identified 10 lineages among 165 isolates in 13 VCGs or VCG complexes. The two largest lineages I and II that contained 65 and 43 isolates respectively and corresponded to the VCG 0124/5/8/12/20 and 0120/15 complexes which were genetically more similar to *F. oxysporum* f. sp. *niveum* a pathogen of watermelon than to each other. The distant relationship provides further support that these major populations of *Foc* might have evolved independently. However, it is also proposed that the pathogen was co-evolved with edible bananas and their wild diploid progenitors (*M. acuminate* and *M. balbisiana*) in Southeast Asia and has been distributed to other countries as infected banana corms and attached soil (Stover 1962). The close relationships between the isolates within each group also suggested that each has a common progenitor. Furthermore, absence of *F. oxysporum* f.sp. *cubense* in areas where highly susceptible cultivars are grown, such as Papua New Guinea and South Pacific Islands, suggested that pathogenic strains do not readily evolve from other populations of *F. oxysporum* (Pegg et al. 1993). The RAPD-PCR analysis of the genetic diversity among isolates from Southeast Asia also supports the co-evolution hypothesis (Pegg et al. 1993). The co-evolution hypothesis has important implications in the selection of resistant genotypes, as these are most likely to be present in regions, where there is great diversity within the host and the pathogen (Vavilow 1922).

Recently, Groenewald (2006) also used AFLP (Amplified fragment length polymorphism) technique to study relationships within and between natural populations of *Foc* isolates at the genome level and also to investigate whether AFLPs could discriminate between VCGs of *Foc*. Distance analysis of AFLP data clearly divided isolates of *Foc* into two major clades. Clade 1 supported by a bootstrap score of 100 %, contained VCGs 0120, 0120/15, 0121, 0122, 0126, 0129, 01213, 01213/16, 01216 and 01219 which were collected from around the world including the South African isolates. Clade 2 supported by a bootstrap score of 74% contained VCGs 0123, 0124, 0125, 01217 and 01218 which contained isolates from Australasia. The *Foc* isolates from the second clade grouped closer to *F. oxysporum* of other formae speciales than to *Foc* isolates of the first clade. Further, the AFLP analysis divided *Foc* isolates into seven genotypic groups. The first genotypic group contained VCG 0120 and VCG complex 0120/15. Genotypic group 2 included VCGs 0126, 0129, 0122 and 01219, and the third genotypic group included the tropical race 4 isolates VCGs 01213, 01216 and 01213/01216. *Foc* VCG 0121 belongs to genotypic group 4, and VCGs 0123 and 01217 were in genotypic group 5. VCG 01218 represented the sixth genotypic group, while group 7 included VCGs 0124 and 0125. It was finally concluded that isolates within each VCG generally produced similar banding patterns, and were therefore, closely related, independent of geographical origin or host source. VCG 0120 isolates are homogeneous despite of their geographical origins (Asia, Australia, South Africa, Central and South America) and different hosts (Cavendish, High Gate and Lady Finger banana cultivars). This is important in terms of future disease management programs (Groenewald 2006).

13.4 Relationship between Races and VCG

The relationship between races and VCGs within *F.oxysporum* f. sp. *cubense* is not well defined (Bentley et al. 1995). Ploetz (1990) found that within a worldwide collection of isolates, multiple VCGs occurred within a race and a single VCG may have representatives from more than one race. In India, many of the *Foc* isolates obtained from the race 1 susceptible cultivars, cross-reacted with nit-M testers of *Foc* isolates obtained from race 2 susceptible cultivars and vice versa. The *nit-1* mutants obtained from race 2 differential cv. Monthan-ABB (44MT, 45 MT, 181MT) have formed heterokaryon with nit-M testers obtained from race-1 differential cv. Rasthali (19RT) (Thangavelu 2008). Similarly, the *nit-1* mutants of race-1 (14RT, 107RT, 132RA, 127RKa, 188RN) obtained from race 1 differential cv. Rasthali (Silk-AAB) has formed heterokaryon with nit-M testers obtained from race 2 differential cv. Monthan (Thangavelu 2008). In certain cases, for example, the *nit-1* mutants recovered from *Foc* isolates obtained from cv. Hill banana (Pome-AAB) formed heterokaryon with nit-M testers recovered from *Foc* isolates of both race 1 and race 2. The pot culture studies conducted to validate the above results also confirmed the cross reaction between various *Foc* isolates of race 1 and race 2 (Fig. 13-3). The *Foc* isolates 5MT (race 2) caused wilt disease in cv. Karpuravalli which is race1 susceptible. Similarly, the race1 *Foc* isolate-19RT produced wilt symptoms in cv. Monthan

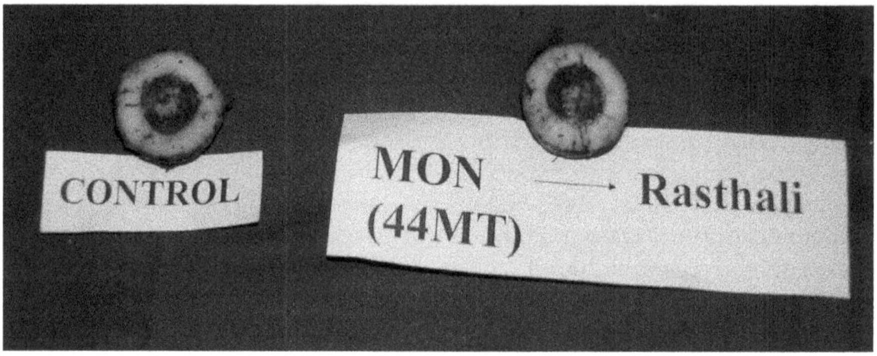

Figure 13-3 Cross reaction of *FOC* of race 2 from cv. Monthan (ABB) on race 1 susceptible cv. Rasthali (Silk-AAB) under pot culture condition. Control- cv. Monthan inoculated with *FOC* from the same cultivar.

which is race 2 susceptible. In contrast, within Australian populations of *F. oxysporum* f. sp. *cubense,* there is good correlation between race and VCG, as only one race has been found to occur within each VCG (Brake et al. 1990; Moore et al. 1993).

13.5 Marker for Diagnosis of *Foc*

Plant diseases can be managed most effectively if control measures are introduced at an early stage of disease development. Disease diagnosis and pathogen identification by traditional methods involving the isolation of the pathogen and characterization by inoculation tests is labor-intensive and time consuming (Alves-Santos et al. 2002). Technological advances in molecular biology and biotechnology continue to be applied to the development of rapid, specific and sensitive tools for the detection of plant pathogens (Miller and Martin 1988).

In addition to being time consuming, identification of the *Foc* pathogen based on morphological characteristics requires expertise in *Fusarium* taxonomy (Jurado et al. 2006). Furthermore, identification of *Fusarium* with microscopy is inefficient and difficult to detect in plant tissues, which may contain various disease complexes (Yergeau et al. 2005). Under these circumstances, a rapid diagnostic assay for *Foc* would certainly help, not only in preventing the entry and spread of the pathogen through planting material but also help the farmers to know the presence of *Foc* pathogen in the field prior to planting. Identification of the disease in growing plants at an early stage of infection using diagnostic tools helps in controlling the disease. The development of diagnostic tools would be useful for basic research in epidemiology and fungal population genetics of *Foc* (Schilling et al. 1996).

Diagnostic methods based on the polymerase chain reaction (PCR) amplification of specific DNA sequences have been utilized successfully and routinely for the detection and identification of a variety of important fungal plant pathogens. Sharon (2003, unpubl. data) sequenced the intergenic spacer (IGS) region of the ribosomal DNA (rDNA) of nine representative isolates of *Foc* of Australia. Sequence analysis of the IGS region of *Foc* isolates, showed a 196bp insertion in a representative isolate of VCG 0124. This insertion sequence was shown to be homologous to a novel family of short interspersed nuclear elements (SINEs) called *Foxy*, characterized in *Fusarium oxysporum* f. sp. *lycopersici* (*Fol*). This insertion sequence was designated as *Foxy-Foc*. Based on these nucleotide polymorphisms, oligonucleotide primer sequences were designed for the specific amplification of the Australian race 1 (R1), sub-tropical race 4 (SR4) and tropical race 4 (TR4) strains of *Foc*. Screening of these primers with a range of isolates of *Foc*, *Fusarium oxysporum* and *Fusarium* species indicated that the primers designed to amplify specifically for race 1, sub-tropical race 4 and tropical race 4 amplified only Australian *Foc* isolates of race 1, sub-tropical race- 4 and tropical race 4. Unfortunately, these primers also amplified the closely related exotic isolates of *Foc* and the reason might be due to lack of sequence variability in the IGS region (Sharon 2003, unpubl.). Similarly, a sequence characterized amplified region (SCAR) as well as ribosomal DNA internal transcribed spacer (rDNA-ITS) region targeted primers were designed by exploiting the variations of rDNA- ITS regions of several *Foc* isolates from India (Fig. 13-4) These markers have amplified only the pathogenic *Foc* of India. These primers have also been validated with field grown plants (Thangavelu 2010 unpubl.). In addition to amplification of specific DNA sequences such as rDNA-IGS/ITS regions, DNA fingerprinting with random amplified polymorphic DNA (RAPD) is also a powerful tool for fungal pathogen identification (Koike et al. 1995; Mes et al. 1999; Fungaro et al. 2004; Lin et al. 2009). Using this RAPD technique, Lin et al. (2008) have developed a molecular method to detect *Foc* race 4 isolates in Taiwan. By PCR amplification, the primer set *Foc*- 1/*Foc*-2 derived from the sequence of a random primer OP-A02 amplified fragment, produced a 242 bp size

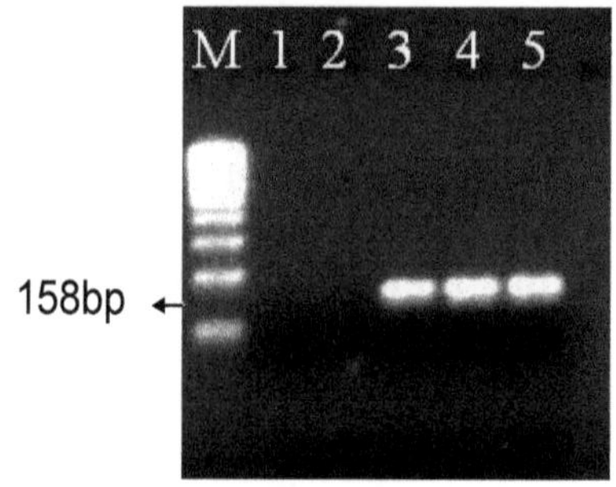

Figure 13-4 PCR amplification of *FOC* pathogen present in *FOC* infested soil and also in the *Foc* infected corm and pseudostem of banana by SCAR marker. M – Marker (100 bp); 1. Control (without infection); 2. Negative control; 3. *Foc* infested soil 4. *Foc* infested Corm; 5. *Foc* infested pseudostem.

DNA fragment which was specific to *Foc* race 4. With the optimized PCR parameters, the molecular method was sensitive and could detect small quantities of *Foc* DNA as low as10 pg in 50 to 2,000 ng host genomic DNA with high efficiency. They also demonstrated that by using this PCR assay with *Foc*-1/*Foc*-2 primer set, *Foc* race 4 could be easily distinguished from other *Foc* races 1 and 2, and separated other *formae speciales* of *F. oxysporum*. This PCR diagnosis protocol can be used to screen naturally *Foc*-infected banana samples and also to rapidly test the presence of *Foc* race 4 in breeding materials of banana for resistance to *Foc* race 4.

13.6 Strategies to Generate Banana Tolerance to Fungal Pathogens

Fungal pathogens cause major yield losses in banana production. Fungal diseases are generally controlled by crop rotation, sanitation, cultivation of disease resistant plants and use of fungicides. Fungicides remain the primary approach for disease control. Several disadvantages are inherent in the use of fungicides: (1) they are expensive and unaffordable by small scale farmers in the developing world; (2) they are potentially or actually harmful to the environment; and (3) extended application of fungicides can accelerate the emergence of new virulent fungal races. Vascular wilts and soil borne diseases like Fusarium wilt in banana cannot be controlled by fungicides. For a number of fungal diseases, plant breeding will not provide a solution as there are simply no natural sources of resistance available to the breeder. Biotechnology offers another scientific approach to create fungus-resistant plant varieties by using antifungal genes from beyond the species barrier. Simultaneously, other agronomically important genes can be introduced into elite banana germplasm.

Banana is attacked by numerous pathogens throughout its life cycle and the cell wall is undoubtly the first line of defense and is very effective against non-host pathogens. In case of host-specific interactions, the pathogen can escape from the plant's defense mechanisms and spread systemically inside the banana host plant (compatible interaction) or can be prevented from diseasing the plant (incompatible interaction). Incompatible interactions often result in tissue necrosis at the site of infection, called hypersensitive response (HR) (Govrin and Levine 2000; Jabs and Slusarenko 2000).

Manipulating the hypersensitive response could pave the way for the production of plants, especially bananas are resistant to a broad range of pathogens. In addition to the hypersensitive response, plants have developed a range of other defense mechanisms against pathogens, which include increased lignification of the cell wall (Lamb and Dixon 1997), the synthesis of small antibiotic molecules or phytoalexins (Lamb and Dixon 1997; Punja 2001), and the production of reactive oxygen species (Govrin and Levine 2000; Liang et al. 2001). However, the biochemical complexity of these mechanisms makes their application in molecular breeding for fungal resistance difficult. One group of the large number of genes activated upon pathogen recognition by a plant is genes that encode pathogenesis-related (PR) proteins (Dixon et al. 1996; Punja 2001). PR proteins have been defined as proteins encoded by the host plant but induced only in pathological or related stress conditions. To date, more than 10 families of PR proteins have been classified, which include β-1, 3-endoglucanases (PR-2 proteins), endochitinases (PR-3 proteins), proteinase inhibitors (PR-6 proteins), proteinases (PR-7 proteins) and extracellular oxidases (PR-9 proteins).

Many plants maintain a long lasting resistance after pathogen attack even in previously uninfected tissues not only against the pathogen that first triggered one or more defense mechanisms but also to a broad spectrum of other pathogens. This kind of induced resistance or systematic acquired resistance (SAR) is best characterized in selected plants and is accompanied by the expression of a large set of genes, among others, those encoding PR proteins (Verberne et al. 2000; Donofrio and Delaney 2001; Lee et al. 2001). A large amount of research is currently devoted to elucidate the mechanisms underlying this SAR response and this approach represents a promising source of potential genes for genetic transformation of banana for resistance against fungal pathogens, especially *Fusarium oxysporum* f. sp. *cubense* (*Foc*). Efforts are mainly directed at finding the signals leading to SAR and subsequently, understanding the induction of a large number of defense genes. Among other molecules, salicylic acid, methyl jasmonate and ethylene seem to be involved in different steps of a cascade of biochemical reactions leading to SAR (Dong 1998; Dempsey et al. 1999). On the basis of signal transduction in resistant plants, development of a reliable method that allows generation of banana plants to be "immune" to a broad spectrum of pathogens using a molecular breeding approach is possible in the future.

13.7 Transgenic Banana for the Control of Fungal Diseases

One of the major steps in genetic engineering of banana for disease resistance is to test the effectiveness of new proteins against the target fungal pathogen especially in *Foc*. The "antifungal protein" strategy relies on the constitutive or pathogen inducible expression of genes encoding proteins, which exhibit a fungitoxic or fungistatic effect (Datta et al. 1999). This strategy exploits single or a few genes, which is/are more amenable for genetic resistance engineering than the more complex defense mechanisms like SAR. The identification of these genes has made it possible to subsequently evaluate their specific roles and importance in disease response pathways using transgenic plants developed with genetic engineering techniques (Pink and Puddephat 1999; Melchers and Stuiver 2000; Echeverria et al. 2008; Sreeramanan et al. 2010).

It is generally accepted that most PR proteins play an important role in the plant defense response to fungal infection because PR proteins show *in vitro* antifungal activity either alone or in combination with other types of PR proteins. PR proteins have already been shown to be good candidates for engineering fungus-resistant crops especially in banana (Datta et al. 2000; Sagi 2000; Punja 2001; Chakrabarti et al. 2003; Pei et al. 2005; Maziah et al. 2007; Sreeramanan et al. 2008; Sreeramanan et al. 2010). Thus, intra or intercellular targeting of an antifungal protein might increase plant resistance, which also implies that broad-spectrum resistance by the antifungal protein strategy will probably require co-expression of several antifungal proteins targeted to different locations within the plant tissue. It is possible indeed to target for instance vacuolar PR proteins extracellularly by removing the short carboxyl-terminal propeptide (Melchers et al. 1993). Interestingly, transgenic tobacco plants expressing the soybean β-1,3-glucanase gene alone, which is involved in the production of a phytoalexin elicitor-releasing factor in soybean, exhibited high resistance to two fungi including *Phytophtora parasitica* cv. *nicotiane* and *Alternaria longipes* (Yoshikawa et al. 1993). Since β-1,3-glycan is a major component of the fungal cell wall and is widely distributed among phytopathogenic fungi, Yoshikawa et al. (1993) believed that this system could lead to broad spectrum resistance. Nevertheless, this would probably require simultaneous intra- and extracellular expression. With respect to this approach, Sreeramanan et al. (2008) developed an efficient *Agrobacterium*-mediated transformation method of banana cultivar "Rasthali" (AAB) with β-1,3-glucanase using tiny meristem tissues and demonstrated increased disease tolerance to *Foc* (race 1; VCG 01217).

In tomato, transgenic plants expressing only a chitinase or a β-1, 3-glucanase transgene were susceptible to *Fusarium oxysporum*, but plants expressing both the genes had significantly higher resistance than the plants expressing only chitinase or β-1,3-glucanase (Jongedijk et al. 1995). Similarly, Jach et al. (1995) demonstrated that tobacco plants expressing a barley β-1,3-glucanase and a chitinase gene had a greatly enhanced resistance compared to plants expressing only one of these two enzymes. Tobacco plants expressing a gene encoding rice basic chitinase and others expressing an alfalfa acidic glucanase were developed separately. Crossing both transgenic lines generated hybrid plants. Evaluation of these hybrids, which were heterozygous for each transgene, and the homozygous selfed progeny, showed that the combined effect of two transgenes in conferring protection against *Cercospora nicotiane* was greater than that afforded by a single transgene. Sreeramanan et al. (2010) successfully transfered the chitinase and β-1, 3-glucanase genes into silk banana (*Musa* spp. AAB group) cv. "Rasthali" using particle bombardment method. Since the cell wall of *Foc* has chitin and β-1,3-glucan, the combined expression of chitinases and β-1,3-glucanases genes have been shown to act synergistically to increase tolerance to a higher level. It was concluded that the combinatorial expression of antifungal genes is an effective approach to enhance tolerance to Fusarium wilt disease (Sreeramanan et al. 2010). These results suggest that the combined expression of PR 2 (β-1,3-glucanase) and PR3 (chitinase) gives effective protection against fungal infection.

Specific attention should be paid to the possible effect of PR-proteins on mycorrhizal fungi, which are important for stimulating the growth of many plants including banana (Declerck et al. 1995). For instance, a delay of colonization by the vesicular-arbuscular mycorrhizal fungus *Glomus mosseae* was observed in tobacco plants expressing a class 11 (extracellular) PR-2 protein, whereas no negative effects were noticed when other PR proteins (PR-1a, PR-3, PR-4 and PR-5) were expressed (Vierheilig et al. 1995). Although such beneficial fungi might have developed a mechanism to avoid the damaging effect of host antifungal proteins, it will be worthwhile to test the activity of each antifungal protein that is expressed in transgenic plants against the known mycorrhizal fungi.

Besides PR-proteins, plants express numerous peptides (shorter than 100 amino acids) with antimicrobial properties including thionins, cysteine-rich defensins, which are structurally related to insect defensins and found in several tissues of more than 20 plant species, non-specific lipid transfer proteins, hevein-type and knottin-type peptides (Cammue et al. 1995; Broekaert et al. 1997; Sagi 2000). In banana, the particle bombardment technique has enabled the production of several hundreds of transgenic banana lines expressing various genes encoding defensin type antimicrobial peptides (AMPs) and nsLTPs alone or in combination (Sagi et al. 1995).

Chakrabarti et al. (2003) reported that transgenic banana expressing a synthetic substitution analog of magainin, a protein from skin secretions of *Xenopus laevis*, was resistant to two fungal pathogens, *Fusarium oxysporum* f.sp. *cubense* and *Mycosphaerella fijiensis*.

Ribosome-inactivating proteins (RIPs) from plants have been shown to interfere with fungal metabolism (Kim et al. 2003). RIPs inhibit translation in eukaryotic target cells by cleavage of the N-glycosidic bond of a specific adenine in the 28S rRNA subunit (Stirpe et al. 1992). They are divided into two groups: single-chain type 1 and type ll proteins containing two subunits, of which one chain (A lectin) can bind to fungal cell surfaces. The expression of non-plant hydrolytic enzymes (bacterial and fungal chitinases, lysozyme) and antimicrobial peptides (insect cecropins) is another strategy to engineer resistance against plant pathogenic fungi (Sagi 1999; Punja 2001; Emani et al. 2003). In addition, Pei et al. (2005) developed transgenic banana plants expressing human lysozyme, which showed a degree of resistance against *Foc*, race 4.

Resistance gene products served as receptors for pathogen Avr factors to recognize the Avr factors indirectly, which could trigger one or more signal transduction pathways that in turn activate deference responses in the plant to prevent pathogen infection (Staskawicz et al. 1995; Punja 2001). There are several examples of the expression of resistance (R)-genes in transgenic plants. Plant R-genes were reported, which encode proteins with cytoplasmic nucleotide binding sites and leucine-rich repeat (NBS-LRR) structure domains that confer resistance to a wide variety of pathogens, especially *Foc* race 4 (Dang and Jones 2001; Echeverria et al. 2008). Echeverria et al. (2008) reported characterization of candidate disease resistance genes of the nucleotide binding site (NBS) type from banana and correlation of a transcriptional polymorphism with resistance to *Foc* race 4. They used degenerate primers to amplify genomic NBS-LRR from wild banana (*Musa acuminate* ssp. *malaccensis*) plants resistant to the *Foc*, race 4. Subsequently, Echeverria et al. (2008) suggested that RNA interference (RNAi) technology could be used to test the function of multiple banana resistance gene candidates (RGC) by silencing their corresponding targets in disease resistant banana genotypes, which would be able to identify a particular potential R-gene.

Banana plant-pathogen interactions are presumed to generate signals that activate nuclear genes involved in plant defense responses leading to the induction of stress-reacted compounds and enzymes such as hydrogen peroxide, phonelic compounds, phenylalanine ammonia lyase, peroxidase and polyphenoloxidase (De Ascensao and Dubery 2000; Sreeramanan et al. 2005). The accumulation of hydrogen peroxide has been determined to be one of the earliest events that occur at host pathogen recognition and it has been postulated to play an important role in plant defense (Baker and Orlandi 1995; Punja 2001; Sreeramanan et al. 2005). Another activator of defense responses that has been engineered is hydrogen peroxide generated through expression of genes encoding for glucose oxidase (Punja 2001). An elevation of the hydrogen peroxide concentration was observed in potato plants transformed with a glucose oxidase gene from *Aspergillus niger* and resulted in enhanced resistance to late-blight fungus *Phytophthora infestants* (Wu et al. 1995). Interestingly, hydrogen peroxide and phenylalanine ammonia lyase activities were also found to be the most sensitive and responsive compounds to *Foc* (race 1, VCG 01217) after spore inoculation in banana transformed with antifungal genes (Sreeramanan et al. 2005). However, a reduction in hydrogen peroxide concentration also seems to increase disease resistance as constitutive expression of a pathogenesis-related peroxidase (*Shpx6a*) from the tropical forage legume *Stylosanthes humili* in transgenic tobacco and canola lines resulted in a slight, but significant resistance against *Phytophthora parasitica* var. *nicotiane* and *Leptospheria maculans*, respectively (Kazan et al. 1998). In banana, transgenic "Rasthali" plantlets, which contain both chitinase and β-1,3-glucanase genes show higher chitinase, β-1,3-glucanase, hydrogen peroxide, phenylalanine ammonia lyase, peroxidase and polyphenoloxidase enzyme activities compared to the untransformed plantlets after inoculation with the *Foc* race 1 (Sreeramanan et al. 2005).

The progress made in genetic engineering of banana for enhanced resistance to fungal pathogens especially *Foc* is an indication of the importance of fungal pathogens in *Musa*. Although greater progress has been made in banana transformation, identification and introduction of potential antifungal genes into banana to reduce losses caused by *Foc* remains a major challenge.

Abbreviations

AFLP	:	Amplified Fragment Length Polymorphism
AMPs	:	Antimicrobial peptides
Avr factor	:	Avirulence factor
DAF	:	DNA amplification fingerprinting
Foc	:	*Fusarium oxysporum* f.sp.*cubense*
Fol	:	*Fusarium xysporum* f.sp.*lycopersici*
HR	:	Hypersentive response
IGS	:	Intergenic Spacer Region
ITS	:	Internally Transcribed Spacer region
NBS-LRR	:	Nucleotide binding site-Leucine-rich repeat
nit	:	nitrate non- utilizing auxotropic mutants
nsLTPs	:	non-specific lipid transfer proteins
PCR	:	Polymerase Chain Reaction

PR proteins	:	Pathogenesis Related Proteins
R gene	:	Resistance gene
RAPD	:	Random amplified Polymorphic DNA
rDNA	:	Ribosomal DNA
RFLP	:	Restriction Fragment Length Polymorphism
RGC	:	Resistance Gene Candidates
RIPs	:	Ribosome Inactivating Proteins
RNAi	:	RNA interference
SAR	:	Systemic Acquired Resistance
SCAR	:	Sequence Characterized Amplified Region
SINES	:	Short Interspersed Nuclear Elements
SR4	:	Subtropical Race 4
TR4	:	Tropical Race 4
VCG	:	Vegetative Compatibility Grouping
vic	:	Vegetatively incompatible

References

Alves-Santos FM, Garcıa-Sanchez RB, Dıaz-Mınguez JM (2002) A DNA-based procedure for *in planta* detection of *Fusarium oxysporum* f. sp. *phaseoli*. Phytopathology 92: 237–244.

Baker CJ, Orlandi EW (1995) Active oxygen in plant pathogenesis. Annu Rev Phytopathol 33: 299–322.

Bentley S, Bassam BJ (1996) A robust DNA amplification fingerprinting system applied to analysis of genetic variation within *Fusarium oxysporum* f. sp. *cubense*. J Phytopathol 144: 207–213.

Bentley S, Pegg KG, Dale JL (1995) Genetic variation among a world-wide collection of isolates of *Fusarium oxysporum* f. sp. *cubense* analyzed by RAPD-PCR fingerprinting. Mycol Res 99: 1378–1384.

Bentley S, Pegg KG, Moore NY, Davis RD, Buddenhagen IW (1998) Genetic variation among vegetative compatibility groups of *Fusarium oxysporum* f. sp. *cubense* analyzed by DNA fingerprinting. Phytopathology 88: 1283–1293.

Blomme G, Eden-Green S, Mustaffa M, Nwauzoma B, Thangavelu R (2011) Major diseases of banana. In: M Pillay, A Tenkouano (eds) Banana Breeding: Progress and Challenges. CRC Press, Boca Raton, FL, USA, pp 85–119.

Boehm EWA, Ploetz RC, Kistler HC (1994) Statistical analysis of electrophoretic karyotype variation among vegetative compatibility groups of *Fusarium oxysporum* f.sp. *cubense*. Mol Plant-Microbe Interact 7: 196–207.

Bosland PW, Williams PH (1987) An evaluation of *Fusarium oxysporum* from crucifers based on pathogenicity, isozyme polymorphism, vegetative compatibility and geographical origin. Can J Bot 65: 2067–2073.

Brake VM, Pegg KG, Irwin JAG, Langdon PW (1990) Vegetative compatibility groups within Australian populations of *Fusarium oxysporum* f. sp. *cubense*, the cause of Fusarium wilt of bananas. Aust J Agric Res 41: 863–870.

Broekaert WF, Cammue BPA, De Bolle MFC, Thevissen K, De Samblanx GW, Osborn RW (1997) Antimicrobial peptides of plants. Crit Rev Plant Sci 16: 297–323.

Buddenhaggen I (2009) Understanding strain diversity in *Fusarium oxysporum* f.sp. *cubense* and history of introduction of 'Tropical Race 4' to better manage banana production. Acta Hort 828: 193–204.

Cammue BP, Thevissen AK, Hendriks M, Eggermont K, Goderis IJ, Proost P, Van Damme J, Osborn RW, Guerbette F, Kader JC, Broekaert WF (1995) A potent antimicrobial protein from onion (*Allium cepa* L.) seeds showing strong sequence homology to plant lipid transfer protein. Plant Physiol 109: 445–455.

Chakrabarti A, Ganapathi TR, Mukherjee PK, Bapat VA (2003) MSI-99, a magainin analogue, imparts enhanced disease resistance in transgenic tobacco and banana. Planta 216: 587–596.

Correll JC (1991) The relationship between formae speciales, races, and vegetative compatibility groups in *Fusarium oxysporum*. Phytopathology 81: 1061–1064.

Correll JC, Puhalla JE, Schneider RW, Kraft JM (1985) Differentiating races of *Fusarium oxysporum* f.sp. *pisi* based on vegetative compatibility (Abstr.) Phytopathology 75: 1347.

Correll JC, Puhalla JE, Schneider RW (1986) Identification of *Fusarium oxysporum* f.sp. *apii* on the basis of colony size, virulence and vegetative compatibility. Phytopathology, 76: 396–400.

Correll JC, Klittich CJR, Leslie JF (1987) Nitrate non-utilizing mutants of *Fusarium oxysporum* and their use in vegetative compatibility tests. Phytopathology 77: 1640–1646.

Croft JH, Jinks JL (1977) Aspects of the population genetics of *Aspergillus nidulens*. In: JE Smith, JA Pateman (eds) Genetics and Physiology of *Aspergillus*. Academic Press, New York, USA, pp 339–360.

Daniells JW (2009) Global banana disease management–Getting serious with sustainability and Food Security. Acta Hort 828: 411–416.

Daniells J, Davis D, Peterson R, Pegg K (1995) Gold finger: Not as resistant to Sigatoka/yellow Sigatoka as first thought. Infomusa 4–6.

Dang IL, Jones J (2001) Plant pathogens and integrated defence responses to infection. Nature 411: 826–833.

Datta K, Velazhahan R, Oliva N, Mew T, Khush GS, Muthukrishnan S, Datta SK (1999) Over expression of the cloned rice thaumatin-like protein (PR-5) gene in transgenic rice plants enhances environmental friendly resistance to *Rhizoctonia solani* causing sheath blight disease. Theor Appl Genet 98: 1138–1145.

Datta K, Koukoliková-Nicola Z, Baisakh N, Oliva N, Datta SK (2000) *Agrobacterium*-mediated engineering for sheath blight resistance of indica rice cultivars from different ecosystems. Theor Appl Genet 100: 832–839.

Davis RI, Moore NY, Bentley S, Gunua TG, Rahamma A (2000) Further records of *Fusarium oxysporum* f.sp. *cubense* from New Guinea. Aust Plant Pathol 29: 224.

De Ascensao ARDCF, Dubery IA (2000) Panama Disease: Cell wall reinforcement in response to elicitors from *Fusarium oxysporum* f. sp. *cubense* race four. Amer Phytopathol Soc 90: 1173–1180.

Declerck S, Plenchette C, Strullu DG (1995) Mycorrhizal dependency of banana (*Musa acuminate*, AAA group) cultivar. Plant Soil 7: 183–187.

Dempsey DA, Shah J, Klessig DF (1999) Salicylic acid and disease resistance in plants. Crit Rev Plant Sci 18: 547–575.

Dixon RA, Lamb CJ, Masoud S, Sewalt VJH, Paiva NL (1996) Metabolic engineering: prospects for crop improvement through the genetic manipulation of phenylpropanoid biosynthesis and defense responses-a review. Gene 179: 61–71.

Dong X (1998) SA, JA, ethylene and disease resistance in plants. Curr Opin Plant Biol 1: 316–323.

Donofrio NM, Delaney TP (2001) Abnormal callose response phenotype and hyper susceptibility to *Peronospora parasitica* in defense-compromised *Arabidopsis nim 1-1* and salicylate hydroxylase-expressing plants. Mol Plant–Microbe Interact 14: 439–450.

Echeverria SP, Dale JL, Harding RM, Smith MK, Collet C (2008) Characterization of disease resistance gene candidates of the nucleotide binding site (NBS) type from banana and correlation of a transcriptional polymorphism with resistance to *Fusarium oxysporum* f.sp. *cubense* race 4. Mol Breed 22: 565–579.

Emani C, Garcia JM, Lopata-Finch E, Pozo MJ, Uribe P, Kim DJ, Sunilkumar G, Cook DR, Kenerley CM, Rathore KS (2003) Enhanced fungal resistance in transgenic cotton expressing an endochitinase gene from *Trichoderma virens* (2003). Plant Biotechnol J 1: 321–326.

Fungaro MHP, Vissotto PC, Sartori D, Vilas-Boas LA., Furlaneto MC, Taniwaki MH (2004) A molecular method for detection of *Aspergillus carbonarius* in coffee beans. Curr Microbiol 49:123–127.

Gorden TR, Correll JC, McCain AH (1986) Host specificity and vegetative compatibility in *Verticillium alboatrum* (Abstr.). Phytopathology 76: 1111.

Govrin EM, Levine A (2000) The hypersensitive response facilitates plant infection by the necrotrophic pathogen *Botrytis cinerea*. Curr Biol 10: 751–757.

Groenewald S, Van den berg N, Marasas WFO, Viljoen A (2006) The application of high-throughput AFLP's in assessing genetic diversity in *Fusarium oxysporum* f. sp. *cubense*. Mycol Res 110: 297–305.

Jabs T, Slusarenko AJ (2000) The hypersensitive response In: AJ. Slusarenko, RSS Fraser, LC van Loon. (eds) Mechanisms of Resistance to Plant Diseases. Kluwer Academic Publishers, Dordrecht, The Netherlands, pp 279–323.

Jach G, Görnhardt B, Mundy J, Logemann J, Pinsdorf E, Leah R, Schell J, Mass C (1995) Enhanced quantitative resistance against fungal disease by combinatorial expression of different barley antifungal proteins in transgenic tobacco. Plant J 8: 97–109.

Jones DR (1999) Introduction to banana, abaca, and enset. In: DR Jones (ed) Diseases of Banana, Abaca, and Enset. CABI Publishing, Wellingford, UK, pp 1–36.

Jones DR (2000) Diseases of Banana, Abacá and Enset. CABI Publishing, Wallingford, UK.

Jongedijk E, Tigelaar H, Van Roekel JSC, Bres-Vloemans SA, Dekker I, Van den Elzen PJM, Cornelissen BJC, Melchers LS (1995) Synergistic activity of chitinases and β-1, 3-glucanases enhances fungal resistance in transgenic tomato plants. Euphytica 85: 173–180.

Jurado M, Vázquez C, Marín S, Sanchis V, González-Jaéna MT (2006) PCR-based strategy to detect contamination with mycotoxigenic *Fusarium* species in maize. Syst Appl Microbiol 29: 681–689.

Kazan K, Goulter KC. Way HM, Manners JM (1998) Expression of a pathogenesis-related peroxidase of *Stylosanthes humilis* in transgenic tobacco and canola and its effect on disease development. Plant Sci 136: 207–217.

Kim JK, Jang IC, Wu R, Zuo WN, Boston RS, Lee YH, Ahn IP, Nahm BH (2003) Co-expression of a modified maize ribosome-inactivating protein and a basic chitinase gene in transgenic rice plants confers enhanced resistance to sheath blight. Transgen Res 12: 475–484.

Kistler HC, Miao V (1992) New modes of genetic change in filamentous fungi. Annu Rev Phytopathol 30: 131–152.

Koenig RL, Ploetz, RC, Kistler HC (1997) *Fusarium oxysporum* f.sp. *cubense* consists of a small number of divergent and globally distributed clonal lineages. Amer Phytopathol Soc 87: 915–923.

Koike M, Watanabe M, Nagao H, Ohshima S (1995) Molecular analysis of Japanese isolates of *Verticillium dahliae* and *V. alboatrum*. Lett Appl Microbiol 21: 75–78.

Lamb R, Dixon RA (1997) The oxidative burst in disease resistance. Annu Rev Plant Physiol & Plant Mol Biol 48: 251–275.

Lee MW, Qia M, Yang Y (2001) A novel jasmonic acid inducible rice *myb* gene associates with fungal infection and host cell death. Mol Plant–Microbe Interact 14: 527–535.

Liang H, Maynard CA, Allen RD, Powell WA (2001) Increased *Septoria musiva* resistance in transgenic hybrid poplar leaves expressing a wheat oxalate oxidase gene. Plant Mol Biol 45: 619–629.

Lin YH, Chang JY, Liu ET, Chao CP, Huang JW, Chang PFL (2009) Development of a molecular marker for specific detection of *Fusarium oxysporum* f.sp. *cubense* race 4. Eur J Plant Pathol 123: 353–365.

Maziah M, Sreeramanan S, Puad A, Sariah M (2007) Production of transgenic banana cultivar, Rasthali (AAB) via *Agrobacterium*-mediated transformation with a rice chitinase gene. J Plant Sci 2: 504–517.

Melchers LS, Stuiver MH (2000) Novel genes for disease resistance breeding. Curr Opin Plant Biol 3: 147–152.

Melchers LS, Sela-Buurlage MB, Vloemans SA, Woloshuk CP, Van Roekel JSC, Pen J, Van den Elzen PJM, Cornelissen BJC (1993) Extracellular targeting of the vacuolar tobacco proteins AP24, chitinase and β-1,3-glucanase in transgenic plants. Plant Mol Bio 21: 583–593.

Mes JJ, Weststeijn EA, Herlaar F, Lambalk JJM, Wijbrandi J, Haring MA (1999) Biological and molecular characterization of *Fusarium oxysporum* f. sp. *lycopersici* divides race 1 isolates into separate virulence groups. Phytopathology 89: 156–160.

Miao VPW (1990) Using karyotype variability to investigate the origins and relatedness of isolates of *Fusarium oxysporum* f.sp. *cubense*. In: RC Ploetz (ed) *Fusarium* Wilt of Banana. American Phytopathological Society, St Paul, MN, USA, pp 55–62.

Miller SA, Martin RR (1988) Molecular diagnosis of plant disease. Annu Rev Phytopathol 26: 409–432.

Moore NY (1994) *Fusarium* wilt of banana: pathogen variability and host pathogen interaction. Ph.D. thesis, University of Queensland, Australia.

Moore NY, Hargreaves PA, Pegg KG, Irwin JAG (1991) Characterization of strains of *Fusarium oxysporum* f.sp *cubense* by production of volatiles. Aust J Bot 39: 161–166.

Moore NY, Pegg KG, Allen RN, Irwin JAG (1993) Vegetative compatibility and distribution of *Fusarium oxysporum* f. sp. *cubense* in Australia. Aust J Exp Agric 33: 797–802.

Moore NY, Bentley S, Pegg KG, Jones DR (1995) *Fusarium* wilt of banana. *Musa* disease fact sheet. INBAP, Montpellier, France, 5: 4

Moore N, Pegg KG, Buddenhagen IW, Bentley S (2001) *Fusarium* wilt of banana: A diverse clonal pathogen of a domesticated clonal host. In: BA Summerell, JF Leslie, D Backhouse, WL Bryden, L Burgess (eds) Paul E Nelson Memorial Symposium. *Fusarium*. The American Phytopathological Society, St. Paul, MN, USA, pp 212–224.

O'Donnell K, Kistler HC, Cigelnik E, Ploetz RC (1998) Multiple evolutionary origins of the fungus causing Panama disease of banana: Concordant evidence from nuclear and mitochondrial gene genealogies. Proc Nat Acad Sci USA, 95: 2044–2049.

Pegg KG, Moore NY, Sorensen S (1993) Fusarium wilt in the Asian Pacific region. In: RV Valmayor, SC Hwang, R Ploetz, SW Lee, NV Roa (eds) Proceedings of International Symposium on Recent Developments in Banana Cultivation Technology (INIBAP/ ASPNET: Los Bafios, Laguna, the Philippines, pp 255–269.

Pegg KG, Moore NY, Sorenson S (1994) Variability in populations of *Fusarium oxysporum* f.sp. *cubense* from the Asia/Pacific region. In: DR Jones (ed) The Improvement and Testing of Musa: A Global Partnership. Proceeding of the First Global Conference of the International Musa Testing Program, FHIA, Honduras. INIBAP, Montpellier, France, pp 70–82.

Pegg KG, Shivas RG, Moore NY, Bentley S (1995) Characterization of a unique population of *Fusarium oxysporum* f. sp. *cubense* causing *Fusarium* wilt in Cavendish bananas at Carnarvon, Western Australia. Aust J Agric Res 46: 167–178.

Pegg KG, Moore NY, Bentley S (1996) *Fusarium* wilt of banana in Australia: a review. Aust J Agric Res 47: 637–650.

Pei X, Chen S, Wen R, Shang YE, Huang J, Zhang Y, Wang B, Wang Z, Jia S (2005) Creation of Transgenic Bananas Expressing Human Lysozyme Gene for Panama Wilt Resistance. J Int Plant Biol 47: 971–977.

Pink D, Puddephat I (1999) Development of disease resistance genes by plant transformation —a 'mix and match' approach. Trends Plant Sci 4: 71–75.

Ploetz RC (1990) *Fusarium* Wilt of Banana. American Phytopathological Society, St. Paul, MN, USA.

Ploetz, RC (1993) *Fusarium* wilt (Panama disease) in Africa: current status and outlook for smallholder agriculture. In: CS Gold, B Gemmill (eds) Biological and Integrated Control of Highland Banana and Plantain Pests and Diseases. IITA, Ibadan, Nigeria, pp 312–323.

Ploetz RC (1994) Panama disease: return of the first banana menace. Int J Pest Manag 40: 326–336.
Ploetz RC (2000) Panama disease: A classic and destructive disease of banana. Online. Plant Health Progress.
Ploetz R (2004) Diseases and pests: A review of their importance and management. Infomusa 13: 11–16.
Ploetz RC (2005) Panama disease, an old enemy rears its ugly head: parts 1 and 2. In: Plant Health Progress, APSnet: Online doi: 10.1094/PHP-2005-1221-01-RV.
Ploetz RC, Correll JC (1988) Vegetative compatibility among races of *Fusarium oxysporum* f. sp. *cubense*. Plant Dis 72: 325–328.
Ploetz RC, Pegg KG (1997) *Fusarium* wilt of banana and Wallace's line: was the disease originally restricted to his Indo-Malayan region? Australasian Plant Pathol 26: 239–249.
Ploetz RC, Pegg KG (2000) Fusarium wilt. In: DR Jones (ed) Diseases of Banana, Abacá and Enset. CABI Publishing, Wallingford, UK, pp 143–159.
Ploetz RC, Herbert J, Sebasigari K, Hernandez JH, Pegg KG, Ventura JA, Mayato LS (1990) Importance of *Fusarium* wilt in different banana-growing regions. In: RC Ploetz (ed) *Fusarium* Wilt of Banana. American Phytopathol Soc. St Paul, MN, USA, pp 9–26.
Puhalla JE (1985) Classification of strains of *Fusarium oxysporum* on the basis of vegetative compatibility. Can J Bot 63: 179–183.
Punja ZK (2001) Genetic engineering of plants to enhance resistance to fungal pathogens-a review of progress and future prospects. Can J Plant Pathol 23: 216–235.
Rishbeth J (1955) *Fusarium* wilt of bananas in Jamaica. Some observations on the epidemiology of the disease. Ann Bot 19: 293–329.
Sagi L (2000) Genetic engineering on banana-future possibilities. In: D Jones (ed) Diseases of Banana, Plantain, Abaca and Enset. CABI, Wallingford, UK, pp 465–515.
Sagi L, Panis B, Remy S, Schoofs H, De Smet K, Swennen R, Cammue BPA (1995) Genetic transformation of banana and plantain (*Musa* spp.) via particle bombardment. Bio/Technology 13: 481–485.
Schilling AG, Möller EM, Geiger HH (1996) Polymerase chain reaction-based assays for species-specific detection of *Fusarium culmorum, F. graminearum* and *F. avenaceum*. Phytopathology 86: 515–523.
Shivas RG, Phileman E (1996) First record of *Fusarium oxyporum* f.sp. *cubense* on banana in Papua New Guinea. Australia Plant Pathol 25: 260.
Simmonds NW, Shepherd K (1955) Taxonomy and origins of cultivated bananas. J Linn Soc Bot (London) 55: 302–312.
Sorensen S, Pegg KG, Dale JL (1993) RAPD-PCR analysis of genetic variation within Australian populations of *Fusarium oxysporum* f.sp. *cubense* In: RV Valmayor, SC Hwang, R Ploetz, SW Lee, VN Rao (eds) Proc. Int Symp Recent Dev Banana Cultivation Technol. INIBAP-ASPNET, Los Banos, Laguna, the Philippines, pp 285–295.
Sreeramanan S, Maziah M, Sariah M, Abdullah MP, Xavier R (2005) Bioassay method for testing *Fusarium* wilt disease tolerance in transgenic banana. Sci Hort 108: 378–389.
Sreeramanan S, Maziah M, Xavier R (2008) Transfer of β-1,3-glucanase gene into banana for tolerance to *Fusarium* wilt race 1 disease using *Agrobacterium*-mediated transformation system. Transgen Plant J 2: 176–185.
Sreeramanan S, Maziah M, Sariah M, Xavier R (2010) Genetic engineering for tolerance to *Fusarium* wilt race 1 in *Musa sapientum* cv Rastali (AAB) using biolistic gun transformation system. In: L Tripathi (ed) Tree and Forestry Science and Biotechnology, Global Science Books, Japan, pp 65–75.
Staskawicz BJ, Ausubel FM, Baker BJ, Ellis JG, Jones JDG (1995) Molecular genetics of plant disease resistance. Science 268: 661–667.
Stirpe F, Barbieri L, Battelli LG, Soria M, Lappi D (1992) Ribosome-inactivating proteins from plants: present status and future prospects. Bio/Technology 10: 405–412.
Stover RH (1962) Fusarial Wilt (Panama Disease) of Bananas and Other *Musa* Species. Commonwealth Mycological Institute, Kew, England.

Su HJ, Hwang SC, Ko WH (1986) Fusarial wilt of Cavendish bananas in Taiwan. Plant Dis 70: 814–818.
Thangavelu R (2008) Banana wilt. Final Technical Report of Network project on wilt of crops with special reference to cultural morphological molecular characterization and pathogenic variability of isolates in India. Indian Institute of Pulses Research, Kanpur, India, pp 266–285.
Thangavelu R, Mustaffa MM (2010) First report on the occurrence of a virulent strain of *Fusarium* wilt pathogen (race 1) infecting Cavendish (AAA) group of Bananas in India. Plant Dis 94: 1379.
Thangavelu R, Sundararaju P, Sathiamoorthy S, Raghuchander T, Velazhahan R, Nakkeeran S, Palaniswami A (2001) Status of *Fusarium* wilt of banana in India. In: AB Molina, NH Nik Masdek, KW Liew (eds) Banana Fusarium Wilt Management: Towards Sustainable Cultivation. INIBAP-ASPNET, Los Banos, Laguna, the Philippines, pp 58–63.
Vavilow NI (1922) The law of holmologous series in variation. J Genet 12: 47.
Verberne MC, Verpoorte R, Bol JF, Mercado-Blanco J, Linthorst HJM (2000) Overproduction of salicylic acid in plants by bacterial transgenes enhances pathogen resistance. Nat Biotechnol 18: 779–783.
Vierheilig H, Alt M, Lange J, Gut-Rella M, Wiemken A, Boller T (1995) Colonization of transgenic tobacco constitutively expressing pathogenesis-related proteins by the vesicular arbuscular mycorrhizal fungus *Glomus mosseae*. Appl Environ Microbiol 61: 3031–3034.
Waite BH (1963) Wilt of *Heliconia* spp. caused by *Fusarium oxysporum* f. sp. *cubense* race 3. Trop Agric (Trinidad) 40: 299–305.
Wardlaw CW (1961) Banana Diseases, including Plantains and Abaca. Longmans, Green and Co. Ltd, London, UK.
Wu G, Shortt BJ, Lawrence EB, Levine EB, Fitzsimmons KC, Shah DM (1995) Disease resistance conferred by expression of a gene encoding H_2O_2-generating glucose oxidase in transgenic potato plants. Plant Cell 7: 1357–1368.
Yergeau E, Filion M, Vujanovic V, St-Arnaud M (2005) A PCR-denaturing gradient gel electrophoresis approach to assess *Fusarium* diversity in asparagus. J Microbiol Meth 60: 143–154.
Ying-Hong LIN, Kan-ShuCHEN, Tsung-DaoLIOU, Jenn-Wen HUANG and Pi-Fang Linda CHANG (2008) Development of a molecular method for rapid differentiation of watermelon lines resistant to Fusarium oxysporum f. sp. niveum. Botanical Studies 50: 273–280.
Yoshikawa M, Tsuda M, Takeuchi Y (1993) Resistance to fungal diseases in transgenic tobacco plants expressing the phytoalexin elicitor-releasing factor, β-1, 3-endoglucanase, from soybean. Naturwissenschaften 80: 417–420.

14

Disease Resistance-Related Genes in Banana: the Cases of *NBS*, *Pto* and *NPR1* Genes

Santy Peraza-Echeverria,[1,a,*] *Virginia Aurora Herrera-Valencia*[1,b] and *Andrew James*[1,c]

ABSTRACT

Knowledge of molecular and physiological basis of resistance to different biotic stresses, such as plant diseases caused by viruses, bacteria, fungi and oomycetes, has gained by map-based cloning, experimental physiology and genetic engineering approaches in model, tractable plants, such as *Arabidopsis* and tomato. This knowledge has allowed the identification and cloning of homologues of key regulators of disease resistance in intractable perennial and partially sterile important crop plants; in this case, banana. In this chapter we provide an overview of the cloning and molecular characterization of two classes of disease resistance (R) gene homologues in banana, the *NBS* (nucleotide-binding site) and the *Pto* (serine/threonine kinase) classes of R genes. Furthermore, we give insights about the recent cloning and characterization of three banana genes showing homology to the non-expresssor of pathogenesis related gene 1 (*NPR1*), which is a master transcriptional regulator of systemic acquired resistance (SAR). Finally, we describe the potential applications of these genes to develop disease resistance in this crop.

[1]Unidad de Biotecnología, Centro de Investigación Científica de Yucatán, Calle 43 No. 130 Colonia Chuburná de Hidalgo, Mérida, Yucatán, México. C.P. 97200.
[a]e-mail: *santype@cicy.mx*
[b]e-mail: *vicky@cicy.mx*
[c]e-mail: *andyj007@cicy.mx*
*Corresponding author

List of abbreviations after the text.

14.1 Introduction

Banana is a staple food in many developing countries and an important export commodity for numerous farmers in the tropics. Several diseases threaten world banana production causing significant yield losses every year and, among these, fungal diseases are of major concern. The foliar fungal disease black Sigatoka, the causal agent of which is the airborne fungus *Mycosphaerella fijiensis*, is considered the most economically important leaf disease of banana (Churchill 2010). For example, in southern Mexico and throughout Central America it has been estimated that the cost of pesticide sprays to control *M. fijiensis* accounts for up to 30% of production costs (Agrios 2005). The intensive fungicide applications are not only expensive for many farmers but also represent a serious health risk to plantation workers and threaten the environment. Another devastating fungal disease is Panama disease caused by the soil-borne fungus *Fusarium oxysporum* f. sp. *cubense* (*FOC*). *FOC* race 1 wiped out the banana industry in the 1950s, when this industry was based on the cultivar "Gros Michel". Now a new race of this pathogen, *FOC* race 4, has emerged as a serious threat to the current commercial banana cultivars of the Cavendish subgroup, which are resistant to *FOC* race 1 but susceptible to *FOC* race 4 (Ploetz and Pegg 2000; Ploetz 2005). Sources of resistance have been found in wild bananas for these two fungal diseases, however the molecular nature of the resistance is currently unknown. Therefore, knowledge about the genes involved in disease resistance in banana is crucial for the genetic improvement and sustainable production of this crop.

Plants have evolved sophisticated strategies to perceive a vast array of potential harmful microbes and to translate this perception into an effective immune response. The primary immune response recognizes common features of microbial pathogens, such as flagellin, chitin, glycoproteins and lipopolysaccharides (Chisholm et al. 2006; Göhre and Robatzek 2008). These microbial molecules are known as pathogen-associated molecular patterns (PAMPs), which are recognized by transmembrane pattern-recognition receptors (PRR) from the host plant. Perception of microorganisms at the cell surface by PR proteins initiates PAMP triggered immunity (PTI), which is a potent basal defense response that usually halts pathogen attack, resulting in non-host resistance. However, many pathogens can overcome PTI by delivering virulence factors (effectors), which suppress basal defense signaling into plant cells. As a counter defensive strategy to PTI suppression, plants deploy resistance (R) proteins that recognize a given effector either directly or indirectly and activate effector-triggered immunity (ETI) resulting in disease resistance (Chisholm et al. 2006). ETI often culminates in a form of programmed cell death at the site of attack, which constrains further spread of the pathogen. This type of cell suicide is also called hypersensitive response (HR). This local response often triggers non-specific resistance throughout the plant, a phenomenon known as systemic acquired resistance (SAR) (Durrant and Dong 2004). Once triggered, SAR provides resistance to a wide range of pathogens. One key regulator of SAR is the *NPR1* (non-expressor of pathogenesis-related gene 1) gene, which encodes a transcriptional regulator (Pieterse and Van Loon 2004). Overexpression of *NPR1* enhances disease resistance levels against a broad variety of pathogens (Cao et al. 1998; Friedrich et al. 2001; Lin et al. 2004; Makandar et al. 2006). Importantly, this resistance is not usually associated with any adverse plant phenotypes.

ETI-mediated resistance has several attractive features for disease control in the field. When induced in a timely manner, the concerted responses can efficiently stop pathogen growth with no co-lateral damage to the plant. No input is required from the farmer and there are no adverse environmental effects. Consequently, ETI-mediated resistance is one of the top priorities in plant breeding and genetic improvement programs. For effective utilization of ETI genes, it is important to clone and characterize them. Several disease resistance-related genes involved in the ETI responses have been cloned from the model plant *Arabidopsis* and numerous crops using a map-based cloning approach (Hammond-Kosack and Parker 2003; Gurr and Rushton 2005; Van Ooijen et al. 2007). However, in the case of banana, a map-based approach has been difficult up to now, because of chromosomal translocations and the small size of segregating populations. Recently it was possible to isolate R-like or *NPR1*-like genes based on sequence homology and degenerate PCR (Pei et al. 2007; Peraza-Echeverria et al. 2007, 2008, 2009; Azhar et al. 2008; Endah et al. 2008; Miller et al. 2008; Zhao et al. 2009). The availability of these sequences has paved the way for further functional analysis and potential application. Regarding this last point, molecular biotechnology offers a broad range of strategies to engineer disease resistance by transferring single or multiple disease resistance-related genes into current banana cultivars using genetic transformation. Overall, genetic improvement of plants via molecular biotechnology has several advantages over traditional breeding strategies, such as overcoming the species barrier and sterility of clonally propagated plants (e.g., banana), the ability to eliminate unwanted genetically linked traits (genetic drag) and also to rapidly transfer genes into commercially elite cultivars. This chapter describes recent advances in cloning disease resistance-related genes in banana and different molecular biotechnology strategies to develop disease resistance in this crop.

14.2 NBS-Resistance Gene Candidates

Most characterized plant R-genes are predicted to encode proteins with cytoplasmic nucleotide-binding site and leucine-rich repeat (NBS-LRR) domains that confer resistance to a wide variety of pathogens and pests including viruses, bacteria, fungi, nematodes and also insects (Dangl and Jones 2001). NBS-LRR genes are abundant in plant genomes with 149 and 480 isolated from *Arabidopsis* and rice, respectively (Meyers et al. 2003; Zhou et al. 2004), mostly organized in clusters (Hulbert et al. 2001). The NBS-LRR proteins are believed to recognize pathogens and respond by activating signal transduction pathways leading to disease resistance (Belkhadir et al. 2004). The NBS-LRR class can be divided in two subclasses, the TIR and the non-TIR, depending on the presence of a domain at the N-terminus with homology to the *Drosophila* Toll and mammalian Interleukin-1 receptors (TIR) (Meyers et al. 1999). Non-TIR-NBS-LRR genes are present in both monocotyledonous (monocot) and dicotyledonous (dicot) plants whereas TIR-NBS-LRR genes appear to be restricted to dicot plants (Meyers et al. 2003; Zhou et al. 2004). The NBS domain comprises three motifs predicted to bind ATP or GTP, and several conserved motifs whose functions are unknown. This region has homology to two activators of apoptosis in animal cells: APAF-1 and CED. By analogy to these well characterized regulators of programmed cell death, the corresponding domain in NBS-LRR proteins might operate as an intramolecular signal transducer (Van der Biezen and Jones 1998). Biochemical evidence has revealed that the NBS domain of the tomato Mi-1 and I2 non-TIR-NBS-LRR resistance proteins can bind and hydrolase ATP *in vitro* (Tameling et al. 2002), which reinforces the idea that the NBS domain functions in signal transduction. The TIR and non-TIR subclasses of NBS-LRR genes can also be distinguished by motifs found within the NBS domain or by a single amino acid residue in the final portion of the NBS kinase-2 motif, which in most cases is an aspartic acid for the TIR subclass and a tryptophan for the non-TIR subclass (Meyers et al. 1999). PCR with degenerate primers targeting the conserved GVGKTT (P-loop), GSRRIIITTRD or GLPLA motifs of the NBS domain of R-genes has resulted in the isolation of numerous NBS-containing genes from a variety of plant species including banana (Pei et al. 2007; Azhar et al. 2008; Miller et al. 2008; Peraza-Echeverria et al. 2008). The R-gene homologs isolated by this PCR approach have been called resistance gene candidates (RGC).

In the case of banana, several NBS-RGC genes have been isolated from both wild and cultivated bananas. Several features of the banana NBS-RGC sequences suggest they belong to the non-TIR-NBS-LRR subclass of disease resistance genes. For example, the characteristic motifs of the NBS domain of known resistance genes (Meyers et al. 1999; Pan et al. 2000) are present in each banana RGC at similar positions (Fig. 14-1).

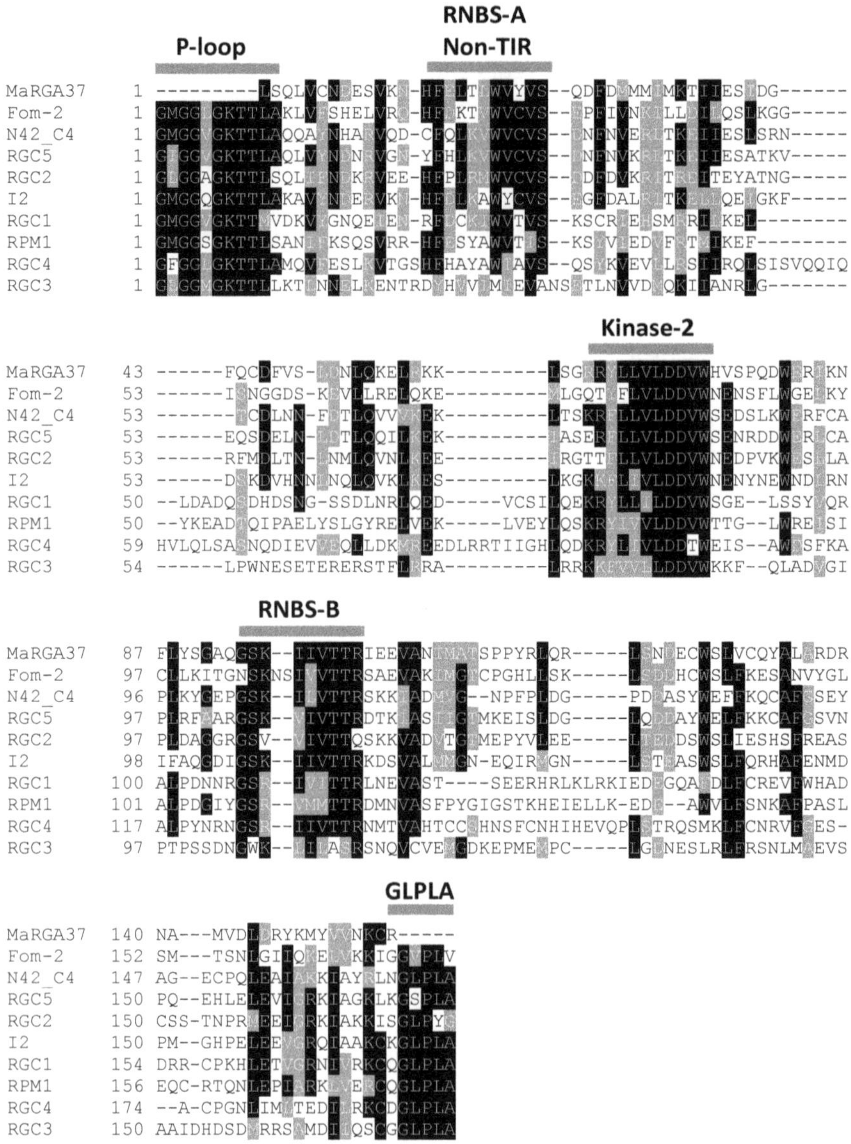

Figure 14-1 ClustalX alignment of the predicted amino acid sequences of some banana NBS-RGCs, RGC1-RGC5 (EU239819-EU239823) (Peraza-Echeverria et al. 2008), MaRGA37 (ER936009) (Miller et al. 2008), N42-C4 (AM931310) (Azhar and Heslop-Harrison 2008), and three R-proteins, I2 (AAD27815), Fom-2 (AAS80152) and RPM1 (A57072) that confer resistance to *Fusarium oxysporum* (I2 and Fom-2) and *Pseudomonas syringae*, respectively. Conserved NBS motifs are indicated above the alignment, while identical amino acids are shaded in black and conservative substitutions are shaded in grey.

One of these motifs, the highly conserved P-loop, has been shown to bind ATP in the NBS-LRR resistance proteins I2 and Mi from tomato (Tameling et al. 2002), suggesting that the banana NBS-RGC proteins may also bind ATP. The non-TIR (nT) motif (Bai et al. 2002), which is associated only with the non-TIR subclass of NBS sequences, is found in the N-terminal region of the banana NBS-RGC sequences (Peraza-Echeverria et al. 2008) and none of the motifs associated with the TIR subclass are found in the corresponding region of the banana NBS-RGC proteins. Moreover, the banana NBS-RGC genes, RGC1, RGC2 and RGC5 showed the presence of a putative coiled-coil (CC) structure in the non-TIR domain, which is another common feature of this region (Peraza- Echeverria et al. 2008). In the rice genome, for example, 174 of the 535 NBS sequences contained a CC motif in the non-TIR domain (Zhou et al. 2004). The finding of non-TIR-NBS-LRR genes in banana is consistent with the structure of R-genes isolated from other monocots where the TIR domain appears to be absent (Meyers et al. 1999; Pan et al. 2000; Bai et al. 2002; Zhou et al. 2004). The non-TIR-NBS-LRR subclass is widespread in plant genomes, with the *Arabidopsis thaliana*, *Vitis vinifera*, *Populus trichocarpa* and *Oryza sativa* genomes containing 55, 196, 233, and 480 non-TIR-NBS-LRR related sequences, respectively (Meyers et al. 2003; Zhou et al. 2004; Tuskan et al. 2006; Velasco et al. 2007). By analogy, this type of sequence is likely to be abundant in the banana genome. Based on an analysis of R-genes isolated from several dicot and monocot species, it was found that the absence of the TIR domain in R-genes could be predicted by the presence of the motif RNBS-A-non-TIR near to the P-loop, and also by the presence of a tryptophan residue (W) at the end of the kinase-2 motif (Meyers et al. 1999). This prediction also applies to the banana NBS-RGC sequences isolated so far (Fig. 14-1) (Pei et al. 2007; Azhar et al. 2008; Miller et al. 2008; Peraza-Echeverria et al. 2008).

To date, the TIR domain has not been found in the structure of monocot NBS-LRR R-genes even in the complete rice genome sequence (Bai et al. 2002; Cannon et al. 2002; Zhou et al. 2004). Phylogenetic analysis supports the classification of the banana NBS-RGC sequences into the non-TIR subclass since they all cluster with other NBS sequences of the non-TIR subclass (Pei et al. 2007; Azhar et al. 2008; Peraza-Echeverria et al. 2008; Miller et al. 2008). It has been hypothesized that the loss of the TIR domain from the NBS-LRR genes in monocot plants may have occurred during the divergence of the monocots and dicots (Pan et al. 2000). Since banana shares a common evolutionary origin with other monocot plants, it is likely that this domain is also absent from the structure of all banana R-genes. Therefore, the fact that only NBS sequences of the non-TIR subclass were isolated in these previous studies is unlikely to be an artifact of the degenerate PCR approach used for the amplification process but instead resulted from the absence of the TIR domain in the Musaceae family. Indeed, the sequencing of the banana genome, which is currently in progress, will shed light on this matter (www.musagenomics.org).

NBS-RGC sequences have also been used to screen bacterial artificial chromosome (BAC) libraries derived from the wild species *M. acuminata* Calcutta IV (AA), *M. balbisiana* "Pisang Klutuk Wulung" (PKW) (BB) and the commercial triploid *M. acuminata* "Grand Naine" (AAA) (Miller et al. 2008). A banana NBS-RGC probe known as MaRGA08 occurred as both a single copy and as multiple copies in validated BACs in the three banana genotypes. In the case of *M. acuminata* Calcutta IV, the BAC clones harbored mainly single-copy NBS-RGC sequences, in contrast to "Grand Naine" and PKW, where BACs contained up to nine and 11 copies, respectively, indicating a clustering pattern of the banana NBS-RGC genes. This finding is consistent with the clustering pattern of NBS-RGC genes from other plants (Hulbert et al. 2001).

14.3 *Pto*-Resistance Gene Candidates

Pto (resistance to *Pseudomonas syringae* pathovar *tomato*) was the first plant R-gene cloned that participates in a gene-for-gene interaction with a pathogen (Martin et al. 1993). It encodes an intracellular serine/threonine protein kinase (STK) of 321 amino acids and it is one of the best-characterized R-genes to date (Oh and Martin 2010). The Pto kinase directly interacts with either of the two pathogen type III effector proteins AvrPto or AvrPtoB (Martin et al. 1993; Kim et al. 2002). In concert with Prf, a NBS-LRR protein, Pto then activates ETI to halt pathogen growth. Overexpression of *Pto* in tomato, under the control of the strong cauliflower mosaic virus (CaMV) 35S promoter, triggers defense responses in the absence of pathogen attack. *Pto*-overexpressing plants showed a resistant phenotype against the bacterium pathogen *Xanthomonas campestris* pv. *vesicatoria* and also to the fungal pathogen *Cladosporium fulvum* (Tang et al. 1999). Hence, the tomato *Pto* gene or its homologs from other plants could be used for engineering broad-spectrum pathogen resistance in other crops. *Pto* resistance gene candidates have been isolated from bean, grapevine, banana and chestnut rose using PCR with degenerate primers targeting the highly conserved subdomains of STK proteins (Vallad et al. 2001; Di Gaspero and Cipriani 2003; Peraza-Echeverria et al. 2007; Xu and Deng 2010). In the case of banana, seven distinctive classes of Pto-RGCs from genomic DNA of the cultivar "Tuu Gia" have been reported (Peraza-Echeverria et al. 2007). These sequences were called Tg-4, Tg-6, Tg-9, Tg-10, Tg-12, Tg-13 and Tg-67 with GenBank accession numbers EF492518, EF492519, EF492520, EF492521, EF492522, EF492523 and EF492524, respectively. The level of identity between the predicted amino acid sequence of Pto-RGCs and the corresponding region of the tomato Pto protein was high and ranged from

58% (Tg-67) to 68% (Tg-13), whereas amino acid identity among the Pto-RGCs ranged from 65.3% (Tg-10 vs Tg-12) to 84.1% (Tg-4 vs Tg-13).

The complete genome sequence of *Arabidopsis* (genome size of 125 Mbp) revealed the presence of 15 Pto-RGCs (Arabidopsis Genome Initiative 2000), while a draft of the rice genome sequence (genome size of 420 Mbp) revealed a similar number of Pto-RGCs with 14 (Goff et al. 2002). These findings indicate that the number of Pto-RGCs in these two plant genomes is conserved. They also indicate that the number of Pto-RGCs in a plant genome is small in comparison to the NBS-LRR class of R-genes, which has a large number of divergent genes in the *Arabidopsis* and rice genomes, with 149 and 480 genes, respectively (Meyers et al. 2003; Zhou et al. 2004). The genome size of *M. acuminata* is estimated at 600 Mbp (Dolezel et al. 1994); assuming that the number of Pto-RGCs in a plant genome does not increase significantly according to the genome size, it is then possible that in banana the number of Pto-RGCs could be similar to that in *Arabidopsis* or rice. Indeed, the complete sequence of the banana genome will provide comprehensive data on the number of Pto-RGC sequences in this Musaceae species (http://www.musagenomics.org).

Alignment of banana Pto-RGC and tomato Pto predicted proteins showed that several regions of the Pto protein such as the STK subdomains (I-IX) and the activation domain, which is involved in the specific binding of AvrPto are highly conserved in the banana Pto-RGCs (Frederick et al. 1998) (Fig. 14-2A).

Moreover, three autophosphorylation sites in the activation domain of Pto (Sessa et al. 2000) are conserved in the corresponding region of all banana Pto-RGCs (Peraza-Echeverria et al. 2007). Of the three Pto autophosphorylation sites (Thr195, Ser198 and Thr199) conserved in the banana Pto-RGCs, Ser198 is required for the AvrPto-Pto-mediated hypersensitive response (Sessa et al. 2000) and it is present in the majority of banana Pto-RGCs with the exception of Tg-12, in which the serine has been replaced by threonine, thereby providing an alternative phosphorylation site. Two other Pto residues, Thr204 and Tyr207, are not autophosphorylated *in vitro* and their mutant versions $pto^{Thr204Ala}$ or $pto^{Tyr207Ala}$ do not interact with AvrPto in yeast (Pedley and Martin 2003). Interestingly, transient expression of the mutants $pto^{Thr204Asp}$ and $pto^{Tyr207Asp}$ in leaf tissue of *Nicotiana benthamiana* produces a cell death response in the absence of AvrPto that resembles a hypersensitive response (Rathjen et al. 1999). The cloning of the full cDNA sequence of Pto-RGCs will allow the possibility to create, through site-directed mutagenesis, mutant versions of the banana Pto-RGCs. These mutants will help in answering some fundamental questions regarding for example, whether the Pto-RGC encoded proteins are autophosphorylated *in vitro*, and whether substitution of tyrosine by aspartate in the corresponding site of Pto (Tyr207) will lead to a HR-like response. All banana Pto-RGC

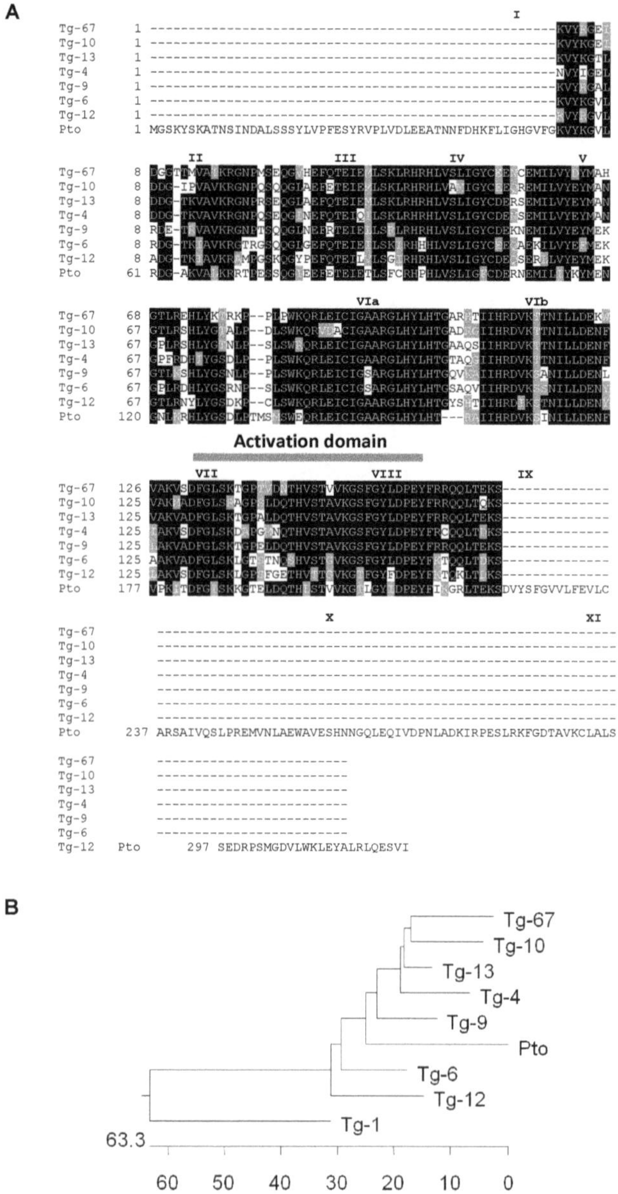

Figure 14-2 Pto-RGCs from the banana cultivar Tuu Gia. (A) ClustalX alignment of the predicted amino acid sequences of banana Pto-RGCs (Peraza-Echeverria et al. 2007) and the full length sequence of the tomato Pto resistance protein (A49332). The subdomains and activation domain of protein kinases (Hanks and Quinn 1991) are indicated. Identical amino acids are shaded in black and conservative substitutions are shaded in gray. (B) Phylogenetic tree of the predicted amino acid sequences of banana Pto-RGCs, Pto R protein and a banana wall-associated receptor-like kinase (Tg-1) (EF492525).

products displayed conserved serine-threonine kinase subdomains (Hanks and Quinn 1991), suggesting that the uncovered genes are likely to encode active kinases. Moreover, most residues of the Pto activation domain involved in pathogen recognition and HR induction (Pedley and Martin 2003) are highly conserved in banana Pto-RGCs suggesting that these residues might play a similar role in banana.

Phylogenetic analyses of Pto and Pto-RGC sequences have suggested that these sequences form a unique group of kinases in plants (Vallad et al. 2001; Vleeshouwers et al. 2001). In agreement with this finding the banana Pto-RGCs formed a group with the Pto disease resistance protein suggesting that these sequences share a common evolutionary origin with the tomato Pto protein and possibly a similar function in disease resistance (Fig. 14-2B) (Peraza-Echeverria et al. 2007). Furthermore, phylogenetic analysis of Pto-RGCs from different *Solanum* species has revealed that *Pto* ortholog genes are more similar than paralogs suggesting that the origin of *Pto* could predate the radiation of *Solanum* species (Vleeshouwers et al. 2001). This ancient origin of *Pto* is further supported by the fact that both *Pto* and a *Pto* orthologue (*LhirPto*) are functional in *N. benthamiana* (Riely and Martin 2001). The phylogenetic analysis shown by Peraza-Echeverria et al. (2007) supports and extends these previous observations since all banana Pto-RGCs were grouped in clades that contained Pto-RGCs from both monocot and dicot species suggesting that the origin of this type of sequence might have predated the divergence of monocot and dicot plants.

14.4 *NPR1*-like Genes

NPR1 (non-expressor of pathogenesis related gene 1) is a key positive regulator of systemic acquired resistance (SAR), which protects plant tissues, distant from the site of pathogen attack, from a broad range of biotrophic /necrotrophic fungi, viruses or insects (Cao et al. 1998; Pieterse and Van Loon 2004). In response to pathogen attack, the plant hormones salicylic acid (SA), jasmonic acid (JA) and ethylene (ET) accumulate in plant tissues and act in signaling networks. In the case of SA accumulation, the signal is transduced by *NPR1* into the nucleus where the expression of pathogenesis-related (PR) genes are activated leading to SAR. Some of the PR proteins have shown to have anti-microbial properties (Durrant and Dong 2004). *NPR1* also, at least partially, regulates induced systemic resistance (ISR), which is activated upon colonization of roots by certain non-pathogenic rhizobacteria (Pieterse et al. 1998; Spoel et al. 2003).

The number of *NPR1* homologs per genome has been found to vary from two in soybean, to six in *Arabidopsis*, although two of these six, *AtNPR3* and *AtNPR4*, have been shown to be negative regulators of defense response (Zhang et al. 2006). Two additional, but distant, *NPR1*-like genes of *Arabidopsis*, *AtNPR5* and *AtNPR6*, have functions in plant development (Hepworth et al. 2005). In the case of non-induced SAR, the NPR1 protein exists as an inactive oligomeric form in the cytosol and has been found to be expressed constitutively at a low level (Henanff et al. 2009). Upon accumulation of SA, a change in redox potential, to a more reducing environment occurs and the NPR1 protein is reduced to an active monomeric state and translocates to the nucleus. This is mediated by a bipartite nuclear localization sequence and by the reduction of intermolecular disulphide bonds between highly conserved cysteine residues of the NPR1 protein (Mou et al. 2003). Additionally, the deduced amino acid profile of all functional NPR1 proteins examined to date have been found to have from four to six ankyrin repeats (a 33 residue motif consisting of two alpha helices separated by loops) as a central domain and an amino-terminal broad–complex tramtrack bric-a-brack poxvirus and zinc finger domain (BTB/POZ). Both of these domains facilitate protein-protein binding.

Upon translocation to the nucleus, the monomeric NPR1 interacts with the transcription-factor family known as TGA, which is a subclass of the basic leucine zipper (bZIP) family of transcription factors. TGAs have been shown to regulate the SA-responsive PR genes *PR1*, *PR2* (β-1,3-glucanase) and the thaumatin-like *PR5*, all of which are associated with the activation of SAR (Despres et al. 2000, 2003; Zhang et al. 1999; Zhou et al. 2000). Additionally, it has been shown that activation of SA suppresses JA signaling and that SA-activated NPR1 negatively regulates JA gene expression. The nuclear localization of NPR1 is not essential for this function (Spoel et al. 2003). Other genes integrate signals from different pathways; these include the *Arabidopsis* WRKY70 transcription factor, which has been shown to be an activator of PR genes and a repressor of JA-induced gene expression, and has been shown to be regulated by NPR1 (Li et al. 2004). The first report of overexpression of the cDNA *NPR1* sequence in *Arabidopsis* by Cao et al. (1998) used the constitutive cauliflower mosaic virus (CaMV35s) promoter for overexpression studies. These authors found that the expression of the *PR-1*, *PR-2* and *PR-5* defense genes was significantly enhanced in transgenic plants in response to artificial inoculation with a bacterial and an oomycete pathogen, and with the SAR inducers SA and 2,6 dichloroisonic acid (INA; a functional analog of SA). They also found that increased disease resistance was correlated with higher expression of the *NPR1* gene. One of the important findings of this pioneering work was the observation that, in the absence of pathogen or inducer (SA, INA), the levels of the PR proteins were the same in wild type and transgenics. These results indicated that NPR1 would probably make an ideal candidate for a genetic engineering approach to disease resistance; transgenic plants overexpressing *NPR1* would not express genes involved in disease resistance until challenged by the pathogen. To isolate the *NPR1* gene this group created a mutant of *Arabidopsis* that was deficient in the induction of SAR, *Atnpr1*, which carries a point mutation in the ankyrin repeats domain (Zhang et al. 1999). This mutant was used to create a genetic map and the *NPR1* gene isolated using positional cloning. The SAR-deficient *npr1* has been used to test the functionality of *NPR1* putative orthologs in heterologous assays; for example in soybean the *NPR1* homologs *GmNPR-1* and *GmNPR-2* when expressed in the *Arabidopsis Atnpr1* mutant, were able to induce PR-1, following induction by INA (Sandu et al. 2009).

With the exception of rice, in all transgenic NPR1 overexpressors to date, the PR proteins have been induced at a higher level only when challenged by a pathogen or after application with chemical SAR inducers such as INA, benzothiadiazole (BTH: a functional analog of SA) or SA. Contrary to these findings, over expression of both *AtNPR1* and the rice homolog *OsNPR1 / NH1* in rice have been found to induce constitutive expression of at least four defense genes. Resistance to the bacterial pathogen *Xanthomonas oryzae* was enhanced in transgenic rice with both *AtNPR1* and the rice homolog. However, the *AtNPR1* overexpressors developed lesion mimic spots in response to both low light and the SAR inducer BTH whilst the *OsNPR1 / NH1* overexpressors developed lesion mimic spots on leaves at the pre-flowering stage and, under low light conditions, were dwarfed (Fitzgerald et al. 2004; Chern et al. 2005). The work of Chern et al. (2005) speculated that the difference between *Arabidopsis* and rice in regulation of defense gene induction may be due to the higher levels of endogenous SA present in rice leaves (Silverman et al. 1995). In conclusion, the successful utilization of over expressing *NPR1* for enhanced disease resistance in genetically modified plants should take into account the possibility of differences in physiology between different plant species.

To date, there are only two peer-reviewed publications dealing with homologs of the *NPR1* gene in banana (Endah et al. 2008; Zhao et al. 2009). In the first work, Endah et al. (2008) isolated two full length *NPR1*–like genes, *MNPR1A* and *MNPR1B*, by using PCR and the rapid amplification of cDNA ends techniques (RACE). The two sequences share 78% similarity in amino acid sequences. Both genes harbor the conserved ankyrin repeat and the BTB/POZ zinc finger domains; however, Endah et al. (2008) found differences between the two genes in both the relative positions of the two domains and in some amino acids (Fig. 14-3).

The expression profiles of these two genes and *PR-1* and *PR-3*, were then analyzed by quantitative RT-PCR, in two banana cultivars differing in their susceptibility to the pathogen *Fusarium oxysporum* f. sp. *cubense*, race 4 (FOC). Both *NPR1*-like and *PR* genes were induced above basal levels by treatment with SA, JA and inoculation with FOC (Figs. 14-4, 14-5 and 14-6).

In response to SA the *MNPR1A* levels in both cultivars were not significantly different from the basal level whereas *MNPR1B* expression

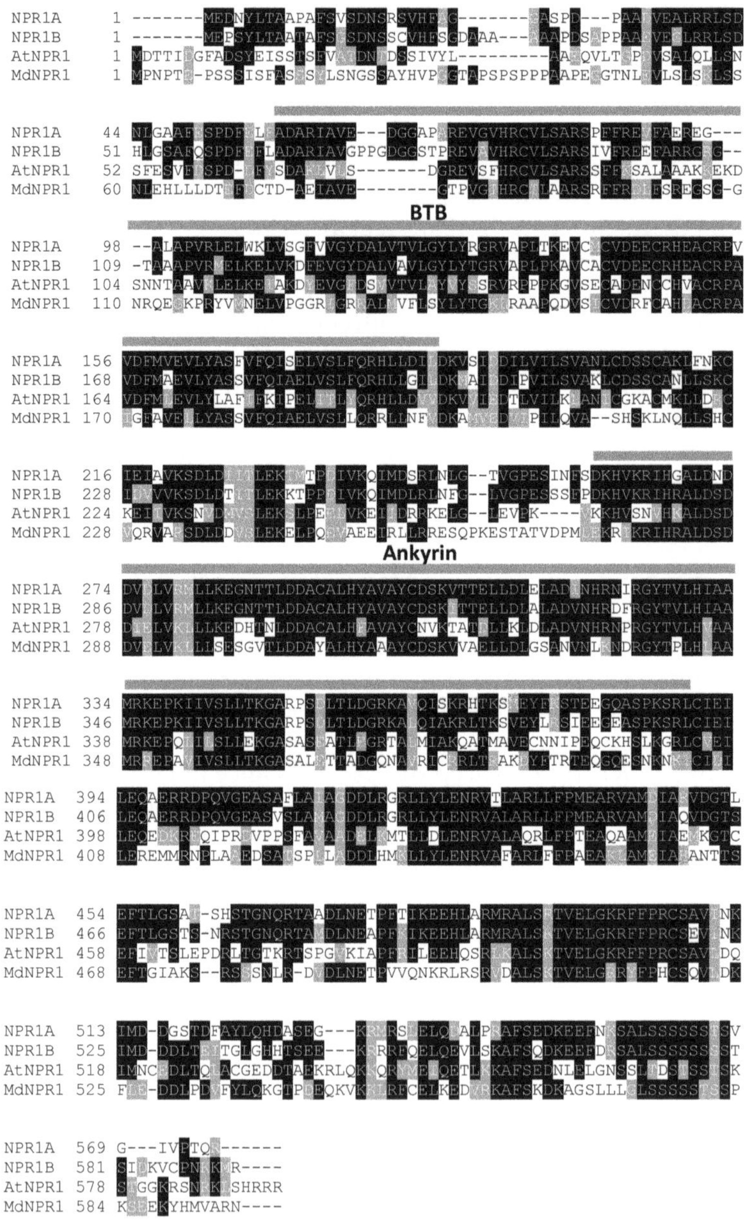

Figure 14-3 ClustalX alignment of three banana NPR1 predicted amino acid sequences, NPR1A (ABI93182), NPR1B (ABL63913) (Endah et al. 2008), MdNPR1 (ACJ04030) (Zhao et al. 2009) and the *Arabidopsis* NPR1 protein (ATIG64280). The BTB and Ankyrin domains are indicated above the sequences. Identical amino acids are highlighted in black and conservative substitutions are marked in gray.

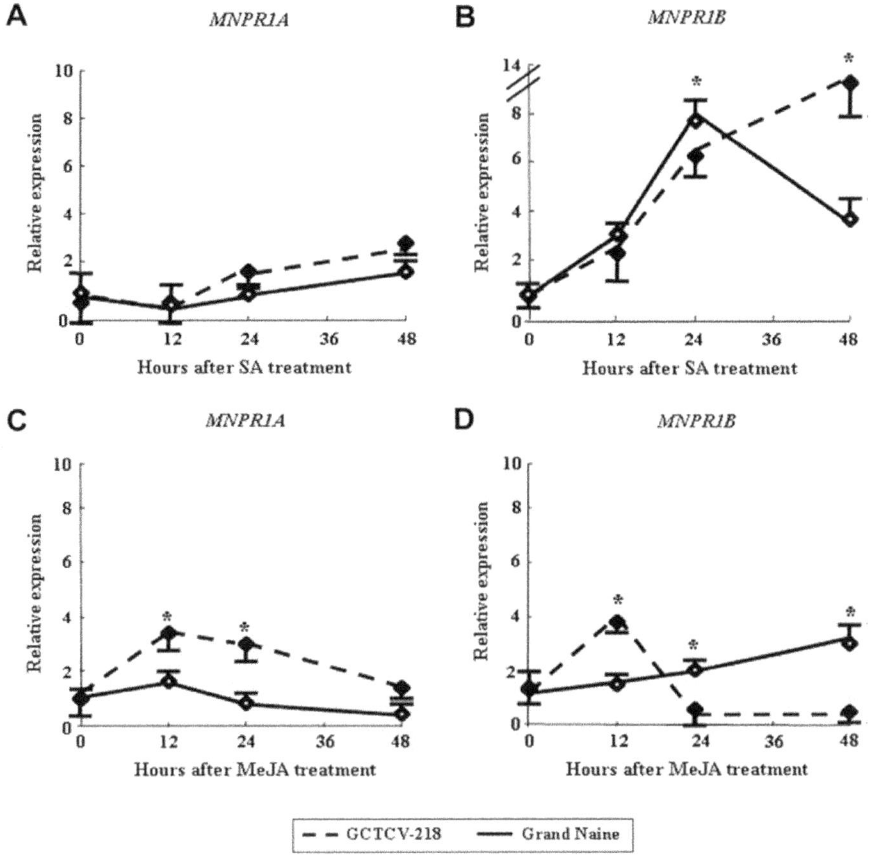

Figure 14-4 The expression of the NPR1-like genes in the roots of FOC race 4- susceptible Grand Naine and the resistant GCTV-218 banana cultivars after treatment with SA (A, B) and JA (C,D) during 48 h post treatment. Results are shown as the standard errors of the mean of 6 plants; * indicates significant differences at $P < 0.05$ (Endah et al. 2008, with permission).

was significantly higher in both cultivars (Fig. 14-4A and B). SA treatment also induced *PR-1* and *PR-3* expression in both the cultivars, however the expression of *PR-1* was higher in the *FOC*-resistant cultivar than in the susceptible (Fig. 14-5A and B). In response to JA, the expression of both *MNNPR1A* and *MNNPR1B* was significantly increased over basal levels in both cultivars; however there was a higher level of expression in the *FOC* resistant cultivar during the first 12 hours post-treatment; subsequently the expression of *MNNPR1B* returned to basal levels in the resistant cultivar but was maintained at above normal levels in the susceptible cultivar (Fig. 14-4C and D). In regard to *PR-1* expression there were significant differences between the cultivars after treatment with SA and JA (Fig. 14-5A and C). Whereas *PR-1* expression was significantly higher in the *FOC* resistant

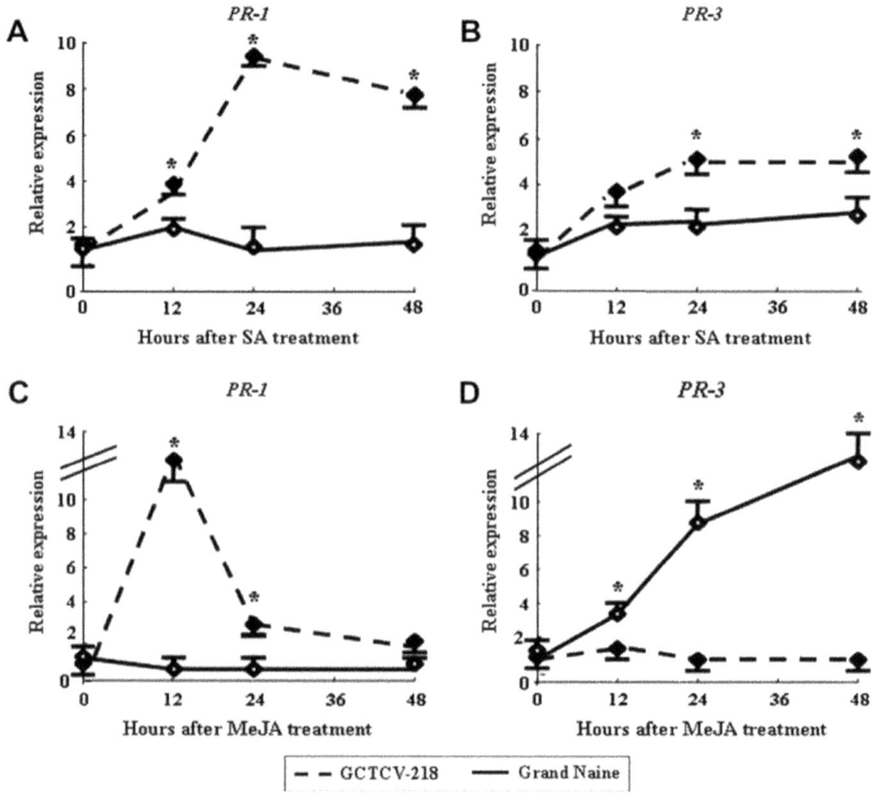

Figure 14-5 The expression of the PR-1 and PR-3 genes in the roots of FOC race 4- susceptible Grand Naine and the resistant GCTV-218 banana cultivars after treatment with SA (A, B) and JA (C,D) during 48 h post treatment. Results are shown as the standard errors of the mean of 6 plants; * indicates significant diffrerences at $P < 0.05$ (Endah et al. 2008, with permission).

cultivar, during the first 24 hours in response to both SA and JA, the case for *PR-3* was contrary where expression was significantly higher in response to JA, in the *FOC*- susceptible cultivar throughout the 48 hours in which expression was measured (Fig. 14-5D). In response to *FOC* inoculation of the roots of both the cultivars there were significant differences: whereas there was little difference in *MNPRIB* expression in the susceptible cultivar there was an increase in expression in *MNPR1A* at 24 hours post-inoculation (Fig. 14-6A). In comparison, there was a highly significant increase in *MNPR1A* expression in the resistant cultivar within 12 hours of inoculation (Fig. 14-6B). With both cultivars it appears that *MNPR1B* expression was not induced by *FOC* and therefore may not be involved in *FOC* resistance. There were large differences in *PR* expression between the two cultivars; in the susceptible cultivar *PR-1* expression was not significantly higher than basal level expression whereas *PR-3* expression at 24 hours post-inoculation

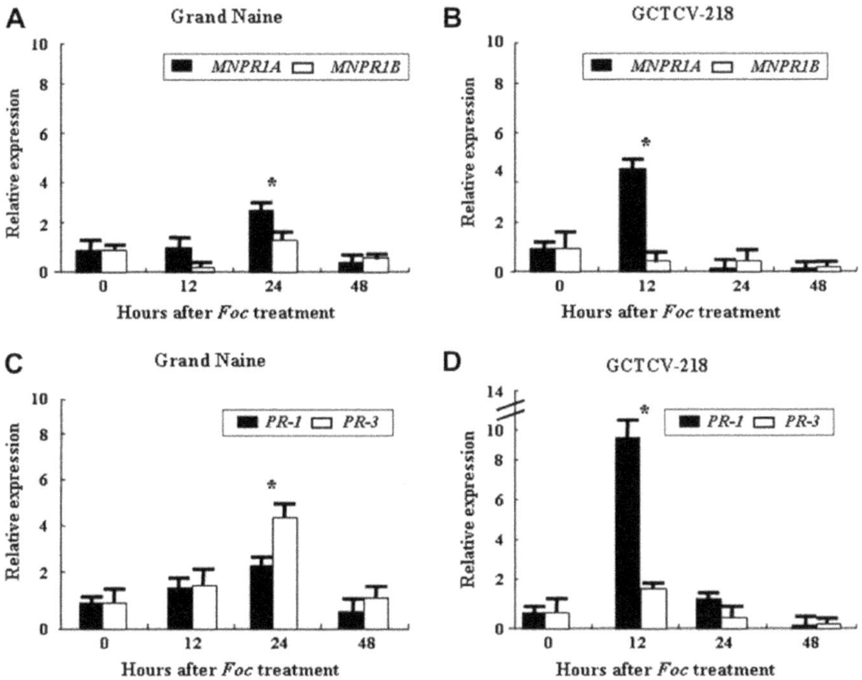

Figure 14-6 The expression of *MNPRA* and *PR-1* (black bars) and *MNPRB* and PR-3 (white bars) genes in the roots of FOC—infected roots of Grand Naine and the resistant GCTV-218 banana cultivars 48 h post treatment. Results are shown as the standard errors of the mean of 6 plants; * indicates significant diffrerences at $P < 0.05$ (Endah et al. 2008, with permission).

was significantly higher (Fig. 14-6C). In contrary to these results, in the resistant cultivar, *PR-1* expression was significantly higher than basal levels at 12 hours post inoculation but declined to near basal levels between 24 and 48 hours post *FOC* inoculation (Fig. 14-6D) whereas *PR-3* expression did not significantly increase during the 48 hours time period in which expression was monitored.

In the absence of an ongoing system for banana transformation at the laboratory of Endah et al. (2008), this group proposed transforming *Arabidopsis* with the two *NPR1*–like genes from banana and monitoring the effect on PR expression and induction of disease resistance. The second report of *NPR1*–like genes in banana comes from the work of Zhao et al. (2009). This group isolated a full-length *NPR1*-like gene designated *MdNPR1* of 4,422 bp in length from a *FOC* race 4 resistant cultivar (cv. Dongguan Dajiao, genome ABB) (Fig. 14-3). A phylogenetic analysis indicated that *MdNPR1* grouped more closely to maize *NPR1* whilst *MNPR1A* and *MNPR1B* grouped more closely to *NPR1* from rice and barley: Zhao et al. (2009) suggested that *MdNPR1* may have a different function to the *NPR1*-

like genes isolated by Endah et al. (2008). Zhao et al. (2009) found that *MdNPR1* was expressed constitutively at low levels in both the susceptible and resistant cultivars. However, induction by SA over basal levels was only observed in the resistant but not in the susceptible cultivar. In response to *FOC* inoculation, *MdNPR1* was expressed at elevated levels (two- to three-fold over basal levels) in the resistant cultivar, within 4–8 weeks of inoculation with *FOC* whilst no such increase of expression was observed in the susceptible cultivar.

14.5 Potential Applications of NBS-RGCs, *Pto*-RGCs and *NPR1*-like Genes for Generating Disease Resistance in Banana

The collection of NBS-RGCs, *Pto*-RGCs and *NPR1*-like genes available today for banana research has a tremendous potential for developing new banana varieties with resistance to the most damaging diseases of this crop such as black Sigatoka and Panama disease. The banana biotechnology tool kit has grown notably over the past two decades allowing the introduction of new genes in banana from reporter genes to antiapoptotic genes (Khanna et al. 2004, 2007). Therefore, the generation of disease resistance in banana using NBS-RGCs, *Pto*-RGCs and *NPR1*-like genes could soon be a reality using different molecular strategies. The banana NBS-RGCs may provide markers tightly linked to R-gene loci that could be used in high-resolution genetic mapping as a tool for map-based cloning. On the other hand, technologies of the post-genomic era, such as RNA interference (RNAi) (Waterhouse and Helliwell 2003) could facilitate testing the function of multiple NBS-RGCs or *Pto*-RGCs in wild banana plants resistant to the most devastating pathogens. Those plants that shift from resistant to susceptible due to the silencing of different NBS-RGCs or *Pto*-RGCs could assist in the identification of a particular R-gene after pathogen challenge. The technology of binary bacterial artificial chromosome (BIBAC) libraries may also facilitate the identification of functional R-genes; this technology can be used to transfer, via *Agrobacterium tumefaciens*, large DNA fragments (up to 120 kb) into the plant genome (He et al. 2003). A BIBAC library has recently been constructed for the cultivar "Tuu Gia" (*Musa acuminata*), which is resistant to the most devastating diseases of banana such as black Sigatoka and Panama disease (Ortiz-Vázquez et al. 2005). This technology coupled with a highly efficient agro-transformation method developed for banana (Khanna et al. 2004) could allow transforming disease-susceptible banana cultivars (e.g., "Grand Naine") with BIBAC clones harboring NBS-RGCs organized either as singletons or clusters. This approach would lead to a NBS-BIBAC collection of banana lines ready to be used for pathogen-resistance screenings.

In the case of the *Pto* gene, previous reports have shown that overexpression in tomato activates a defense response leading to bacterial and fungal disease resistance (Tang et al. 1999). A similar approach applied in banana may lead to the generation of resistance to the same type of pathogens. Interestingly, expression of tomato *Pto* mutants, such as $pto^{Thr204Asp}$ or $pto^{Tyr207Asp}$, can constitutively activate a HR-like response in the absence of *P. syringae* (Rathjen et al. 1999). Expression of these engineered *Pto* mutant genes under the control of a defined inducible promoter could be a strategy to protect crops against pathogens through the hypersensitive response (Rathjen et al. 1999). The cloning of the full-length cDNA sequences of the banana *Pto*-RGCs will permit assessing their potential to confer disease resistance. Regarding the *NPR1* gene from *Arabidopsis* there are reports showing that the overexpression of *NPR1* in this plant (Cao et al. 1998; Friedrich et al. 2001), and in rice (Chern et al. 2001), tomato (Lin et al. 2004), wheat (Makandar et al. 2006), tobacco (Meur et al. 2008) and carrot (Wally et al. 2009) induces enhanced bacterial and fungal resistance. Likewise, overexpression of endogenous *NPR1* orthologs in rice (Chern et al. 2005) and apple (Malnoy et al. 2007) has led to enhanced bacterial and fungal resistance. These findings make *NPR1* an attractive candidate for engineering disease resistance in tropical plants such as banana. Thus, overexpression of banana *NPR1*-like genes in cultivars such as Grand Naine has the potential of activating the SAR response leading to resistance against a broad range of pathogens. The application of these technologies in banana foresees a promising future to unravel the function of NBS-RGCs, *Pto*-RGCs and *NPR1*-like genes in this crop and develop pathogen resistance using a cisgenic approach (genes from the same or closely related species).

Acknowledgments

The authors wish to thank the research grants 106289 and 24579 from the National Council on Science and Technology, México.

Abbreviations

APAF-1	:	Apoptotic protease activating factor 1
ATP	:	Adenosine Triphosphate
AVR	:	Avirulence
BAC	:	Bacterial Artificial Chromosome
BIBAC	:	Binary Bacterial Artificial Chromosome
BTB/POZ	:	Broad –Complex Tramtrack Bric-a-Brack Poxvirus and Zinc Finger Domain
BTH	:	Benzothiadiazole
CaMV35S	:	Cauliflower Mosaic Virus 35S Promoter

ETI	:	Effector-Triggered Immunity
ET	:	Ethylene
FOC	:	*Fusarium oxysporum* f. sp. *cubense*
GTP	:	Guanosine Triphosphate
HR	:	Hypersensitive Response
ISR	:	Induced Systemic Resistance
JA	:	Jasmonic Acid
NBS	:	Nucleotide Binding Site
NBS-LRR	:	Nucleotide-Binding Site and Leucine-Rich Repeat
NPR1	:	Non-Expressor of Pathogenesis-Related Gene 1
PAMPs	:	Pathogen-Associated Molecular Patterns
PCR	:	Polymerase Chain Reaction
PR	:	Pathogenesis-Related Genes
PRR	:	Pattern Recognition Receptors
PTI	:	PAMP-Triggered Immunity
R	:	Resistance Gene
RACE	:	Rapid Amplification of cDNA Ends
RGC	:	Resistance Gene Candidate
RNAi	:	RNA interference
SA	:	Salicylic Acid
SAR	:	Systemic Acquired Resistance
STK	:	Serine/Threonine Protein Kinase
TIR	:	Homology to the *Drosophila* Toll and Mammalian Interleukin-1 Receptors

References

Agrios GN (2005) Plant Pathology. Academic Press, San Diego, California, USA.

Arabidopsis Genome Initiative (2000) Analysis of the genome sequence of the flowering plant *Arabidopsis thaliana*. Nature 408: 796–815.

Azhar M, Heslop-Harrison JS (2008) Genomes, diversity and resistance gene analogues in *Musa* species. Cytogenetics 121: 59–66.

Bai J, Pennill LA, Ning J, Lee SW, Ramalingam J, Leach JE, Hulbert SH (2002) Diversity in nucleotide binding site-leucine rich repeat genes in cereals. Genome Res 12: 1871–1884.

Belkhadir Y, Subramaniam R, Dangl JL (2004) Plant disease resistance protein signalling: NBS-LRR proteins and their partners. Curr Opin Plant Biol 7: 391–399.

Cannon SB, Zhu H, Baumgarten AM, Spangler R, May G, Cook DR, Young ND (2002) Diversity, distribution and ancient taxonomic relationships within the TIR and non-TIR NBS-LRR resistance gene subfamilies. J Mol Evol 54: 548–562.

Cao H, Li X, Dong X (1998) Generation of broad-spectrum disease resistance by over-expression of an essential regulatory gene in systemic acquired resistance. Proc Natl Acad Sci USA, 95: 6531–6536.

Chern M, Fitzgerald HA, Canlas PE, Duroy A, Navarre A, Ronald PC (2005) Overexpression of a rice NPR1 homolog leads to constitutive activation of defense response and hypersensitivity to light. Mol Plant Microbe Interact 18: 511–520.

Chisholm ST, Coaker G, Day B, Staskawicz BJ (2006) Host-microbe interactions: shaping the evolution of the plant immune response. Cell 124: 803–814.

Churchill ACL (2010) *Mycosphaerella fijiensis*, the black leaf streak pathogen of banana: progress towards understanding pathogen biology and detection, disease development, and the challenges of control. Mol Plant Pathol DOI:10.1111/j.1364-3703.2010.00672.X.

Dangl L, Jones J (2001) Plant pathogens and integrated defence responses to infection. Nature 411: 826–833.

Despres C, DeLong C, Glaze S, Liu E, Fobert PR (2000) The *Arabidopsis* NPR1/NIM1 protein enhances the DNA binding activity of a subgroup of the TGA family of bZIP transcription factors. Plant Cell 12: 279–290.

Despres C, Chubak C, Rochon A, Clark R, Bethune T, Desveau D, Fobert PR (2003) The *Arabidopsis* NPR1 disease resistance protein is a novel cofactor that confers redox regulation of DNA binding activity to the basic domain/leucine zipper transcription factor TGA1. Plant Cell 15: 2181–2191.

Di Gaspero G, Cipriani G (2003) Nucleotide binding site/leucine-rich repeats, Pto-like and receptor-like kinases related to disease resistance in grapevine. Mol Gen Genom 269: 612–623.

Dolezel J, Dolezelova M, Novak FJ (1994) Flow cytometric estimation of nuclear DNA amount in diploid bananas (*Musa acuminata* and *M. balbisiana*). Biol Plant 36: 351–357.

Durrant WE, Dong X (2004) Systemic acquired resistance. Ann Rev Phytopath 42: 185–209.

Endah R, Beyene G, Kiggundu A, van den Berg N, Schluter U, Kunert K, Chikwamba R (2008) Elicitor and Fusarium-induced expression of NPR1-like genes in banana. Plant Physiol Biochem 46: 1007–1014.

Fitzgerald HA, Chen MS, Navarre R, Ronald PC (2004) Overexpression of (At) NPR1 in rice leads to a BTH- and environment-induced lesion-mimic/cell death phenotype. Mol Plant Microbe Interact 17: 140–151.

Frederick RD, Thilmony RL, Sessa G, Martin GB (1998) Recognition specificity for the bacterial avirulence protein AvrPto is determined by Thr-204 in the activation loop of the tomato Pto kinase. Mol Cell 2: 241–245.

Friedrich L, Lawton K, Dietrich R, Willits M, Cade R, Ryals J (2001) *NIM1* overexpression in *Arabidopsis* potentiates plant disease resistance and results in enhanced effectiveness of fungicides. Mol Plant Microbe Interact 14: 1114–1124.

Hammond-Kosack K, Parker JE (2003) Deciphering plant-pathogen communication: fresh perspectives for molecular resistance breeding. Curr Opin Biotech 14: 177–193.

Goff SA, Ricke D, Lan TH, Presting G, Wang R, Dunn M, Glazebrook J, Sessions A, Oeller P, Varma H, Hadley D, Hutchison D, Martin C, Katagiri F, Lange BM, Moughamer T, Xia Y, Budworth P, Zhong J, Miguel T, Paszkowski U, Zhang S, Colbert M, Sun WL, Chen L, Cooper B, Park S, Wood TC, Mao L, Quail P, Wing R, Dean R, Yu Y, Zharkikh A, Shen R, Sahasrabudhe S, Thomas A, Cannings R, Gutin A, Pruss D, Reid J, Taytigian S, Mitchell J, Eldredge G, Scholl T, Miller RM, Bhatnagar S, Adey N, Rubano T, Tusneem N, Robinson R, Feldhaus J, Macalma T, Oliphant A, Briggs S (2002) A draft sequence of the rice genome (*Oryza sativa* L. ssp. *japonica*). Science 296: 92–100.

Göehre V, Robatzek S (2008) Breaking the barriers: microbial effector molecules subvert plant immunity. Annu Rev Phytopathol 46: 189–215

Gurr SJ, Rushton PJ (2005) Engineering plants with increased disease resistance: what are going to express? Trends Biotechnol 23: 275–282.

Hanks SK, Quinn AM (1991) Protein kinase catalytic domain sequence database: identification of conserved features of primary structure and classification of family members. Method Enzymol 200: 38–62.

He RF, Wang Y, Shi Z, Ren X, Zhu L, Weng Q, He GC (2003) Construction of a genomic library of wild rice and *Agrobacterium*-mediated transformation of large insert DNA linked to *BPH* resistance locus. Gene 321: 113–121.

Henanff GL, Heitz T, Mestre P, Mutterer J, Walter B, Chong J (2009) Characterization of *Vitis vinifera* NPR1 homologs involved in the regulation of *Pathogenesis-Related* gene expression. BMC Plant Biol 9: 54.
Hepworth SR, Zhang Y, McKim S, Li X, Haughn GW (2005) BLADE –ON PETIOLE-dependent signaling controls leaf and floral patterning in Arabidopsis. Plant Cell 17: 1434–1448.
Hulbert SH, Webb CA, Smith M, Sun Q (2001) Resistance gene complexes: evolution and utilization. Annu Rev Phytopathol 39: 285–312.
Khanna HK, Becker D, Kleidon J, Dale JL (2004) Centrifugation assisted *Agrobacterium*-mediated transformation (CAAT) of embryogenic cell suspensions of banana (*Musa* spp. Cavendish AAA and Lady finger AAB). Mol Breed 14: 239–252.
Khanna HK, Paul JY, Harding RM, Dickman MB, Dale JL (2007) Inhibition of *Agrobacterium*-induced cell death by antiapoptotic gene expression leads to very high transformation efficiency of banana. Mol Plant Pathol Interact 20: 1048–1054.
Kim YJ, Lin NC, Martin GB (2002) Two distinct *Pseudomonas* effector proteins interact with the Pto kinase and activate plant immunity. Cell 109: 589–598.
Li J, Brader G, Palva ET (2004) The WRKY70 transcription factor: a node of convergence for jasmonate -mediated and salicylate-mediated signals in plant defense. Plant Cell 16: 319–331.
Lin WC, Lu CF, Wu JW, Cheng ML, Lin YM, Yang NS, Black L, Green SK, Wang JF, Cheng CP (2004) Transgenic tomato lines expressing the *Arabidopsis NPR1* gene display enhanced resistance to a spectrum of fungal and bacterial diseases. Transgen Res 13: 567–581.
Makandar R, Essig JS, Schapaugh MA, Trick HN, Jyoti S (2006) Genetically Engineered Resistance to Fusarium Head Blight in Wheat by Expression of *Arabidopsis NPR1.* Mol Plant Microbe Interact 19: 123–129.
Malnoy M, Jin Q, Borejsza-Wysocka EE, He SY, Aldwinckle HS (2007) Overexpression of the Apple *MpNPR1* Gene Confers Increased Disease Resistance in *Malus* × *domestica.* Mol Plant Microbe Interact 20: 1568–1580.
Martin GB, Brommonschenkel SH, Chunwongse J, Frary A, Ganal MW, Spivey R, Wu T, Earle ED, Tanksley SD (1993) Map-based cloning of a protein kinase gene conferring disease resistance in tomato. Science 262: 1432–1435.
Meur G, Budatha M, Srinivasan T, Kumar KRR, Gupta AD, Kirti PB (2008) Constitutive expression of *Arabidopsis* NPR1 confers enhanced resistance to the early instars of *Spodoptera litura* in transgenic tobacco. Physiol Plant 133: 765–775.
Meyers B, Dickerman A, Michelmore R, Sivaramakrishnan S, Sobral B, Young N (1999) Plant disease resistance genes encode members of an ancient and diverse protein family within the nucleotide-binding superfamily. Plant J 20: 317–332.
Meyers BC, Kozik A, Griego A, Kuang H, Michelmore RW (2003) Genome-wide analysis of NBS-LRR-encoding genes in *Arabidopsis*. Plant Cell 15: 809–834.
Miller RN, Bertioli DJ, Baurens FC, Santos CM, Alves PC, Martins NF, Togawa RC, Souza MT, Pappas GJ (2008) Analysis of non-TIR NBS-LRR resistance gene analogs in *Musa acuminata* Colla: isolation, RFLP marker development, and physical mapping. BMC Plant Biol 8: 15.
Mou Z, Fan W, Dong X (2003) Inducers of plant systemic acquired resistance regulate NPR1 function through redox changes. Cell 113: 935–944.
Oh CS, Martin GB (2010) Effector-triggered immunity mediated by the Pto kinase. Trends Plant Sci 16: 132–140.
Ortiz-Vázquez, E, Kaemmer D, Zhang HB, Muth J, Rodríguez-Mendiola M, Arias-Castro C, James A (2005) Construction and characterization of a plant transformation-competent BIBAC library of the black Sigatoka resistant banana *Musa acuminata* cv. Tuu Gia (AA). Theor Appl Genet 110: 706–713.
Pan Q, Wendel J, Fluhr R (2000) Divergent evolution of plant NBS-LRR resistance gene homologues in dicot and cereal genomes. J Mol Evol 50: 203–213.
Pedley KF, Martin GB (2003) Molecular basis of *Pto*-mediated resistance to bacterial speck disease in tomato. Annu Rev Phytopathol 41: 215–243.

Pei X, Li S, Jiang Y, Zhang Y, Wang Z, Jia S (2007) Isolation, characterization and phylogenetic analysis of the resistance gene analogues (RGAs) in banana (*Musa* spp.). Plant Sci 172: 1166–1174.

Peraza-Echeverria S, James-Kay A, Canto-Canché B, Castillo-Castro E (2007) Structural and phylogenetic analysis of Pto-type disease resistance gene candidates in banana. Mol Genet Genomics 278: 443–453.

Peraza-Echeverria S, Dale JL, Harding RM, Smith MK, Collet C (2008) Characterization of disease resistance gene candidates of the nucleotide binding site (NBS) type from banana and correlation of a transcriptional polymorphism with resistance to *Fusarium oxysporum* f. sp. *cubense* race 4. Mol Breeding 22: 565–579.

Peraza-Echeverria S, Dale JL, Harding RM, Collet C (2009) Molecular cloning and *in silico* analysis of potential *Fusarium* resistance genes in banana. Mol Breeding 23: 431–443.

Pieterse CMJ, Van Loon LC (2004) NPR1: the spider in the web of induced resistance signaling pathways. Curr Opin Plant Biol 7: 456–464.

Pieterse CMJ, van Wees SCM, van Pelt JA, Knoester M, Laan R, Gerrits H, Weisbeek PJ, van Loon LC (1998) A novel signaling pathway controlling induced systemic resistance in Arabidopsis. Plant Cell 10: 1571–1580.

Ploetz RC (2005) Panama disease, an old nemesis rears its ugly head: part 1, the beginnings of the banana export trades. Plant Health Progr doi:10.1094/PHP-2005-1221-01-RV.

Ploetz RC, Pegg KG (2000) Fungal diseases of the root, corm and pseudostem. In: DR Jones (ed). Diseases of Banana, Abaca and Enset. CAB International, Wallingford, UK, pp 143–172.

Rathjen JP, Chang JH, Staskawicz BJ, Michelmore RW (1999) Constitutively active *Pto* induces a Prf-dependent hypersensitive response in the absence of *avrPto*. EMBO J 18: 3232–3240.

Riely BK, Marti GB (2001) Ancient origin of pathogen recognition specificity conferred by the tomato disease resistance gene *Pto*. Proc Natl Acad Sci USA, 98: 2059–2064.

Sandu D, Tasma IM, Frasch R, Bhattacharyya KM (2009) Systemic acquired resistance in soybean is regulated by two proteins, Orthologous to *Arabidopsis* NPR1. BMC Plant Biol 9: 105 doi:10.1186/1471-2229-9-105.

Sessa G, D'Ascenzo M, Martin GB (2000) Thr38 and Ser198 are Pto autophosphorylation sites required for the AvrPto-Pto-mediated hypersensitive response. EMBO J 19: 2257–2269.

Silverman P, Seskar M, Kanter D, Schweizer P, Metraux JP, Raskin I (1995) Salicylic Acid in Rice (Biosynthesis, Conjugation, and Possible Role). Plant Physiol 108: 633–639.

Spoel SH, Koornneef A, Claessens SMC, Korzelius JP, Van Pelt JA, Mueller MJ, Buchala AJ, Metraux JP, Brown K, Kazan K (2003) NPR1 modulates cross-talk between salicylate- and Jasmonate -dependent defense pathways through a novel function in the cytosol. Plant Cell 15: 760–770.

Tameling W, Elzinga S, Darmin P, Vossen J, Takken F, Haring M, Cornelissen B (2002) The tomato R gene products I2 and Mi-1 are functional ATP binding proteins with ATPase activity. Plant Cell 14: 2929–2939.

Tang X, Xie M, Kim YJ, Zhou J, Klessig DF, Martin GB (1999) Overexpression of *Pto* activates defense responses and confers broad resistance. Plant Cell 11: 15–29.

Tuskan GA, DiFazio S, Bohlmann J, Grigoriev I, Hellsten U, Putman N, Ralph S, Rombauts S, Salamov A, Schein J, Sterck L, Aerts A, Bhalerao RR, Bhalerao RP, Blaudez D, Boerjan W, Brun A, Brunner A, Busov V, Campbell M, Carlson J, Chalot M, Chapman J, Chen GI, Cooper D, Coutinho PM, Couturier J, Covert S, Cronk Q, Cunningham R, Davis J, Degroeve S, Dejardin A, dePamphilis C, Detter J, Dirks B, Dubchak I, Duplessis S, Ehlting J, Ellis B, Gendler K, Goodstein D, Gribskov M, Grimwood J, Groover A, Gunter L, Hamberger B, Heinze B, Helariutta Y, Henrissat B, Holligan D, Holt R, Huang W, Islam-Faridi N, Jones S, Jones-Rhoades M, Jorgensen R, Joshi C, Kangasjarvi J, Karlsson J, Kelleher C, Kirkpatrick R, Kirst M, Kohler A, Kalluri U, Larimer F, Leebens-Mack J, Leplé JC, Locascio P, Lou Y, Lucas S, Martin F, Montanini B, Napoli C, Nelson DR, Nelson C, Nieminem K, Nilsson O, Pereda V, Peter G, Philippe R, Pilate G, Poliakov

A, Razumovskaya J, Richardson P, Rinaldi C, Ritland K, Rouzé P, Ryaboy D, Schmutz J, Schrader J, Segerman B, Shin H, Siddiqui A, Sterky F, Terry A, Tsai CJ, Uberbacher E, Unneberg P, Vahala J, Wall K, Wessler S, Yang G, Yin T, Douglas C, Marra M, Sandberg G, Van de Peer Y, Rokhsar D (2006) The genome of black cottonwood *Populus trichocarpa* (Torr. and Gray). Science 313: 1596–1604.

Vallad G, Rivkin M, Vallejos C, McClean P (2001) Cloning and homology modelling of a Pto-like protein kinase family of common bean (*Phaseolus vulgaris* L.). Theor Appl Genet 103: 1046–1058.

Van der Biezen E, Jones JDG (1998) Plant disease-resistance proteins and the gene-for-gene concept. Trends Biochem Sci 23: 454–456.

Van Ooijen G, Van den Burg HA, Cornelissenn BJC, Takken FLW (2007) Structure and function of resistance proteins in Solanaceous plants. Annu Rev Phytopathol 45: 43–72.

Velasco R, Zharkikh A, Troggio M, Cartwright DA, Cestaro A, Pruss D, Pindo M, FitzGerald LM, Vezzuli S, Reid J, Malacarne G, LLiev D, Coppola G, Wardell B, Micheletti D, Macalma T, Facci M, Mitchell JT, Perazzolli M, Eldredge G, Gatto P, Oyzerski R, Moretto M, Gutin N, Stefanini M, Chen Y, Segala C, Davenport C, Dematte L, Mraz A, Battilana J, Stormo K, Costa F, Tao Q, Si-Ammour A, Harkins T, Lackey A, Perbost C, Taillon B, Stella A, Solovyev V, Fawcett JA, Sterck L, Vandepoele K, Grando SM, Toppo S, Moser C, Lanchbury J, Bogden R, Skolnick M, Sgaramella V, Bhatnagar S, Fontana P, Gutin A, Van der Peer Y, Salamini F, Viola R (2007) A high quality draft consensus sequence of the genome of a heterozygous grapevine variety. PLoS One 12: 1–18.

Vleeshouwers VGAA, Martens A, van Dooijeweert W, Colon LT, Govers F, Kamoun S (2001) Ancient diversification of the *Pto* kinase family preceded speciation in *Solanum*. Mol Plant Microbe Interact 14: 996–1005.

Wally O Jayaraj J, Punja ZK (2009) Broad-spectrum disease resistance to necrotrophic and biotrophic pathogens in transgenic carrots (*Daucus carota* L.) expressing an *Arabidopsis NPR1* gene. Planta doi 10.1007/s00425-009-1031-2.

Waterhouse PM, Helliwell CA (2003) Exploring plant genomes by RNA-induced gene silencing. Nature Rev 4: 29–38.

Xu Q, Deng X (2010) Cloning and phylogenetic analysis of serine/threonine kinase class defense-related genes in a wild fruit crop 'chestnut rose'. BMC Res Notes 3: 202.

Zhang Y, Fan W, Kinkema M, Li X, Dong X (1999) Interaction of NPR1with basic leucine zipper protein transcription factors that bind sequences required for salicylic acid induction of the *PR-1*gene. Proc Natl Acad Sci USA, 96: 6523–6528.

Zhang Y, Cheng YT, Qu N, Zhao Q, Bi D, Li X (2006) Negative regulation of defense responses in *Arabidopsis* by two NPR1 paralogs. Plant J 48: 647–656.

Zhao JT, Huang X, Chen YP, Chen YF, Huang XL (2009) Molecular Cloning and Characterization of an Ortholog of *NPR1* Gene from Dongguan Dajiao (*Musa* spp. ABB). Plant Mol Biol Rep 27: 243–249.

Zhou JM, Trifa Y, Silva H, Pontier D, Lam E, Shah J, Klessig DF (2000) NPR1 differentially interacts with members of the TGA/OBF family of transcription factors that bind an element of the *PR-1* gene required for induction by salicylic acid. Mol Plant Microbe Interact 13: 191–202.

Zhou T, Wang T, Chen JQ, Araki H, Jing Z, Jiang K, Shen J, Tian D (2004) Genome-wide identification of NBS genes in *japonica* rice reveals significant expansion of divergent non-TIR NBS-LRR genes. Mol Genet Genomics 271: 402–415.

15

A Case for Molecular Breeding in *Musa*

Michael Pillay,[1,*] *Kaliyaperumal Ashokkumar,*[2,a] *Arun Siva Kumar Shunmugam*[2,b] and *Sivalingam Elayabalan*[3]

ABSTRACT

Conventional breeding of *Musa* that is based mainly on mass phenotypic recurrent selection is handicapped by a number of factors. The rapid development of molecular biology techniques and their application to plant breeding has resulted in significant genetic gains in agricultural crops. Marker assisted breeding will be very useful for a crop like banana that has a relatively long life cycle. DNA markers are being sought for several characters of importance in *Musa* including resistance to pests and diseases. Achievements and prospects of molecular breeding for black Sigatoka, Fusarium, Banana bunchy top virus (BBTV), nematodes and Xanthomonas wilt resistance are discussed. In addition gains made in nutritional enhancement of banana are described. The development of modern plant molecular and quantitative genetics in the last two decades has the potential to revolutionize what has mostly been experienced-based empirical plant breeding. This chapter outlines the value of modern molecular tools for molecular breeding of banana.

Keywords: breeding challenges, molecular markers, molecular breeding achievements

[1]Vaal University of Technology, Private Bag X021, Vanderbijlpark 1900, Gauteng, South Africa.
[2]Department of Plant Sciences, University of Saskatchewan, 51 Campus Drive, Saskatoon, SK, Canada, S7N 5A8.
[a]e-mail: *biotech.ashok@gmail.com*
[b]e-mail: *anbeshivam481@gmail.com*
[3]Crop Tech Ltd, P.O. Box 1367 Arusha, Tanzania; e-mail: *balabiotech@gmail.com*
*Corresponding author

List of abbreviations after the text.

15.1 Introduction

Currently the production of improved banana and plantain cultivars that are nutritionally acceptable to consumers, with resistance or tolerance to biotic and abiotic stresses, and reduced post-harvest losses, has been met largely through conventional breeding that has made steady progress over the years producing a large number of hybrids (Rowe 1984; Vuylsteke et al. 1995). The pressure of an increasing population and consequent increase in demand for food on the one hand and the depletion of arable land on the other have placed new emphases on conventional plant breeding (Pillay et al. 2011). However, conventional breeding of *Musa* is handicapped by sterility and a number of other factors that are discussed in Pillay et al. (2002), and Pillay and Tripathi (2006, 2007).

The rapid development of molecular techniques and their application to plant breeding has resulted in significant genetic gains in agricultural crops, some of which have already entered the market (Newell-McGloughlin 2008). Molecular and biotechnological tools such as marker-assisted breeding, tissue culture, *in vitro* mutagenesis and genetic transformation can contribute to solving or reducing some of the constraints of conventional banana breeding. This chapter examines some aspects of molecular breeding in *Musa*.

15.1.1 Breeding Challenges in Musa

Musa is a polyploid crop with ploidy ranging from diploid ($2n = 2x = 22$) to tetraploids ($2n = 4x = 44$). Most cultivated bananas are triploids ($2n = 3x = 33$) and sterile harboring various combinations of either one, two or three A, B, S, or T genomes. New banana cultivars are exceptionally cumbersome to develop. Selection for desirable characters is time consuming and it may take up to 12 years to develop a new cultivar. *Musa* breeding is based mainly on phenotypic mass recurrent selection. The high levels of heterozygosity make identification of ideal parental material difficult and very large populations are required for selection of individual clones with good agronomic traits. This is virtually impossible to attain due to the low seed set in crosses. Generally, few seeds are obtained (an average of 1 to 1.5) and acquiring large numbers of seeds is a labor intensive and tedious process (Ortiz and Vuylsteke 1995; Ssebuliba et al. 2006a, b, 2009). The genes for resistance to diseases and pests are introgressed from wild diploid species. Wild species also carry many undesirable traits, e.g., low yield and non-parthenocary. The process of eliminating the unwanted traits requires several backcrosses that lengthen the breeding process. The multigenic nature and low heritability of some traits also slow down the breeding process. *Musa* breeding is also problematic due to the narrow genetic diversity of the germplasm (Pillay et al. 2001; Nyine and Pillay 2011) and the lack of information about wild species that carry useful agronomic traits. Only a few wild diploids have been used so far and mostly as male parents by majority of the breeding programs. As cultivated banana is propagated asexually, its genetic base is narrow with diversity dependent on somatic mutation. Limited genetic variation has resulted in a crop lacking resistance to fungal, bacterial and viral pathogens and numerous pests (Miller et al. 2009). Very little knowledge exists on the genetics of important agronomic traits in *Musa* and precise genetic control is known for relatively few traits (see Chapters 6, 7).

15.1.2 Production Constraints

The production constraints of *Musa* have been well documented (Pillay et al. 2002; Pillay and Tripathi 2006, 2007; Tenkouano et al. 2011). Briefly the production of bananas worldwide is threatened by a complex of foliar diseases, nematodes, viruses and pests. The use of resistant varieties is considered to be the most effective, economical and environmentally friendly approach to controlling diseases and pests. Two of the most important fungal diseases include black Sigatoka (*Mycosphaerella fijiensis* Morelet) and fusarium wilt (*Fusarium oxysporum* Schlect. f.sp. *cubense* (E.F. Smith). The main pests include a complex of nematodes (*Radopholus similis, Pratylenchus* spp. *Helicotylenchus*) and the banana weevil (*Cosmopolites sordidus* Germar). New diseases such as banana *Xanthomonas* wilt (BXW) have been recently identified in East Africa.

15.1.3 Breeding Objectives in **Musa**

The most important objectives of *Musa* breeding include:

- increased bunch size and yield
- host plant resistance against the major pathogens including those causing Sigatoka, Fusarium and Xanthomonas wilts, and viruses
- host plant resistance against nematodes and insect pests
- fruit quality traits, e.g., increased vitamin A, iron and zinc levels
- better adaptation to abiotic stresses such as drought, heat and other stresses that may be enforced by predictions in climate change.

Breeding for yield is a major target followed by breeding for host plant resistance to pathogens and pests that impact on yield.

15.2 Molecular Breeding

Molecular breeding (MB) is the generic term used to describe several modern breeding methods including, (i) marker-assisted selection (MAS) —the selection of specific alleles for traits conditioned by a few loci, (ii) marker-assisted backcrossing (MABC)—the transfer of a limited number of loci from one genetic background to another, including transgenes, (iii) marker-assisted recurrent selection (MARS)—the identification and selection of several genomic regions involved in the expression of complex traits to "assemble" the best-performing genotype within a single, or across related, populations, and (iv) genome wide selection (GWS)—selection based on markers without significant testing and without identifying *a priori* a subset of markers associated with the trait (Ribaut et al. 2010). MABC is one of the most anticipated and frequently cited benefits of molecular markers as indirect selection tools in breeding programs (Semagn et al. 2006).

Routine use of MAS in ongoing plant breeding programs has not been achieved as yet. The implementation of MAS has been slow due to the high relative cost compared to conventional phenotypic selection. To be useful to plant breeders, gains made from MAS must be more cost-effective than gains through traditional breeding or MABC must generate significant time savings to justify the additional cost involved (Semagn et al. 2006).

Since a large number of traits in plants are polygenic, MABC of traits controlled by single genes is the most effective way of using DNA markers effectively. The improvement of quantitative trait loci (QTL) through MABC has produced variable results ranging from limited success and/ or even a failure to a few highly successful stories (Semagn et al. 2006). Marker-assisted breeding will be very useful for a crop like banana that has a relatively long life cycle. The use of molecular markers for the indirect selection of improved cultivars speeds up the selection process by alleviating time-consuming approaches of direct screening under greenhouse or field conditions. Some of the most important characters to the *Musa* breeder have been reported to have an oligogenic epistatic basis (Ortiz 1995).

15.2.1 Molecular Markers in Musa

It is now generally accepted that molecular markers represent the most significant advance in breeding technology in the last few decades and are currently the most important application of molecular biology to plant breeding. There appears to be no resistance to the use of molecular marker technology in breeding as there is for genetically-modified organisms (Pillay et al. 2011). DNA markers are being sought for several characters of importance in *Musa* including resistance to pests and diseases. Fruit quality (color, texture, ripening) are other candidate traits for selecting with

DNA markers. Most of these traits are expressed only late in the life cycle of the plants or are difficult to screen. Identification of markers linked to loci governing important traits will facilitate gene introgression and other MAS applications. Accessing genes from various genomes, including the S (*M. schizocarpa*) and T (*M. textilis*) genomes will increasingly become important for *Musa* breeding. To date very few markers have been linked to traits of interest in *Musa* and are limited to markers for disease resistance and the main genomes.

Methylation-sensitive amplification polymorphism (MSAP) markers were used to identify molecular markers associated with resistance to *Mycosphaerella fijiensis* toxins (black Sigatoka) with a set of reference cultivars and somaclonal variants (Gimenez et al. 2006). The study identified four MSAP markers that were associated with resistance to *M. fijiensis* toxins. The MSAP markers showed a high degree of sequence similarity with resistance gene analogs and with retrotransposon sequences. These markers were cited as being useful as molecular indicators of tolerance to *M. fijiensis* toxins and resistance to black Sigatoka.

A reliable molecular method to detect *Fusarium oxysporum* f.sp. *cubense* (*Foc*) race 4 isolates in Taiwan was developed by (Lin et al. 2010). By PCR amplification, the primer set *Foc*-1/*Foc*-2 derived from the sequence of a random primer OP-A02 amplified fragment produced a 242 bp size DNA fragment, which was specific to *Foc* race 4. With the optimized PCR parameters, the molecular method was sensitive and could detect small quantities of *Foc* DNA as low as 10 pg in 50 to 2,000 ng host genomic DNA with high efficiency.

A putative RAPD marker for Sigatoka resistance has been identified at the National Research Center for Banana (NRCB), India. The marker has been cloned, sequenced and converted into a sequence characterized amplified region (SCAR) marker and is being validated using contrasting parents for expression of Sigatoka (*M. musicola*) disease resistance and their progenies. Parallel studies have led to the identification of a putative random amplified polymorphic DNA (RAPD) marker for nematode resistance (S. Uma, pers. comm.) An RAPD marker has been identified for salt tolerance among clones of cv. "Dwarf Cavendish" that were obtained through induced mutagenesis (Miri et al. 2009).

A banana somatic embryogenesis receptor-like kinase (*SERK*) gene, designated as *MaSERK1*, isolated from *Musa acuminata* cv. "Mas" (AA) was associated with somatic embryogenic competence and disease resistance response in *Musa* (Xia et al. 2010). The gene encoded a protein of 628 amino acids with identities of above 82% to *SERK* genes in coconut, rice, maize, *Arabidopsis*, carrot, and *Medicago truncatula*. *MaSERK1* was expressed weakly in male flower clusters, but not in male flower-derived non-embryogenic calli. It was highly expressed in male flower-derived embryogenic calli and embryogenic cell suspensions (ECS). The frequency of somatic embryogenesis of ECS positively correlated with *MaSERK1* transcript levels. *MaSERK1* expression in leaves of cultivar "Dongguan Dajiao" (ABB), known to be resistant to *FOC* race 4, was induced by exogenous salicylic acid (SA) or inoculation with *FOC* race 4. However, *MaSERK1* expression levels in leaves of "Pisang awak" (ABB), known to be susceptible to *FOC* race 4, did not change following either treatment (Xia et al. 2010). It was suggested that *MaSERK1* gene expression not only could serve as a molecular marker for banana somatic embryogenesis, but could also play a role in host plant resistance response to banana pathogens.

15.2.2 Marker-Assisted Introgression

Marker-assisted breeding takes advantage of the association between agronomic traits and allelic variants of genetic markers, mostly molecular markers (Stam 2003). Generally these associations are the result of genetic linkages between markers and gene loci underlying the trait(s) of interest. These associations are also known as linkage disequilibrium. Linkage disequilibrium arise in experimental populations used for linkage mapping, e.g., backcross generations (BC), F_2 segregating populations, recombinant inbred lines (RILs) or doubled haploids (DHs) (Stam 2003). In cross-fertilizing plant species such as *Musa* a mapping population usually consists of a large full-sib family resulting from a cross between single plants of divergent genotypes. Before a plant breeder can utilize linkage-based associations between traits and markers, the associations have to be assessed with a certain degree of accuracy, such that it can be safely relied on, and thus marker genotypes can be used as indicators or predictors of trait genotypes and phenotypes (Stam 2003). For monogenic traits with a clear qualitative contrast between genotypes, such as a single gene-based host plant resistance to pathogens, the assessment of association is straightforward: mapping a monogenic trait goes along with the mapping of markers. For quantitative, multigenic traits, however, a reliable assessment of trait-marker association requires large-scale field experiments as well as statistical techniques, known as QTL mapping (Stam 2003). Progress in the breeding of plantain and banana has been restricted by the complex genetic structure and behavior of cultivated polyploid *Musa*. Mapping in *Musa* has been hampered by the low levels of male and female fertility and seed viability and the absence of large segregating populations. The key to successfully integrating marker-aided breeding into breeding programs will lie in identifying applications in which markers offer real advantages over conventional breeding methods or complement them in novel ways (Semagn et al. 2006). Marker-aided breeding offers significant advantages in the following cases.

1) When phenotypic screening is expensive, difficult or impossible.
2) When the trait is of low heritability (incorporating genes that are highly affected by environment).
3) When the selected trait is expressed late in plant development, like fruit and flower features or adult characters in species with a juvenile period.
4) For incorporating genes for host plant resistance to pathogens or pests that cannot be easily screened for due to special requirement for the gene to be expressed.
5) When the expression of the target gene is recessive.
6) To accumulate multiple genes for one or more traits within the same cultivar, a process used is called gene pyramiding (Sharma et al. 2004; Barone et al. 2005; Yang et al. 2005).

Highly precise MAS approaches require the development of high density linkage maps. Improved molecular markers systems are required to enhance the adoption of MAS. Several factors are important when considering MAS including ease of use, robustness, cost and linkage to trait of interest (de Koeyer et al. 2010). The ideal marker systems for polyploid crops should be dosage sensitive and have the ability to distinguish heterozygous genotypes with multiple haplotypes within the target genomic region by the marker (de Koeyer et al. 2010).

Currently, several molecular marker methods have been used in *Musa* and these differ from each other in their technical requirements, sensitivity and reliability (see Chapter 4). New markers systems are available and have not been exploited in *Musa* as yet.

15.2.3 Gene Pyramiding

Gene pyramiding is the accumulation of multiple genes for one or more traits within the same cultivar (Barone et al. 2005; XiangYan et al. 2005). Genetic stocks produced from gene pyramiding can be used in breeding programs. Gene pyramiding is a very useful approach for the introgression of genes controlling different agronomic traits into one cultivar to ensure that the cultivar has acquired several traits simultaneously (Semagn et al. 2006). For example, genes leading to host plant resistance to different races or biotypes of a pathogen or an insect pest can be pyramided together to make a line with multi-race or multi-biotype resistances, which could be more durable than any single-race or single-biotype resistance (Jiang et al. 2004). The joint expression of pyramided genes was found to provide numerical increases or a broader spectrum of host plant resistance over that conferred by single genes through gene interaction and quantitative complementation (Yoshimura et al. 1995; Singh et al. 2001). Gene pyramiding has been successfully applied in several crop breeding programs, and many cultivars and lines possessing multiple attributes have been produced (Porter et al. 2000; Wang et al. 2001; Samis et al. 2002; Jiang et al. 2004).

Traits which are traditionally regarded as quantitative and not targeted by gene pyramiding programs can be improved using gene pyramiding if major genes affecting the trait are identified (Ashikari and Matsuoka 2006). Gene pyramiding is, however, difficult using conventional breeding methods due to the dominance and epistasis effects of genes governing disease resistance (the stronger resistance genes will always mask the less strong, which cannot be revealed without screening using a virulent strain on the former—in itself undesirable) (Semagn et al. 2006). Moreover, genes with similar reactions to two or more races—so called race-non specific or partial resistance—are difficult to identify and transfer through conventional approaches (Singh et al. 2001), and virtually impossible if stronger race-specific genes are present.

Gene pyramiding programs are also thought to be highly cost intensive. While the breeder's work will be made easier by using a single donor, phenotyping is still required to select the desired segregants in field experiments. It is also highly likely that whatever genes are being stacked into one cultivar might lose their usefulness by the time they are pyramided and subsequently by the time they are used by the breeder.

15.2.4 Marker Systems and Germplasm Characterization

The most frequent use of molecular marker methods in *Musa* has been limited to germplasm characterization and diversity analysis. There are about 1,500 to 3,000 *Musa* accessions with a wide range of morphological variation and genome constitutions (Heslop-Harrison and Schwarzacher 2007) within the germplasm. About 1,000 *Musa* cultivars and 180 wild species, are maintained in tissue culture at the International Transit Centre (ITC) in the Catholic University of Leuven (KULeuven) in Belgium, and these provide a valuable reference collection that is mostly in the public domain and freely accessible for research and breeding. Numerous banana researchers in Asia have developed field-based germplasm collections and well-curated internet databases are now disseminating information about these collections (Pollefeys et al. 2004). Although diversity can be assessed by morphology and flow cytometry, these analyses have limitations and there remain questions about the presence of multiple genotypes with a single name or a single genotype with multiple names (Heslop-Harrison and Schwarzacher 2007). Therefore, DNA-based molecular diversity studies will help to direct plant breeders towards appropriate germplasm to test and select, and to focus germplasm collections towards representing the full range of diversity present in the genus at all ploidy levels (Heslop-Harrison and Schwarzacher 2007). The various techniques used to assess *Musa* diversity and phylogenetic relationships have been addressed in Chapter 3.

15.2.5 Transgenic Breeding

Conventional breeding of bananas is hindered by a number of factors including the long-generation time, triploidy and sterility of most edible cultivars (Pillay et al. 2002). Sources of resistance to many of the major pests and diseases have been identified in a few wild diploid species. However, most landraces are often sterile and cannot be used in breeding, while crosses involving wild species result in the transfer of many unwanted traits together with the desired resistance genes. Furthermore, there are certain diseases such as banana bunchy top virus (BBTV) for which sources of resistance are not known (Sagi et al. 1998).

Although conventional breeding of *Musa* is faced with difficulties, currently available transformation methods may not solve all these difficulties. Increased understanding of responses in *Musa* to biotic and abiotic stresses may provide new opportunities for genetic improvement.

Genetic transformation provides an opportunity for single genes or gene combinations, such as those associated with host plant resistance to pathogens, to be extracted from the genome of the source organism and transferred directly into the desired cultivar, which allows to retain all the original characteristics of the cultivar and adding the desired trait. Furthermore, since most banana cultivars do not produce seeds under natural conditions, crosses with other cultivars or species will seldom occur. In these cases, the introduced gene remains confined to the cultivar in which it has been introduced (Sagi et al. 1998).

Relative success in genetic engineering of bananas and plantains has been achieved enabling the transfer of foreign genes into some cultivars (Sagi et al. 2007). But the scarcity of useful genes, factors that affect transgene expression such as RNA interference, interactions between transgenes and those already present in the plant and the quantitative nature of some traits are still problems that must be considered before accepting that genetic transformation is the only choice for *Musa* improvement. Protocols for the introduction of genes, including the efficient regeneration of shoots in tissue cultures, and transformation methods still remain as major bottlenecks in genetic engineering (Sharma et al. 2005).

15.3 Achievements and Prospects of Transgenic Breeding in *Musa*

15.3.1 Resistance to Black Sigatoka and Fusarium Wilt Disease

Transformation of banana and plantains for fungal diseases started in the 1990s (Sagi et al. 1995). Various transformation techniques have been used to produce transgenic bananas for the cultivars "Williams" (AAA) export banana, "Gros Michel" (AAA) fruit banana, "Bluggoe" (ABB) cooking banana and "Three Hand Planty" (AAB) plantain with antifungal peptides which are highly active *in vitro* against major pathogenic fungi such as black Sigatoka and Fusarium wilt of bananas (Remy et al. 2000). The transgenics showed resistance to black Sigatoka under laboratory conditions. Antimicrobial proteins (AMPs) which are stable, cysteine-rich small peptides isolated from seeds of diverse plant species were also used in developing transgenics for fungal diseases (Sagi et al. 1998).

Improved resistance to Sigatoka was obtained when banana was transformed with the endochitinase gene *ThEn-42* from *Trichoderma harzianum* and the grape stilbene synthase (*StSy*) gene (Vishnevetsky et al. 2010). The superoxide dismutase gene *Cu,Zn-SOD* from tomato, under control of the ubiquitin promoter, was also added to this cassette to improve scavenging of free radicals generated during fungal attack. A 4-year field trial demonstrated several transgenic banana lines with improved tolerance to Sigatoka. Since the genes conferring Sigatoka tolerance may have a wide range of antifungal activities the regenerated banana plants were also inoculated with the fungus *Botrytis cinerea*. The best transgenic lines exhibiting Sigatoka tolerance were also found to have tolerance to *B. cinerea* in laboratory assays (Vishnevetsky et al. 2010). Gene discovery via analysis of EST data from cDNA libraries produced from *Mycosphaerella fijiensis*-infected leaf material from *M. acuminata* ssp. *burmannicoides* "Calcutta 4" (resistant) and "Grande Naine" (AAA genome, susceptible) is ongoing in Brazil (Miller et al. 2009)

15.3.2 Resistance to Banana Bunchy Top Virus (BBTV)

Transgenic research to develop resistance to BBTV has been in progress in Australia and Hawaii. The replication initiation protein (Rep) of nanoviruses is the only viral protein essential for viral replication and represents an ideal target for pathogen derived resistance. In Australia, a Rep-encoded protein (DNA-S1) was identified that suppressed the replication of BBTV. Different constructs of the *Rep* gene were shown to significantly suppress replication of BBTV in banana embryogenic cell suspensions (Tsao and Tsun-Hui 2008). Using such constructs, transgenic bananas with resistance to BBTV have been developed in Australia and Hawaii. The resistant lines have been field tested in Hawaii.

15.3.3 Resistance to Nematodes

The approach adopted for nematode resistance in *Musa* relies on introducing an additional plant gene coding for a protein called cystatin that prevents the digestion in parasitic nematodes. The cystatin suppresses the nematode's ability to grow, lay eggs and build to population levels that damage crops. The advantage of using cystatins is that they are part of the human diet (e.g., present in cereal seeds or eggs) and have no effect on our digestion or health. This approach has been already been used in developing a transgenic Cavendish bananas (AAA) that showed resistance to *Radopholous similis* one of the major nematodes of *Musa* (Atkinson et al. 2004).

15.3.4 Resistance to Banana Xanthomonas Wilt

Banana Xanthomonas or bacterial wilt disease caused by infection with *Xanthomonas campestris* pv. *musacearum* (BXW) has reached epidemic proportions in the Great Lakes region of East and Central Africa (Biruma et al. 2007). The lack of banana germplasm exhibiting resistance to the disease makes it an ideal target for transformation. Transgenic technologies may hold the key for developing bananas that are resistant to the BXW pandemic. The ferredoxin-like amphipathic protein (*pflp*) and hypersensitive response assisting protein (*hrap*), isolated from sweet pepper (*Capsicum annuum*) are novel proteins that can intensify the harpinPSS-mediated hypersensitive response (Chen et al. 2000). Transgenic rice carrying the *pflp* gene showed enhanced resistance to *Xanthomonas oryzae* pv. *oryzae* (Tang et al. 2001). The *pflp* has also been shown to enhance resistance in transgenic orchids against *E. carotovora* (Liau et al. 2003). The elicitor-induced resistance is not specific against particular pathogens, so it could be a very useful strategy for developing broad spectrum resistance. This strategy has been used for developing transgenic banana with resistance to Xanthomonas wilt. Transgenic lines with *pflp* or *hrap* genes have been developed using a protocol based on the *Agrobacterium tumefaciens* technology (Tripathi et al. 2009). These transformed lines of various cultivars have been validated via PCR assay and Southern blot analysis. They have been tested for disease resistance under laboratory conditions and transgenic bananas showed complete resistance. The transgenic lines are now under field trials in Uganda (Tripathi et al. 2010).

15.3.5 Nutritional Enhancement

Banana is rich in natural antioxidants such as vitamin C and vitamin E (Someya et al. 2002; Amorim et al. 2009a, b). High levels of vitamin A deficiency leads to serious health problems, especially in children in low-income regions of the world, such as parts of Asia, Africa and Latin America (Bloem et al. 2005). Micronutrient deficiencies of iron and zinc also results in serious health problems such as mental and physical retardation, reduced resistance to infections and hypogonadism (Whittaker 1998). The genetic enhancement of micronutrient content (i.e., biofortification) of banana by conventional breeding combined with the use of biotechnological tools has the potential to increase the concentrations of micronutrients (Fe, Zn) and vitamin A in new cultivars (Amorim et al. 2011). Improving the nutritional content of *Musa* would have a significant impact on vitamin and nutrient intake for millions of people who depend on the crop for food.

Researchers in Australia are transforming bananas for increased vitamin A, vitamin E or iron. A large suite of both fruit-specific and constitutive promoters that drive pro-vitamin A, vitamin E, or iron accumulation genes, have been cloned into vectors. Four cultivars "Nakinyika", "Mpologoma", "Nakasabira", and "Sukalindizi" have been selected for this study (Dale and Tushemeirewe 2008). A study by Fungo and Pillay (2011) showed that cultivars "Nakitembe", "Entukura" and "Nakhaki" had the highest levels of vitamin A among 10 East African Highland bananas in Uganda and these cultivars may be suited for transformation studies for micronutrients.

15.4 Limitations and Prospects of MAS

One of the major limitations in the use of MAS is the high costs associated with the identification and verification of genetic markers, development of genetic maps, etc. Economics is the key determinant for the application of molecular markers in genetic improvement programs (Dekkers and Hospital 2002). Other factors that influence the cost of utilizing marker-aided breeding include inheritance of the trait, method of phenotypic evaluation, and high costs. The main factors that slow down using molecular breeding technologies in most developing countries include poor infrastructure; inadequate capacity and operational support; and lack of an enabling policy, statutory and regulatory framework at country level, which in turn affects research institutions. Despite these difficulties some developing countries are making progress in using biotechnology for *Musa* improvement.

15.5 Conclusion

Although conventional breeding programs have their limitations, they have shown over time that they can be highly successful. Genome manipulations and interspecific crosses have allowed considerable genetic progress in *Musa* breeding but much remains to be done in the identification of parental combinations that are likely to produce progenies with both high mean and genetic variability (Tenkouano 2001). The development of modern plant molecular and quantitative genetics in the last two decades has the potential to revolutionize what has mostly been experience-based empirical plant breeding (Ye and Smith 2008). Molecular breeding is expected to improve the efficiency of crop breeding by selecting and stacking favorable alleles at target loci (Ribaut et al. 2010). New developments and improvements in marker technology, the integration of functional genomics with QTL mapping, and the availability of more high-density maps are the other factors that will greatly affect the efficiency and effectiveness of QTL mapping and marker-aided breeding in the future (Collard et al. 2005). The development of high-density maps that incorporate new marker types, such as single nucleotide polymorphisms (SNPs) and expressed sequence tags (EST) will provide researchers with a greater arsenal of tools for QTL mapping and marker-aided breeding (Semagn et al. 2010). The number of EST and genomic sequences available in databases is growing rapidly (especially from genome sequencing projects), and the accumulation of these sequences will be extremely useful for the discovery of SNPs and data mining for new markers in the future (Gupta et al. 2001). The potential genetic and economic benefits of using molecular breeding need to be critically compared to those achieved or expected from any existing conventional breeding programs.

Abbreviations

AMPs	:	antimicrobial proteins
BBTV	:	banana bunchy top virus
GWS	:	genome wide selection
MABC	:	marker-assisted backcrossing
MARS	:	marker assisted recurrent selection
MAS	:	marker-assisted selection
MB	:	molecular breeding
MSAP	:	methylation-sensitive amplification polymorphism

PCR	:	polymerase chain reaction
pflp	:	ferredoxin-like amphipathic protein
QTL	:	quantitative trait loci
RAPD	:	random amplified polymorphic DNA

References

Amorim EP, Cohen KO, Amorim VBO, Santos-Serejo JA, Silva SO, Vilarinhos AD, Monte DC, Paes NS, Reis RV (2009a) The genetic diversity of carotenoid-rich bananas measured by Diversity Arrays Technology (DArT). Genet Mol Biol 32: 96–103.

Amorim EP, Lessa LS, Ledo CAS, Amorim VBO, Reis RV, Santos-Serejo JA, Silva SO (2009b) Caracterização agronômica e molecular de genótipos diplóides melhorados de bananeira. Rev Bras Frutic 31: 154–161.

Amorim EP, de Oliveira e Silva S, de Oliveira Amorim VB, Pillay M (2011) Quality improvement of cultivated *Musa.* In: M Pillay, A Tenkouano (eds) Banana Breeding: Progress and Challenges. CRC Press, Boca Raton, FL, USA, pp 251–267.

Ashikari M, Matsuoka M (2006) Identification, isolation and pyramiding of quantitative trait loci for rice breeding. Trends Plant Sci. 11: 344–350.

Atkinson HJ, Grimwood S, Johnston K, Green J (2004) Prototype demonstration of transgenic resistance to the nematode *Radopholous simils* conferred on banana by a cystatin. Transgen Res 13: 135–142.

Barone A, Ercolano MR, Langella R, Monti L, Frusciante L (2005) Molecular marker-assisted selection for pyramiding resistance genes in tomato. Adv Hort Sci 19: 147–152.

Biruma M, Pillay M, Tripathi L, Blomme G, Abele S, Mwangi M, Bandyopadhyay R, Muchunguzi P, Kassim S, Nyine M, Eden-Green S (2007) Banana *Xanthomonas* wilt: a review of the disease, management strategies and future research directions. Afr J Biotechnol 6: 953–962.

Bloem MW, de Pee S, Darnton-Hill I (2005) Micronutrient deficiencies and maternal thinness: first link in the chain of nutritional and health events in economic crises. In: A Bendich, RJ Deckelbaum (eds) Primary and Secondary nutrition. 2nd edn. Humana Press, Totowa, NJ, USA, pp 357–373.

Chen CH, Lin HJ, Ger MJ, Chow D, Feng TY (2000) The cloning and characterization of a hypersensitive response assisting protein that may be associated with the harpin-mediated hypersensitive response. Plant Mol Biol 43: 429–438.

Collard BCY, Jahufer MZZ, Brouwer JB, Pang ECK (2005) An introduction to markers, quantitative trait loci (QTL) mapping and marker-assisted selection for crop improvement: The basic concepts. Euphytica 142: 169–196.

Dale JL, Tushemeirewe W (2008) Biofortification of banana: a grand challenge in global health. Conference on Banana and Plantain in Africa: Harnessing International Partnerships to Increase Research Impact, Mombasa, Kenya.

De Koeyer D, Douglass K, Murphy A, Whitney S, Nolan L, Song Y, De Jong W (2010) Application of high-resolution DNA melting for genotyping and variant scanning of diploid and autotetraploid potato. Mol Breed 25: 67–90.

Dekkers JCM, Hospital F (2002) The use of molecular genetics in the improvement of agricultural populations. Nat Rev Genet 3: 22–32.

Dita MA, Rispail N, Prats E, Rubiales D, Singh KB (2006) Biotechnology approaches to overcome biotic and abiotic stress constraints in legumes. Euphytica 147: 1–24.

Fungo R, Pillay M (2011) ß-carotene content of selected banana genotypes from Uganda. Afr J Biotechnol (in press).

Gimenez C, Palacios G, Colmenares M (2006) *Musa* methylated sequences associated with tolerance to *Mycosphaerella fijiensis* toxins. Plant Mol Biol Rep 24: 33–43.

Gupta PK, Roy JK, Prasad M (2001) Single nucleotide polymorphisms: a new paradigm for molecular marker technology and DNA polymorphism detection with emphasis on their use in plants. Curr Sci 80: 524–535.
Heslop-Harrison JS, Schwarzacher T (2007) Domestication, genomics and the future for banana. Ann Bot 100: 1073–1084.
Jiang GH, Xu CG, Tu JM, Li XH, He YQ, Zhang QF (2004) Pyramiding of insect- and disease-resistance genes into an elite indica, cytoplasm. male sterile restorer line of rice, 'Minghui 63'. Plant Breed 123: 112–116.
Liau CH, Lu JC, Prasad V, Lee JT, Hsiao HH, You SJ, Lee JT, Yang NS, Huang HE, Feng TY, Chen WH, Chan MT (2003) The sweet pepper ferredoxin-like protein (*pflp*) conferred resistance against soft rot disease in *Oncidium* orchid. Transgen Res 12: 329–336.
Lin YH, Chang JY, Liu ET, Chao CP, Huang JW, Chang PFL (2010) Development of a molecular marker for specific detection of *Fusarium oxysporum f.sp cubense* race 4. Eur J Plant Pathol 123: 353–365.
Miller RNG, Bertioli DG, Baurens FC, Quirino BF, Ciampi AY, Santos CMR, Martins NY F, Souza Jr. MT, Pappas Jr. GJ (2009) Understanding plant responses to biotic stress: ongoing research in *Musa*. Acta Hort 828: 255–272.
Miri SM, Mousavi A, Naghavi MR, Mirzaii M, Talaei AR, Khiabani BN (2009) Analysis of induced mutants of salinity resistant banana (*Musa acuminata* cv. Dwarf Cavendish) using morphological and molecular markers. Iran J Biotechnol 7: 86–92.
Newell-McGloughlin M (2008) Nutritionally Improved Agricultural Crops. Plant Physiol 147(3): 939–953
Nwakanma DC, Pillay M, Okoli BE, Tenkouano A (2003) PCR-RFLP of the ribosomal DNA internal transcribed spacers (ITS) provides markers for the A and B genomes in *Musa* L. Theor Appl Genet 108: 154–159.
Nyine M, Pillay M (2011) Breeding increases diversity of bananas. Acta Hort (in press).
Ortiz R (1995) *Musa* genetics. In: S Gowen (ed) Bananas and Plantains. Chapman and Hall, London, UK, pp 84–109.
Ortiz R, Vuylsteke D (1995) Effect of the parthenocarpy gene P1 and ploidy in bunch and fruit traits of plantain and banana hybrids. Heredity 75: 460–465.
Pillay M, Tripathi L (2006) Banana: An overview of breeding and genomics research in *Musa*. In: C Kole (ed) Genome Mapping and Molecular Breeding in Plants, vol. 4: Fruits and Nuts. Springer-Verlag, Heidelberg, Germany, pp 282–301.
Pillay M, Tripathi L (2007) Banana breeding. In: MS Kang, PM Priyadarshan (eds) Breeding Major Food Staples. Blackwell Publishing, Boston, MA, USA, pp 393–428.
Pillay M, Nwakanma DC, Tenkouano A (2000) Identification of RAPD markers linked To A and B genome sequences in *Musa.* Genome 43: 763–767.
Pillay M, Ogundiwin E, Nwakanma DC, Ude G, Tenkouano A (2001) Analysis of genetic diversity and relationships in East African banana germplasm. Theor Appl Genet 102: 965–970.
Pillay M, Tenkouano A, Hartman J (2002) Future challenges in *Musa* breeding. In: Crop Improvement: Challenges in the twenty-first century. MS Kang (ed) Food Products Press, Inc New York, pp 223–252.
Pillay M, Ogundiwin E, Tenkouano A, Dolezel J (2006) Ploidy and genome composition of *Musa* germplasm at the International Institute of Tropical Agriculture (IITA). Afr J Biotechnol 5: 1224–1232.
Pillay M, Tenkouano A, Ortiz R (2011) Molecular breeding of other vegetatively propagated crops: lessons for banana. In: M Pillay, A Tenkouano (eds) Banana Breeding: Progress and Challenges. CRC Press, Boca Raton, FL, USA, pp 321–350.
Pollefeys P, Sharrock S, Arnaud E (2004) Preliminary analysis of the literature on the distribution of wild *Musa* species using MGIS and DIVA-GIS. International Network for the Improvement of Banana and Plantain, Montepellier, France.
Porter DR, Burd JD, Shufran KA, Webster JA (2000) Efficacy of pyramiding greenbug (Homoptera: Aphididae) resistance genes in wheat. J Econ Entomol 93: 1315–1318.

Remy S, Buyens A, Cammue BPA, Swennen R, Sagi L (2000) Production of transgenic banana plants expressing antifungal proteins. Acta Hort 490: 219–277.

Ribaut JM, de Vicente MC, Delannay X (2010) Molecular breeding in developing countries: challenges and perspectives. Curr Opinion Plant Biol 13: 1–6.

Rowe PR (1984) Breeding Bananas and Plantains. Plant Breed Rev 2: 135–155.

Samis K, Bowley S, McKersie B (2002) Pyramiding Mn-superoxide dismutase transgenes to improve persistence and biomass production in alfalfa. J Exp Bot 53: 1343–1350.

Sagi L, Panis B, Remy S, Schoofs H, De Smet K, Swennen R, Cammue B (1995) Genetic transformation of banana (*Musa* spp.) via particle bombardment. Bio/Technology 13: 481–485.

Sagi L, Gregory DM, Remy S, Swennen R (1998) Recent developments in biotechnological research on bananas (*Musa* spp.). Biotechnol Genet Eng Rev 15: 313–317.

Sagi L, Remy S, Swennen R (2007) Transgenic and (trans) genomic research in banana (*Musa* spp.). Afr Crop Sci Conf Proc 8: 592–595.

Semagn K, Bjornstad A, Ndjiondjop MN (2006) Progress and prospects of marker assisted backcrossing as a tool in crop breeding programs. Afr J Biotechnol 5: 2588–2603.

Semagn K, Bjørnstad A, Xu Y (2010) The genetic dissection of quantitative traits in crops. Electronic Journal of Biotechnology, North America, 1325 10 2010.

Sharma KK, Bhatnagar-Mathur P, Thorpe TA (2005) Genetic transformation technology: status and problems. *In Vitro* Cell Dev Biol 41: 102–112.

Sharma PN, Torii A, Takumi S, Mori N, Nakamura C (2004) Marker assisted pyramiding of brown planthopper (*Nilaparvata lugens* Stål) resistance genes *Bph1* and *Bph2* on rice chromosome 12. Hereditas 140: 61–69.

Singh S, Sidhu JS, Huang N, Vikal Y, Li Z, Brar DS, Dhaliwal HS, Khush GS (2001) Pyramiding three bacterial blight resistance genes (*xa5, xa13* and *Xa21*) using marker-assisted selection into indica rice cultivar PR106. Theor Appl Genet 102: 1011–1015.

Someya S, Yumiko Y, Okubob K (2002) Antioxidant compounds from bananas (*Musa Cavendish*). Food Chem 79: 351–354.

Ssebuliba R, Magambo M, Makumbi D, Tenkouano A, Rubahaihayo P, Pillay M (2006a) Biological factors affecting seed production in East African Highland bananas. J Crop Imp 16: 67–77.

Ssebuliba R, Talengera D, Makumbi D, Tenkouano A, Pillay M (2006b) Reproductive efficiency and breeding potential of East African highland banana. Field Crops Res 95: 250–255.

Ssebuliba R, Makumbi D, Pillay M (2009) Patterns of seed set in East African Highland Banana (*Musa* sp.) hybrids. J New Seeds 10: 160–170.

Stam P (2003) Marker-assisted introgression: Speed at any cost? In: Th. JL Van Hintum, A Lebeda, D Pinl, JW Schut (eds) Eucarpia Leafy Vegetables, Valencia, Spain, pp 117–124.

Tang K, Sun X, Hu Q, Wu A, Lin CH, Lin HJ, Twyman RM, Christou P, Feng TY (2001) Transgenic rice plants expressing the ferredoxin-like protein (AP1) from sweet pepper show enhanced resistance to *Xanthomonas oryzae* pv. *Oryzae.* Plant Sci 160: 1035–1042.

Tenkouano A (2001) Current issues and future directions for *Musa* genetic improvement research at the International Institute of Tropical Agriculture. Advancing Banana and Plantain R & D in Asia and the Pacific 10: 11–23.

Tenkouano A, Pillay M, Ortiz R (2011) Breeding techniques. In: M Pillay, A Tenkouano (eds) Banana Breeding: Progress and Challenges. CRC Press, Boca Raton, FL, USA, pp 181–202.

Tripathi L, Mwangi M, Abele S, Aritua V, Tushemereirwe WK, Bandyopadhyay R (2009) *Xanthomonas* Wilt: A Threat to Banana Production in East and Central Africa. Plant Dis 93: 440–451.

Tripathi L, Mwaka H, Tripathi JN, Tushemereirwe WK (2010) Expression of sweet pepper *Hrap* gene in banana enhances resistance to *Xanthomonas campestris* pv. *musacearum*. Mol Plant Pathol 11: 721–731.

Tsao T, Tsun-Hui (2008) Towards the development of transgenic banana bunchy top virus (BBTV)-resistant banana plants: interference with replication. PhD thesis, Queensland University of Technology, Australia.

Vishnevetsky J, White TL, Palmateer AJ, Flaishman M, Cohen Y, Elad Y, Velcheva M, Hanania U, Sahar N, Dgani O, Perl A (2010). Improved tolerance toward fungal diseases in transgenic Cavendish banana (*Musa* spp. AAA group) cv. Grand Nain Transgen Res. 20: 61–72.

Vuylsteke D, Ortiz R, Ferris S, Swennen R (1995) 'PITA-9': A black sigatoka-resistant hybrid from the 'False Horn' plantain gene pool. HortScience 30: 395–397.

Wang XY, Chen PD, Zhang SZ (2001) Pyramiding and marker-assisted selection for powdery mildew resistance genes in common wheat. Acta Genet Sin 28: 640–646.

Whittaker P (1998) Iron and zinc interactions in humans. Amer J Clinical Nutr 68: 442S–446S.

Xia H, Lu XY, Zhao JT, Chen JK, Dai XM, Xiao W, Chen YP, Chen YF, Huang XL (2010) *MaSERK1* gene expression associated with somatic embryogenic competence and disease resistance response in banana (*Musa* spp.). Plant Mol Biol Rep 28: 309–316.

XiangYan Z, ZengYan Z, LiPu D, ZhiYong X, Xiao C (2005) Development of wheat germplasm with multi-resistance to powdery mildew, stripe rust and yellow dwarf virus by molecular marker assisted selection. Sci Agric Sin 38: 2380–2386.

Yang WC, Sacks EJ, Ivey MLL, Miller SA, Francis DM (2005) Resistance in *Lycopersicum esculentum* intraspecific crosses to race T1 strains of *Xanthomonas campestris* pv. *vesicatoria* causing bacterial spot of tomato. Phytopathology 95: 519–527.

Yoshimura S, Yoshimura A, Nelson RJ, Mew TW, Iwata N (1995) Tagging *Xa-1*, the bacterial blight resistance gene in rice, by using RAPD markers. J Breed 45: 81–85.

Ye G, Smith KF (2008) Marker-assisted gene pyramiding for inbred line development: basic principles and practical guidelines. Int J Plant Breed 1: 1–10.

16

Conclusion and Prospects in *Musa* Research

Michael Pillay

ABSTRACT

Despite the advances made in genomic studies in banana many challenges remain. There is need for better genetic resistance to pathogens and pests, greater yields and more efficient production for improving food security, greater yield stability, better use of water, less toxic chemical input, and higher quality. This chapter examines some of the issues facing the genetic improvement of *Musa* and makes proposals for future research in the crop. New data is making it possible to exploit comparative genomics to gain an understanding of *Musa* genetics. The initiation of the *Musa* Genomics Consortium seems to be bearing fruit as new initiatives in genomics research in *Musa* is growing. The use of molecular markers, marker-assisted breeding, gene pyramiding and linkage maps is briefly summarized. The chapter suggests that more comprehensive approaches, including quantitative and qualitative analyses of gene expression products, are necessary at the transcriptomics, proteomics, and metabolomic levels in *Musa*.

The chapter highlights how new approaches such as RNA interference, TILLING, proteomics and metabolomics could revolutionize *Musa* research. A summary of the health aspects of banana is provided and reasons are given why bananas could be considered as an important source of pharmaceuticals and as a crop for biofortification. The scope for research in the biotic stresses that affect banana is still wide and the incidence of new diseases such as Xanthomonas wilt has opened new avenues for research. Transgenic plants have been developed for black Sigatoka, Xanthomonas wilt and nematode resistance but their efficacy

Vaal University of Technology, Private Bag X021, Vanderbijlpark 1900, South Africa; e-mail: *mpillay@vut.ac.za*

List of abbreviations after the text.

in the fields has not been proven as yet. While much research has been done in the biotic stresses that affect banana including fungal diseases, nematodes, insects, viruses, and bacteria, little has been done with regards to abiotic stresses which may include drought, flooding, salinity, metal toxicity, mineral deficiency, adverse pH, adverse temperature, and air pollution. The chapter addresses the progress being made in addressing drought, saline and cold stress. Progress in crop genome sequencing, high resolution genetic mapping and precise phenotyping will accelerate the discovery of functional alleles and allelic variation that are associated with traits of interest for plant breeding.

Keywords: genomics, gene functional analysis, nutritional value, biofortification

16.1 Introduction

Despite the advances made for genetically improving this vegetatively propagated crop, many challenges remain. Banana production faces major challenges from biotic and abiotic stresses. This is aptly summarized by Heslop-Harrison and Schwarzacher (2007) that in addition to the critical need for better genetic resistance to pathogens and pests, there is the need for banana to contribute to better food security through greater yields and more efficient production, greater yield stability, better use of water, less toxic chemical input, and higher quality. In this chapter we examine some of the issues facing the genetic improvement of *Musa* and suggest proposals for future research areas in the crop.

16.2 Advanced Tools for *Musa* Research

16.2.1 Genomics

Effective plant breeding requires knowledge of the *Musa* genome, its evolution and structure (Dolezel and Bartos 2005), complemented with functional genomics, allowing bioinformatics and gene discovery. Understanding the evolution of the banana genome through past DNA changes, hybridization and polyploidy will enable prediction of how genomes will change in the future naturally in the evolutionary processes, or in the accelerated processes of plant breeding (Heslop-Harrison and Schwarzacher 2007). New data is making it possible to exploit comparative genomics to gain an understanding of *Musa* genetics. The complete DNA sequence of a *Musa* species enabling access to all the genes and their control sequences would be valuable for both fundamental research and for application in the future improvement of the crop. The genetic understanding and development of tools means that there are prospects for "new generations" of super-domestication through breeding technology involving sexual hybridization, mutation breeding and targeted transformation approaches (Heslop-Harrison and Schwarzacher 2007). The initiation of the *Musa* Genomics Consortium was to stimulate genomics research in *Musa* as a collective effort. Recent publications in *Musa* genomics suggest that this initiative is bearing fruit. However, much more work is required in this field.

16.2.2 Molecular Markers

Molecular markers have been widely used in *Musa* for detecting genetic variation and genetic relationships in the germplasm, identification of duplicate accessions in field and tissue culture germplasm banks, monitoring genetic stability of tissue culture material, analyzing irradiated banana genotypes and identification of trait markers for use in breeding programs. While the term "molecular marker" in the true sense is used to "flag" the position of a particular gene or the inheritance of a particular characteristic, it has been used loosely in scientific literature to mean a fragment of DNA sequence that is associated to a part of the genome. While molecular markers have been widely used in *Musa* in the latter sense, very few markers have been linked to agronomic traits considering the true sense of the term. Although there is an urgent need for molecular markers for a number of traits in *Musa* such as parthenocarpy, apical dominance, suckering behavior or earliness, very few genetic markers have been developed to date. This may be due to the apparent lack of large segregating populations for these traits and perhaps the absence of large breeding programs for the crop.

New marker techniques (see Chapter 4) such as diversity arrays technology (DArT)—which can detect simultaneously variation at numerous genomic loci without sequence information (Kilian et al. 2005), high resolution DNA melting (HRM) analysis—which has several advantages over other genotyping methods (Montgomery et al. 2007; Reed et al. 2007; Erali et al. 2008)—single nucleotide polymorphism (SNPs) (Rafalski 2002), and targeted-induced local lesion in genome (TILLING) (Colbert et al. 2001) should be further investigated in *Musa*. Marker-assisted selection cannot become a reality in *Musa* if genetic markers are not available.

16.3 Molecular Marker-assisted Breeding

Molecular markers speed up the selection process in crop breeding by alleviating time-consuming approaches of direct screening under greenhouse and field conditions. Molecular markers are particularly useful when targeting characters controlled by several genes. The potential to map different quantitative trait loci (QTL) contributing to an agronomical trait and to identify linked molecular markers opens up the possibility to transfer simultaneously several QTLs and to pyramid QTLs for several agronomical traits in one improved cultivar (Dita et al. 2006). The numerous molecular marker-related techniques used in *Musa* have been addressed in Chapter 4. Although the use of marker-assisted selection may be helpful for *Musa* improvement, its practical application has not been realized mainly due to lack of investment and the lack of knowledge of most agronomics traits.

16.4 Gene Pyramiding

Breeding durable host plant resistance to pathogens and pests is a major task for plant breeders and pyramiding different host plant resistance genes into a genotype is one way of achieving this. There are many examples of introgression and pyramiding of favorable alleles and QTLs in crop plants (Richardson et al. 2006; Bernado 2008; Semagn et al. 2006, 2010). However, gene pyramiding has not been reported in *Musa* as yet. In order to develop resistant cultivars to multiple stresses it is necessary to combine several biotechnological approaches such as transgenesis or mutagenesis and marker-assisted selection to pyramid multiple resistance genes into a single cultivar.

16.5 Linkage Maps

Despite the advantages of molecular maps to sustain banana genetic improvement at diploid and triploid levels, mapping studies in *Musa* have not kept pace with the advances made in other crop species. The first partial molecular linkage map in *Musa*, which exhibited 15 linkage groups with 77 markers of which 36% significantly deviated from Mendelian segregation ($P < 0.05$), was developed by Faure et al. (1993). The second map was developed from 89 individuals coming from a selfed *M. acuminata* diploid "M53". It displayed 11 linkage groups and also showed distorted markers (Noyer et al. 1997). The third map featured 14 linkage groups (Vilarhinos 2004). A fourth (high density) map using DArTs and SSRs markers has been developed (Hippolyte et al. 2010). Despite these advances, mapping in *Musa* can still be regarded as being way behind other crops and remains as an open area for further research.

The importance of cytogenetic maps was highlighted in Chapter 2. Further research is needed to identify the pachytene chromosomes and develop a pachytene karyotype for *Musa*. This would pave the way for detailed cytological characterization of the *Musa* genomes and integrate the DNA sequence based maps with cytological maps.

16.6 Gene Functional Analysis

As already mentioned, the efficiency of both MAS and transgenic approaches will be improved by using information from gene expression studies. Understanding the mechanisms employed by plants to defend themselves against stresses and a more complete knowledge about the genes involved, will allow more precise use of marker-aided breeding and transgenics. Sequence information alone, while valuable and a necessary starting point, is insufficient to answer questions concerning gene function, regulatory networks and the biochemical pathways activated in response to stresses.

The *Musa* genome is now being sequenced. The challenge will be to decipher the function of the thousands of genes identified by sequencing and other genome projects. A number of powerful tools are now available to determine the function of a gene. These include anti-sense RNA suppression, targeted gene replacement, insertional mutagenesis, gene silencing and TILLING approaches. More comprehensive approaches, including quantitative and qualitative analyses of gene expression products, are necessary at the transcriptomic, proteomic, and metabolomic levels in *Musa*.

16.7 RNA Interference Studies in *Musa*: Potential for Research

RNA interference (RNAi) is a homology-dependent gene silencing technology that is initiated by double stranded RNA (dsRNA) (Fire et al. 1998). Since then the expression of dsRNA directed against suitable eukaryotic pathogen target genes in transgenic plants has been shown to give protection against harmful nematodes, herbivorous insects, parasitic weeds and fungi (Niu et al. 2010). Although no RNA interference studies have been reported in *Musa,* the technique has potential to be used in a number of areas such as nematode, viral and fungal resistance in the crop.

16.7.1 Nematode Resistance

RNAi was first demonstrated in the nematode *Caenorhabditis elegans* (Fire et al. 1998). Nematodes are serious pests in bananas that destroy the roots of plants as well as in other crop species causing an estimated damage to world agriculture of around US$125 billion a year (Chitwood 2003). A number of genes, expressed in a range of different tissues and cell types can now be targeted for silencing in different plant parasitic nematodes (Rosso et al. 2009). Transgenic plants expressing dsRNA hairpin structures, which target essential genes of parasitic nematodes, have been developed in soybean significantly reducing the reproductive potential of this nematode with 68% reduction in eggs per gram root tissue (Steeves et al. 2006). These results have shown that RNAi is a powerful technique for investigating the function of nematode genes and identifying potential targets for parasite control (Niu et al. 2010).

16.7.2 Insect Resistance

Transgenic plants have also been developed for engineering insect resistance by RNAi. Transgenic cotton plants were shown to have sufficient levels of dsRNA to suppress gene expression in the cotton bollworm midgut and stunt its growth (Mao et al. 2007). This type of research should pave the way for similar experiments targeting the banana weevil.

16.7.3 Fungal Diseases

RNAi had been used to explore gene function in a wide range of fungal species and fungus-like organisms (Nakayashiki and Nguyen 2008). These studies demonstrate that RNAi mechanism may be broadly existent, even though RNA silencing pathways appear to have diversified significantly (Nakayashiki et al. 2006). To date, there is no existing published paper on RNAi-mediated gene suppression in fungi where the dsRNA molecules are taken up from artificial growth media or plant tissue. As a relatively new technology RNAi mediated plants resistance has both advantages and disadvantages over conventional bio-engineering crops resistance.

(i) multiple pathogens' resistance can be achieved by silencing conserved and essential genes.

(ii) the resistance has the potential to be more durable. The RNAi-mediated resistance is based on RNA hybridization rather than protein-protein interaction; the molecular hybridization cannot be inhibited by minority nucleic acid mutation (Escobar et al. 2001). Therefore, the possibility of the pests overcoming the resistance are likely more difficult. In addition, this biotechnology represents a flexible means of developing pest resistant crops. In theory, all the pests' genes showing detrimental knockdown phenotype can be considered as potential targets. So this strategy would not be limited by the scarcity of resistance genes. A key challenge and essential step for the RNAi-based crop-protection strategies is identification of the right targets. The optimal candidates of pathogenicity—related genes were involved in parasitism, detoxification or the essential genes knockout which lead to a lethal phenotype.

The main concern about RNAi is "off-target" effects. RNAi has been considered to be highly gene specific, but cross-hybridization with transcripts containing partial identity to the introduced dsRNA sequence can induce knockdown of unintended genes, this may result in unexpected mutant phenotypes in addition to the target gene (Jackson et al. 2003; Ma et al. 2006). Despite the successes of this technique in several species, RNA silencing has several drawbacks, i.e., phenotypic instability in later generations (Hannon 2002) and the requirement for a reliable plant transformation system.

16.8 TILLING

TILLING (targeting induced local lesions in genomes) is a general reverse-genetics strategy utilizing traditional mutagenesis and high-throughput mutation discovery that has been applied to many species (McCallum et al. 2000; Colbert et al. 2001). TILLING provides a method to combine the power of induced mutations with expanding sequence information for functional genomics and crop improvement projects. Challenges exist for the efficient adaptation of TILLING for banana, and several approaches are being evaluated. While the development of large mutant populations in banana can be considered a bottleneck, reverse-genetics may provide a practical means for genetic studies and plant improvement (Jain et al. 2011). Tilling in banana appears to be a robust and accurate platform for the discovery of polymorphisms in homologous and homeologous gene targets (Till et al. 2010).

16.9 Proteomics

Analogous to genomics, the systematic analysis and documentation of all protein species of an organism or a specific type of tissue is termed proteomics (Wasinger et al. 1995; Cahill et al. 2000; Anderson et al. 2001). Proteomics addresses analytical questions about the abundance and distribution of proteins in the organism, the expression profiles of different tissues and the identification and localization of individual proteins of interest (Kersten et al. 2002). This concept implies the use of high-throughput analytical techniques. Recent technological advancements have allowed the establishment of valuable methods for quantitative and qualitative protein profiling (Canovas et al. 2004). This approach is very important in evaluating stress-responses because mRNA levels do not always correlate with protein accumulation (Gygi et al. 1999). It is known that large differences in protein turnover and post-translational modifications may lead to large variations between transcriptomic and proteomic data. Therefore protein studies are needed to provide information on their levels and activities (Zivy and de

Vienne 2000). Proteomic based techniques that allow large-scale protein profiling are powerful tools for the identification of proteins involved in biotic and abiotic stress-responses in plants (Gygi and Aebersold 2000). Extensive studies have evaluated changes in protein levels in plant tissues in response to stresses (Canovas et al. 2004; Kim et al. 2003). Most of these studies have been conducted on model plants such *Arabidopsis* and rice (Canovas et al. 2004).

Few studies in proteomics have been reported in *Musa*. Proteins that were significantly up- or down-regulated due to the high-sucrose treatment were studied in banana shoot meristems (Carpentier et al. 2007). Functional proteome analysis of the banana plant (*Musa* spp.) using *de novo* sequence analysis of derivatized peptides was also reported by Samyn et al. (2007). The advantages and pitfalls of proteomics and a transcriptomics to study a model plant such as *Musa* that has very little sequence data was addressed by Carpentier et al. (2008). The conclusion was that each technique has its advantages and disadvantages. When a species is poorly characterized, a 2DE (two dimensional electrophoresis) approach will produce better characterization. However, it was suggested that the advanced derivative of serial analysis of gene expression (SuperSAGE) approach is advisable when one is interested in low-abundant or hydrophobic proteins.

Various proteins have been identified from the *Musa* genome. For example, genes that encode ripening traits in banana, including the genotypes AAA (Cavendish), AAB ("Rasthali" and "Poovan"), AB ("Kanthali"), ABB ("Monthan") (Choudhury et al. 2008).

A proteome database has been set up as a collaboration between the Laboratory for Tropical Crop Improvement, Division of Crop Bio-engineering at the Catholic University of Leuven (KULeuven, Belgium) and the Laboratory of Plant Biochemistry and Physiology at the University of Antwerp (Belgium). The aim of this database is to facilitate data integration, interrelation and querying, and to provide interested researchers of the datasets on which the proteome research papers of this project were based. A number of tools are currently under development that will assist in querying and analyzing data. Clearly more research is needed in this area to understand the various interactions that occur between the host and biotic and abiotic factors that affect *Musa*.

16.10 Metabolomics

Transcriptomic and proteomic data are important steps in deciphering a complex biological process, but they are still insufficient to understand them fully since most biological processes are ultimately mediated by cell metabolites (Dita et al. 2006). Alternative mRNA splicing, protein turnover rates and post-translational modifications that modulate protein activity imply that changes in the transcriptome or proteome do not always correspond to alterations in the cell metabolome (Sumner et al. 2003). Therefore, the only way to obtain a complete understanding of both gene function and molecular events controlling complex plant processes is to analyze in parallel the transcriptome, proteome and the metabolome in an integrative manner (Dixon 2001). Large-scale metabolomic studies are important to assess the involvement of metabolites in various stresses (Dita et al. 2006).

Musa metabolomics has great potential to assist in functional genomics and gene function annotations. However, for further applications of metabolomics, strategies to control and standardize technical processes and infrastructure of this science are necessary (DHL Chi and P. Schenk, unpubl.).

Metabolomics plays an important role in designing better crops because molecular signatures associated with key traits identified by this technology help scientists to rearrange metabolic interactions as well as redesign/select crops with the desired phenotype and genotype (Hall et al. 2008). Metabolite profiling also supports genomics for future breeding programs in plants. For example, a metabolomic study on the phenylpropanoid pathway of banana and a comparative genomic study to plants in the family Haemodoraceae reveal the close relationship between these two species (Kamo et al. 2000). Scientists can use information from the biosynthetic phenylpropanoid pathway retrieved from Haemodoraceae model plants for modifying the genetic basis of phytoalexins in banana fruits with implications for ripening traits. Research on plant metabolites significantly expands our knowledge in both plant proteins and genetics; the possible changes in metabolite profiles explain related changes in proteins and genes that control the metabolic pathways. In comparison with proteomics, metabolomics presents higher throughput and more economical directions (Hollywood et al. 2006). Finally the integration of transcriptomics, proteomics, metabolomics and bioinformatics will provide comprehensive data for *Musa* genetic improvement.

16.11 Transcription Factors

Transcription factors (TFs) are proteins that play an important role in controlling the expression of genes in most biochemical pathways (Kasuga et al. 1999; Eulgem 2005). Genomics studies over the last few years have identified numerous TFs (mainly in *Arabidopsis*) and revealed a high degree of complexity and overlap in the transcriptional regulation of gene expression in response to many stresses (Shinozaki and Yamaguchi-Shinozaki 2000). The understanding of the role of TFs may open new avenues for improving resistance or tolerance to stresses (Singh et al. 2002). A given TF can mediate the response to various stresses (Eulgem 2005; Yamaguchi-Shinozaki and Shinozaki 2005). This characteristic makes the TFs especially attractive for genetic transformation, because a single TF gene can result in resistance or tolerance to various stresses. Following this principle, over-expression of a TF that regulates an ABA-responsive gene conferred multiple stress tolerance in rice (Kim et al. 2003). However, different TFs are also known to respond to the same stress with different but overlapping kinetics (Onate-Sanchez and Singh 2002). The use of TFs for genetic improvement requires a comprehensive knowledge of their biological functions.

16.12 Bananas and Edible Vaccines

With the advent of recombinant DNA technology researchers have been trying to develop genetically modified plants that could provide immunity to infectious diseases. Studies have already shown that genetically engineered plants can act as a vaccine. Plants acting as vaccines would offer the advantage of being inexpensive to produce, and thus they could more easily be made available to developing countries. Edible plant vaccines offer significant new opportunities for making safe and effective oral vaccines. Recently, banana has been labeled as an ideal vehicle for the development of vaccines, in view of the easy digestion, pleasant taste and great acceptability of banana by children. Studies have been ongoing to develop different vaccines in banana (Sala et al. 2003; Arntzen et al. 2005; Kumar et al. 2005).

Genes that express potential vaccine proteins for typhoid fever, rabies, and HIV have been introduced into banana plants (Miller 2006). The products are awaiting clinical trials. The future of plant-made vaccines lies in the researcher's ability to develop vaccines against diseases of importance and to test these vaccines for their efficacy in the host. Advantages of the plant expression system over conventional systems will continue to be elucidated as experimental data amass in this new and growing field of vaccinology (Rice et al. 2005).

16.13 Somaclonal Variation and *In Vitro* Mutagenesis

The capacity of tissue culture to generate genetic variations not been fully exploited in *Musa* genetic improvement. Tissue culture generates a wide range of genetic variation in plants, which can be incorporated in plant breeding programs (Jain 2001). It is well known that somaclonal variation involving callus cultivation and somatic embryogenesis has the capacity to generate genetic variation (Larkin and Scowcroft 1981). The possibility of producing agronomically useful somaclones via organogenesis and somatic embryogenesis has already been reported in *Musa* (Jain et al. 2011). *In vitro* mutagenesis strategies such as treatment with ethyl–methane–sulfonate (EMS), fast neutron radiation and insertional mutagenesis have been applied in plant breeding. These methods induce point mutations, deletions, or insertions, respectively and have been useful in breeding for biotic (Bhagwat and Duncan 1998; Kowalski and Cassells 1999) and abiotic (Khan et al. 2001; Fuller and Eed 2003) stress in crops.

The major difficulty with these techniques is the high quantity of individuals required to find the desired trait. Nevertheless, by using *in vitro* selection systems this disadvantage can be minimized.

Putative stress-resistant lines derived from both conventional breeding and transgenic approaches could be screened using *in vitro* selection. This is particularly attractive for some abiotic stresses, where appropriate screening methods are unavailable or have low efficiency. Although the advantages of the recent high-throughput technologies, coupled with genetic transformation, are emerging as attractive approaches, somaclonal variation and *in vitro* mutagenesis followed by *in vitro* selection offers an alternative way for breeding (Dita et al. 2006).

16.14 Nutrition and Biofortification

16.14.1 Nutritional Value of Banana and Production of Pharmaceuticals

Chapter 4 shows that bananas are an important source of health-promoting phytochemicals. The useful compounds in banana have been reported in many studies (Wall 2006; Davey et al. 2007, 2009; Arora et al. 2008; Cohen et al. 2009a, b; Amorim et al. 2011; Fungo and Pillay 2011). Bananas are rich carotenoids and its fractions, vitamin C, total polyphenols, flavonoids, and compounds with antioxidant activity (Amorim et al. 2011) and could therefore be a potential source of raw materials for the pharmaceutical industry. For example, Englberger et al. (2003a, b, c) identified banana genotypes, commonly known as the Fe'i bananas (section *Australimusa*) with β-carotene values concentrations ranging from 56 μg $100g^{-1}$ to 6,360 μg $100g^{-1}$. The genotypes (*Karat* and *Uht en Yap, Musa troglodytarum*) from the islands of Micronesia that contain 275 times more carotenoids than "Cavendish" may be potentially useful in reducing health problems related to deficiency of vitamin A and/or other micronutrients (Amorim et al. 2011). These bananas could become an ideal source of β-carotene for the pharmaceutical industry. Banana contains high concentrations of serotonin and dopamine, two neurotransmitters that act directly in controlling the release of some hormones and in regulating the circadian rhythms of sleep and appetite (Adao et al. 2005). Kanazawa and Sakakibara (2000) found that the dopamine content, a strong natural antioxidant in banana varied from

80–560 mg $100g^{-1}$ in the peel and 2.5-10 mg $100g^{-1}$ in the pulp in Cavendish bananas. Dopamine is easily absorbed by the body and is considered to play an important role in the control of Parkinson's disease. Someya et al. (2002) found higher levels of the antioxidant, gallocatechin, in the peel (158 mg/100 g dry wt) than in the pulp (29.6 mg/100g dry wt) in "Cavendish" banana, whereas González-Montelongo et al. (2010) quantified the antioxidant activity in banana and observed that the peel contains a large amount of dopamine and L-dopamine, two catecholamines with considerable antioxidant activity. Together, these results suggest that bananas could be considered as an important source of pharmaceutical products.

16.14.2 Biofortification

Biofortification is a means of providing naturally fortified foods to malnourished populations in relatively remote areas where commercially marketed fortified foods are not readily available or are unaffordable (Bouis 2003). There is strong emphasis on identifying foods that can provide the recommended daily requirements of micronutrients especially vitamin A, iron and zinc. Diets deficient in these elements are responsible for major health problem in many low income countries within sub-Saharan Africa (Fungo and Pillay 2011). Vitamin A is considered to have many beneficial health effects such as the prevention of cancer, cardiovascular disease, cataracts, and macular diseases as well as neurologic, inflammatory and immune disorders (Arora et al. 2008). The majority of people in low income countries derive most of their vitamin A from plant foods (Van de Berg et al. 2000). Thus provitamin A carotenoids, especially β-carotene, from plant foods are very important in alleviating vitamin A deficiency in developing countries (Kidmose et al. 2007). Increasing the nutritional value of banana through breeding or genetic engineering can have positive effects on the health and well-being of millions of people that depend on banana as a staple food. Biofortification of bananas can be achieved by conventional breeding or via the transgenic approach if suitable genes for micronutrients are identified. A few studies assessing the mineral composition of banana have been published. A single study to transform banana to improve its nutrient status is underway (Chapter 4). More research will be needed if biofortification of banana is to become a reality.

16.14.3 Post-harvest Products and Dietary Fiber from Banana

The development of processed products from bananas has been slow compared with that of other crops. Processing of bananas has been stimulated by the high volumes of bananas that are lost after harvest especially those that are rejected for the export market in the main banana growing countries of South America (Narayana and Pillay 2011). There is no exact data on post-harvest banana losses in different countries but in general 40 to 50% of fruits and vegetables are lost in the tropics (Mejia 2003) due to spoilage of various kinds.

The production of flour from ripe and unripe fruit and to incorporate the flour into various innovative products such as slowly digestible cookies (Aparicio-Saguilan et al. 2007), high-fiber bread (Juarez-Garcia et al. 2006) and edible films (Sothornvit and Pitak 2007) has been used to increase the utilization of banana and plantain. Bakery products with banana flour have low glycemic index and could be used as a dietary aid for people with low caloric requirements (Juarez-Garcia et al. 2006). Unripe banana is also a good source of indigestible carbohydrate, due to the starch content of the pulp, high cellulose, hemicelluloses and lignin levels.

Advances in biotechnology are now being applied to food commodities to increase the production of new products. The development of processing industries could see an increase in the production of processed products (chips, flour, dried pulp, jam and beverages) from banana. Although there are many innovative laboratory studies of processed banana products, the industrial application of these products is still not exploited (Aurore et al. 2009). The fact that alcohol can be produced from banana also suggests that banana could be used for the production of biofuel.

The large genetic diversity of banana is an asset for any program attempting to explore the various uses of this genus. Better use of bananas and plantains could be achieved by investigating their suitability for different types of processing (Aurore et al. 2009). Bananas represent a great potential raw material for food and non-food processing industries.

16.15 Biotic and Abiotic Stresses

16.15.1 Biotic Stresses

The scope for research in the biotic stresses that affect banana is still wide and the incidence of new diseases such as Xanthomonas wilt has opened new avenues for research. Transgenic plants have been developed for black Sigatoka, Xanthomonas wilt and nematode resistance but their efficacy in the fields has not been proven as yet.

16.15.2 Abiotic Stresses

While much research has been done in the biotic stresses that affect banana including fungal diseases, nematodes, insects, viruses, and bacteria, little has been done with regard to abiotic stresses, which include drought, flooding, salinity, metal toxicity, mineral deficiency, adverse pH, adverse temperature, and air pollution.

16.15.3 Climate Change

It is predicted that climate change will cause temperatures and rainfall to rise in some areas and decrease in others. The general consensus in current literature is that climate change will adversely affect most crops. Banana is a tropical/subtropical crop that usually thrives best in warm climates with high rainfall. Bananas are very sensitive to climatic conditions (Robinson 1996). Such change, which affects soil temperature and moisture levels, also determines the vitality of both beneficial organisms and pests. Rainfall is the major limiting factor in the growth and production of bananas. Adequate moisture is critical for plants, especially during germination and fruit development. Among the abiotic stresses affecting banana productivity, drought is probably the most damaging. Drought and salinity are widespread in many regions and are expected to cause, by 2050, serious salinization of more than 50% of all arable lands (Vinocur and Altman 2005). In view of these factors it will be necessary to research in *Musa* traits for drought and tolerance to temperature extremes that could be influenced by climate change.

16.15.4 Development of Drought Resistance in Musa

Drought is one of the greatest environmental constraints worldwide for agriculture (Boyer 1982). Therefore, developing drought tolerance in plants has huge economical importance. Conventional breeding studies for drought tolerance in banana are virtually non-existent. A trial to identify drought tolerant cultivars in banana was set up in Uganda in 2004 (M. Pillay, unpubl.).

The identification of drought-tolerance traits and genes for *Musa* would facilitate breeding for yield stability under water-limiting conditions. To develop novel strategies for drought tolerance in banana, one of the major research goals will be to understand the molecular mechanisms underlying drought tolerance in the crop (Umezawa et al. 2006). Drought triggers a wide range of plant responses, including alterations in gene expression, accumulation of metabolites such as the phytohormone abscisic acid (ABA) or osmotically active compounds, and synthesis of specific proteins (e.g., largely hydrophilic proteins, proteins that function to scavenge oxygen radicals, chaperone proteins; Reddy et al. 2004). The basic strategy of genetic engineering for drought tolerance is to introduce functional genes that are directly involved in these events. Biotechnology has increased our understanding of the processes underlying plant responses to drought at the molecular and whole plant level and made it possible to identify hundreds of genes that are induced under drought and some of these have been cloned. With the use DNA microarray technology, which allows high-throughput analysis of differential mRNA expression, several hundred stress-induced genes have been identified as candidate genes for genetic engineering (Umezawa 2006).

A range of tools from gene expression patterns to transgenic plants has now become available to better understand drought tolerance mechanisms. New techniques, such as genome-wide tools, proteomics, stable isotopes, and thermal or fluorescence imaging, may help bridge the genotype–phenotype gap. The two main biotechnological approaches, which are being followed to develop drought-tolerant germplasm in plants, are genetic engineering and molecular-marker technology.

Modern bananas are hybrids of two ancestral genomes designated A and B. There is evidence that drought tolerance is related to the B-genome of *Musa*. Banana has a superficial root system and a big leaf area index and needs vast amounts of water to ensure a high yield. The yield loss due to drought is estimated at 40% for an area with an annual yearly rainfall lower than 1,200 mm (R. Swennen, KULeuven, pers. comm.). Therefore, a better knowledge on drought tolerance and drought tolerant cultivars is essential in *Musa*.

Studies are underway to identify drought tolerant *Musa* cultivars (*in vitro*) using different osmotic. Preliminary investigations showed that "Cachaco" (ABB) and "Lep Chang Kut" (ABB) were drought tolerant while "Mbwazirume" (AAA) is sensitive to drought. This characterization of water use efficiency of different banana cultivars will be linked to a high throughput proteome analysis, to validate those results at the transcriptome level and to correlate the phenotype to the genomic constitution (R. Swennen, KULeuven, pers. comm.). The large biodiversity in the genus needs to be investigated for drought tolerance. By linking multiple domains of physiology, proteomics, transcriptomics and genomics it is envisaged that the complex mechanisms of drought tolerance in *Musa* will be elucidated. Identification of drought tolerant cultivars could be used *per se* or in crosses to breed plants with enhanced adaptation to drought-prone environments.

16.15.5 Cold Tolerance

Experiments to determine cold-tolerant banana cultivars have been reported (Ram et al. 2008). The former study showed that a number of banana cultivars displayed frost tolerance under field conditions. Among the dessert accessions, four cultivars were found to be tolerant and six were moderately tolerant. Among the cooking bananas, seven accessions were identified as being tolerant while 12 accessions were moderately tolerant. The study also found that cultivars with a higher proportion of the A genome were more susceptible to frost injury, while accessions with high proportion of the B genome were more tolerant to frost injury.

16.15.6 Salinity Stress

Research on salinity stress in *Musa* has been reported in banana cultivars (Gomes et al. 2002; Willadino et al. 2011), in the cultivar "Pacovan" infected with arbuscular mycorrhizal fungi (Yano-Melo et al. 2003), and in induced mutants (Miri et al. 2009). Although these studies showed that some banana cultivars are able to withstand a certain amount of salt stress, further work is essential to gain more information about salinity stress in the crop.

16.16 *Musa* Germplasm Conservation and Exchange

The wild relatives of *Musa* are at risk due to climate change. Wild relatives are a valuable source of genes that are necessary to boost the ability of cultivated crops to resist biotic and abiotic stresses. Current banana breeding programs have used wild relatives as sources of resistance against black Sigatoka, nematodes, banana weevil, and Fusarium wilt, and large bunch sizes, etc. Many studies have shown that domestication has narrowed down the genetic variation in banana (Pillay et al. 2001; Nyine and Pillay 2011). Climate change may make reliance on wild relatives of banana more acute since it may make areas too hot, too cold, too wet or too dry for banana production. There is an urgent need to collect and characterize the wild relatives of banana before they disappear. At the moment, existing collections are conserving only a fraction of the diversity of wild species that are out there.

16.17 Doubled Haploids

Doubled haploid (DH) technology refers to the use of the microspore or anther cultures to obtain haploid embryos. This technology offers breeders a tool for the rapid production of homozygous lines. These homozygous lines can be multiplied and released as cultivars, or used similarly as recombinant inbred lines for molecular mapping and/or in breeding programs (Martinez et al. 2002). An efficient DH production technology can greatly reduce the time and cost of cultivar development (Liu et al. 2002).

16.18 Conclusion

In addition to conventional breeding a variety of biotechnological approaches are now available for genetic improvement of *Musa*. Over the past two decades biotechnology has emerged as a promising tool to overcome pathogens and pests in crop plants. While progress in some crops has been dramatic, limited progress has been achieved in *Musa*. The current advances in tissue culture-based techniques, genetic transformation and MAS, together with the advances in powerful new "omics" technologies offer great potential to improve this situation (Dita et al. 2006). Genomics-assisted breeding approaches along with bioinformatics capacity and metabolomics resources are becoming essential components of crop improvement programs worldwide. Progress in crop genome sequencing, high resolution genetic mapping and precise phenotyping will accelerate the discovery of functional alleles and allelic variation that are associated with traits of interest for plant breeding (Ortiz 2011). This chapter has highlighted some potential new areas of research and some emerging technologies that could greatly benefit the genetic enhancement of *Musa*.

Abbreviations

DArT	:	diversity arrays technology
HRM	:	high resolution melting
RNAi	:	RNA interference
TFs	:	Transcription factors
TILLING	:	targeted-induced local lesion in genome

References

Adao RC, Beatriz A. Glória M (2005) Bioactive amines and carbohydrate changes during ripening of 'Prata' banana (*Musa acuminata* × *M. balbisiana*). Food Chem 90: 705–711.

Amorim EP, de Oliveira e Silva S, de Oliveira Amorim VB, Pillay M (2011) Quality improvement of cultivated *Musa*. In: M Pillay, A Tenkouano (eds) Banana Breeding: Progress and Challenges. CRC Press, Boca Raton, FL, USA, pp 251–267.

Anderson NG, Matheson A, Anderson NL (2001) Back to the future: the human protein index (HPI) and the agenda for post proteomic biology. Proteomics 1: 3–12.

Aparicio-Saguilan A, Sayago-Ayerdi SG, Vargas-Torres A, Tovar J, Ascencio-Otero TE, Bello-Perez LA (2007) Slowly digestible cookies prepared from resistant starch-rich lintnerized banana starch. J Food Compo. Anal 20: 175–181.

Arora A, Choudary D, Agarwal G, Singh VP (2008) Compositional variation in β-carotene content, carbohydrate and antioxidant enzymes in selected banana cultivars. Int J Food Sci Technol 43: 1913–1921.

Arntzen C, Plotkin S, Dodet B (2005) Plant-derived vaccines and antibodies: potential and limitations. Vaccine 23: 1753–1756.

Aurore G, Parfait B, Fahrasmane L (2009) Bananas, raw materials for making processed food products. Trends Food Sci Technol 20: 78–91.

Bernardo R (2008) Molecular Markers and Selection for Complex Traits in Plants: Learning from the Last 20 Years. Crop Sci 48: 1649–1664.
Bhagwat B, Duncan EJ (1998) Mutation breeding of banana cv Highgate (*Musa* spp., AAA Group) for tolerance to *Fusarium oxysporum* f. sp. *cubense* using chemical mutagens. Sci Hort 73: 11–22.
Bouis HE (2003) Micronutrient fortification of plants through plant breeding: can it improve nutrition in man at low cost. Proc Nutr Soc 62: 403–411.
Boyer JS (1982) Plant productivity and environment. Science 218: 443–448.
Cahill DJ, Nordhoff E, O'Brien J, Klose J, Eickhoff H, Lehrach H (2000) Bridging genomics and proteomics. In: S Pennington, M Dunn (eds) 2000 Proteomics. BIOS Scientific Publishers, New York, USA, pp 1–17.
Canovas F, Dumas-Gaudot E, Recorbet G, Jorrin J, Mock H-P, Rossignol M (2004) Plant proteome analysis. Proteomics 4: 285–298.
Carpentier SC, Witters E, Laukens K, Van Onckelen H, Swennen R, Panis B (2007) Banana (*Musa* spp.) as a model to study the meristem proteome: acclimation to osmotic stress. Proteomics 7: 92–105.
Carpentier SC, Coemans B, Podevin N, Laukens K, Witters E, Matsumura H, Terauchi R, Swennen R, Panis B (2008) Functional genomics in a non-model crop: transcriptomics or proteomics? Physiol Planta 133: 117–130.
Chitwood DJ (2003) Research on plant-parasitic nematode biology conducted by the United States Department of Agriculture-Agricultural Research Service. Pest Manag Sci 59: 48–753.
Choudhury SR, Roy S, Sengupta DN (2008) Characterization of cultivar differences in β *1–3* glucanase gene expression, glucanase activity and fruit pulp softening rates during fruit ripening in three naturally occurring banana cultivars. Plant Cell Rep 28: 1641–1653.
Cohen KO, Amorim EP, Paes NS, Silva SO, Santos-Serejo JA, Costa AM, Souza HN, Baiocchi MV (2009a) Teores de flavonóides e polifenóis totais em genótipos diplóides de bananeira. Comunicado Técnico. Embrapa Mandioca e Fruticultura 1: 1–4.
Cohen KO, Amorim EP, Paes NS, Silva SO, Baiocchi MV, Souza HN (2009b) Vitamina C, carotenóides, compostos fenólicos e atividade antioxidante em genótipos de banana. Comunicado Técnico. Embrapa Mandioca e Fruticultura 2: 1–4.
Colbert T, Till BJ, Tompa R, Reynolds S, Steine MN, Yeung AT, McCallum CM, Comai L, Henikoff S (2001) High-throughput screening for induced point mutations. Plant Physiol 126: 480–484.
Davey MW, Stals W, Ngoh-Newilah G, Tomekpe K, Lutsy C, Markham R, Swennen R, Keulemans J (2007) Sampling strategies and variability in fruit pulp micronutrient contents of West and Central African bananas and plantains (*Musa* sp.). J Agric Food Chem 55: 2633–2644.
Davey MW, Van den Berg I, Markham R, Swennen R, Keulemans J (2009) Genetic variability in *Musa* fruit provitamin A carotenoids, lutein and mineral micronutrient contents. Food Chem 115: 806–813.
Dita MA, Rispail N, Prats E, Rubiales D, Singh AB (2006) Biotechnology approaches to overcome biotic and abiotic stress in legumes. Euphytica 147: 1–24.
Dixon RA (2001) Natural products and plant disease resistance. Nature 411: 843–847.
Dolezel J, Bartos J (2005) Plant DNA flow cytometry and estimation of nuclear genome size. Ann Bot 95: 99–110.
Englberger L, Darnton-Hill I, Coyne T, Fitzgerald MH, Marks GC (2003a) Carotenoid-rich bananas: a potential food source for alleviating vitamin A deficiency. Food Nutr Bull 24: 303–318.
Englberger L, Schierle J, Marks GC, Fitzgerald MH (2003b) Micronesian banana, taro, and other foods: newly recognized sources of provitamin A and others carotenoids. J Food Compos Anal 16: 3–19.

Englberger L, Aalbersberg W, Ravi P, Bonnin E, Marks GC, Fitzgerald MH, Elymore J (2003c) Further analyses on Micronesian banana, taro, breadfruit and other foods for provitamin A carotenoids and minerals. J Food Compos Anal 16: 219–236.
Erali M, Voelkerding KV, Wittwer CT (2008) High resolution melting applications for clinical laboratory medicine. Exp Mol Pathol 85: 50–58.
Escobar MA, Civerolo EL, Summerfelt KR, Dandekar AM (2001) RNAi-mediated oncogene silencing confers resistance to crown gall tumorigenesis. Proc Natl Acad Sci USA, 98: 13437–13442.
Eulgem T (2005) Regulation of the Arabidopsis defense transcriptome. Trends Plant Sci 10: 71–78.
Faure S, Noyer JL, Horry JP, Bakry F, Lanaud C, Gonzalez de Leon D (1993) A molecular marker-based linkage map of diploid bananas (*Musa acuminata*). Theor Appl Genet 87: 517–526.
Fire A, Xu S, Montgomery MK, Kostas SA, Driver SE, Mello CC (1998) Potent and specific genetic interference by double-stranded RNA in *Caenorhabditis elegans*. Nature 391: 806–811.
Fuller MP, Eed MHI (2003) The development of multiple stress resistant cauliflower using mutagenesis in conjunction with a microshoot tissue culture technique. Acta Hort 618: 71–76.
Fungo R, Pillay M (2011) ß-carotene content of selected banana genotypes from Uganda. Afr J Biotechnol (in press).
Gomes EWF, Willadino L, Martins LSS, Camara TR (2002) The effects of salinity on five banana genotypes (*Musa* spp). Dev Plant Soil Sci 92: 410–411.
González-Montelongo R, Lobo MG, González M (2010) Antioxidant activity in banana peel extracts: Testing extraction conditions and related bioactive compounds. Food Chem 119: 1030–1039.
Gosal SG, Wani SH, Kang MS (2009) Biotechnology and drought tolerance. J Crop Improv 23: 19–54.
Gygi SP, Aebersold R (2000) Mass spectrometry and proteomics. Curr Opin Chem Biol 4: 489–494.
Gygi SP,Rochon Y, Franza BR, Aebersold R (1999) Correlation between protein and mRNA abundance in yeast. Mol Cell Biol 19: 1720–1730.
Hall RD, Brouwerand ID, Fitzgerald MA (2008) Plant metabolomics and its potential application for human nutrition. Physiol Planta 132: 162–175.
Hannon GJ 2002. RNA interference. Nature 418: 244–251.
Heslop-Harrison JS, Schwarzacher T (2007) Domestication, genomics and the future for banana. Ann Bot 100: 1073–1084.
Hippolyte I, Bakry V, Seguin M, Gardes L, Rivallan R, Risterucci AM, Jenny C, Perrier X, Carreel F, Argout X, Piffanelli P, Khan IA, Miller RNG, Pappas GJ, Mbéguié DM, Matsumoto T, Bernardinis VD, Huttner E, Kilian A, Baurens FC, D'Hont A, Cote F, Courtois B, Glaszmann JC (2010) A saturated SSR/DArT linkage map of *Musa acuminata* addressing genome rearrangements among bananas. BMC Plant Biol 10: 65.
Hollywood K, Brison DR, Goodacre R (2006) Metabolomics: current technologies and future trends. Proteomics 6: 4716–4723.
Jackson AL, Bartz SR, Schelter J, Kobayashi SV, Burchard J, Mao M, Li B, Cavet G, Linsley PS (2003) Expression profiling reveals off-target gene regulation by RNAi. Nat Biotechnol 21: 635–637.
Jain SM (2001) Tissue culture-derived variation in crop improvement Euphytica 118: 153–166.
Jain SM, Till J, Suprasanna P, Roux N (2011) Mutations and cultivar development of banana. In: M Pillay, A Tenkouano (eds) Banana Breeding: Progress and Challenges. CRC Press, Boca Raton, FL, USA, pp 203–217.

Juarez-Garcia E, Agama-Acevedo E, Sayago-Ayerdi SG, Rodriguez-Ambriz SL, Bello-Perez LA (2006) Composition, digestibility and application in breadmaking of banana flour. Plant Foods Human Nutr 61: 131–137.

Kamo T, Hirai N, Tsuda M, Fujioka D, Ohigashi H (2000) Changes in the content and biosynthesis of phytoalexins in banana fruits. Biosci Biotechnol Biochem 64: 2089–2098.

Kanazawa K, Sakakibara H (2000) High content of dopamine, a strong antioxidant, in Cavendish banana. J Agric Food Chem 48: 844–848.

Kasuga M, Liu Q, Miura S, Yamaguchi-Shinozaki K, Shinozaki K (1999) Improving plant drought, salt, and freezing tolerance by gene transfer of a single stress-inducible transcription factor. Nat Biotechnol 17: 287–291.

Kersten B, Bürkle L, Kuhn EJ, Giavalisco P, Konthur Z, Lueking A, Walter G, Eickhoff H, Schneider U (2002) Large-scale plant proteomics. Plant Mol Biol 48: 133–141.

Khan AJ, Hassan S, Tariq M, Khan T (2001) Haploidy breeding and mutagenesis for drought tolerance in wheat. Euphytica 120: 409–414.

Kidmose U, Christensen LP, Agili SM, Thilsted SH (2007) Effect of home preparation practices on the content of provitamin A carotenoids in coloured sweet potato varieties (*Ipomoea batatas* Lam.) from Kenya. Innov Food Sci Emerg Technnol 8: 399–406.

Kilian A, Huttner E, Wenzl P, Jaccoud D, Carling J, Caig V, Evers M, Heller-Uszynska K, Cayla C, Patarapuwadol S, Xia L, Yang S, Thomson B (2005) The fast and the cheap: SNP and DArT-based whole genome profiling for crop improvement. In: Proceedings of the International Congress "In the wake of the double helix: from the green revolution to the gene revolution", 27–31 May 2003, 443–461. Avenue Media, Bologna, Italy.

Kim ST, Cho KS, Yu S, Kim SG, Hong JC, Han CD, Bae DW, Nam MH, Kang KY (2003) Proteomic analysis of differentially expressed proteins induced by rice blast fungus and elicitor in suspension-cultured rice cells. Proteomics 3: 2368–2378.

Kowalski B, Cassells AC (1999) Mutation breeding for yield and *Phytophthora infestans* (Mont.) de Bary foliar resistance in potato (*Solanum tuberosum* L-cv. Golden Wonder) using computerized image analysis in selection. Potato Res 42: 121–130.

Kumar GB, Ganapathi TR, Revathi CJ, Srivinas L, Bapat V (2005) Expression of hepatitis B surface antigen in transgenic banana plants. Planta 222: 484–493.

Larkin PJ, Scowcroft WR (1981) Somaclonal variation—a novel source of variability from cell-cultures for plant improvement. Theor Appl Genet 60: 197–214.

Liu JH, Xu XY, Deng XX (2005) Intergeneric somatic hybridization and its application to crop genetic improvement. Plant Cell Tiss Org Cult 82: 19–44.

Liu WG, Zheng MY, Polle EA, Konzak CF (2002) Highly efficient doubled-haploid production in wheat (*Triticum aestivum* L.) via induced microspore embryogenesis. Crop Sci 42: 686–692.

Ma Y, Creanga A, Lum L, Beachy PA (2006) Prevalence of off-target effects in *Drosophila* RNA interference screens. Nature 443: 359–363.

Mao YB, Cai WJ, Wang JW, Hong GJ, Tao XY, Wang LJ, Huang YP, Chen XY (2007) Silencing a cotton bollworm *P450* monooxygenase gene by plant-mediated RNAi impairs larval tolerance of gossypol. Nat Biotechnol 25: 1307–1313.

Martinez VA, Hill WG, Knott S (2002) On the use of double haploids for detecting QTL in outbred populations. Heredity 88: 423–431.

McCallum CM, Comai L, Greene EA, Henikoff S (2000) Targeted screening for induced mutations. Nat Biotechnol 18: 455–457.

Mejia DJ (2003) Preservation of fruits and vegetables by combined methods Technologies. International Tropical Fruits Network, pp 1–2.

Miller HI (2006) Biotech production is blooming in the tropics. Point of View 26(19): 1.

Miri SM, Mousavi A, Naghavi MR, Mirzaii M, Talaei AR, Khiabani BN (2009) Analysis of induced mutants of salinity resistant banana (*Musa acuminata* cv. Dwarf Cavendish) using morphological and molecular markers. Iran J Biotechnol 7: 86–92.

Montgomery J, Wittwer CT, Palais R, Zhou L (2007) Simultaneous mutation scanning and genotyping by high-resolution DNA melting. Nat Protoc 2: 59–66.

Nakayashiki H, Kadotani N, Mayama S (2006) Evolution and =diversification of RNA silencing proteins in fungi. J Mol Evol 63: 127–135.
Nakayashiki H, Nguyen QB (2008) RNA interference: roles in fungal biology. Curr Opin Microbiol 11: 494–502.
Narayana CK, Pillay M (2011) Postharvest processed products from banana. In: M Pillay, A Tenkouano (eds) Banana Breeding: Progress and Challenges. CRC Press, Boca Raton, FL, USA, pp 269–284.
Niu JH, Jian H, Xu JM, Guo YD, Liu Q (2010) RNAi technology extends its reach: engineering plant resistance against harmful eukaryotes. Afr J Biotechnol 9: 7573–7582.
Noyer JL, Dambier D, Lanaud C, Lagoda PJL (1997) The saturated map of diploid banana (*Musa acuminata*). Plant and Animal Genome Conference V, January 12–16, San Diego, CA, USA.
Nyine M, Pillay M (2011) Breeding increases diversity of bananas. Acta Hort (in press).
Ortiz R (2011) Revisiting the green revolution: seeking innovations for a changing world. Chron Hort 51(1): 6–11.
Onate-Sanchez L, Singh KB (2002) Identification of Arabidopsis ethylene-responsive element binding factors with distinct induction kinetics after pathogen infection. Plant Physiol 128: 1313–1322.
Pillay M, Tripathi L (2006) Banana: An overview of breeding and genomics research in *Musa*. In: C Kole (ed) Genome Mapping and Molecular Breeding in Plants, vol. 4: Fruits and Nuts Springer-Verlag, Heidelberg, Germany, pp 282–301.
Pillay M, Tripathi L (2007) Banana breeding. In: MS Kang, PM Priyadarshan (eds) Breeding Major Food Staples. Blackwell Publishing, Boston, MA, USA, pp 393–428.
Pillay M, Ogundiwin E, Nwakanma DC, Ude G, Tenkouano A (2001) Analysis of genetic diversity and relationships in East African banana germplasm. Theor Appl Genet 102: 965–970.
Pillay M, Tenkouano A, Ortiz R (2011) Molecular breeding of other vegetatively propagated crops: lessons for banana. In: M Pillay, A Tenkouano (eds) Banana Breeding: Progress and Challenges. CRC Press, Boca Raton, FL, USA, pp 321–350.
Rafalski JA (2002) Novel genetic mapping tools in plants: SNPs and LD-based approaches. Plant Sci 162: 329–333.
Ram K, Rajan S, Singh A (2008) Evaluation of banana germplasm (*Musa* spp.) for frost tolerance under field conditions in subtropics. Indian J Hort 65: 35–39.
Reddy AR, Chaitanya KV, Vivekanandan M (2004) Drought-induced responses of photosynthesis and antioxidant metabolism in higher plants. J Plant Physiol 161: 1189–1202.
Reed GH, Kent JO, Wittwer CT (2007) High-resolution DNA melting analysis for simple and efficient molecular diagnostics. Pharmacogenomics 8: 597–608.
Rice J, Ainley WM, Shewen P (2005) Plant-made vaccines: biotechnology and immunology in animal health. Anim Health Res Rev 6: 199–209.
Richardson KL, Vales MI, Kling JG, Mundt CC, Hayes PM (2006) Pyramiding and dissecting disease resistance QTL to barley stripe rust. Theor Appl Genet 113: 485–495.
Robinson JC (1996) Bananas and Plantains. Institute of Tropical and Subtropical Crops, South Africa—University Press, Cambridge, UK.
Rosso MN, Jones JT, Abad P 2009. RNAi and functional genomics in plant parasitic nematodes. Annu Rev Phytopathol 47: 207–232.
Sagi L, Remy S, Swennen R (2007) Transgenic and (trans) genomic research in banana (*Musa* spp.). Afr Crop Sci Conf Proc 8: 592–595.
Sala F, Rigano MM, Barbante A, Basso B, Walmsley AM, Castiglione S (2003) Vaccine antigen production in transgenic plants: Strategies, gene constructs and perspectives. Vaccine 21: 803–808.
Samyn B, Sergeant K, Carpentier S, Debyser G, Panis B, Swennen R, Van Beeumen J (2007) Functional Proteome Analysis of the Banana Plant (*Musa* spp.) Using de Novo Sequence Analysis of Derivatized Peptides. J Proteome Res 6: 70–80.

Semagn K, Bjornstad A, Ndjiondjop MN (2006) Progress and prospects of marker assisted backcrossing as a tool in crop breeding programs. Afr J Biotechnol 5: 2588–2603.
Semagn K, Bjørnstad A, Xu Y (2010) The genetic dissection of quantitative traits in crops. Electronic Journal of Biotechnology, North America, 1325 10 2010.
Shinozaki K, Yamaguchi-Shinozaki K (2000) Molecular responses to dehydration and low temperature: differences and cross-talk between two stress signaling pathways. Curr Opin Plant Biol 3: 217–223.
Someya S, Yoshiki Y, Okubo K (2002) Antioxidant compounds from bananas (*Musa* Cavendish). Food Chem 79: 351–354.
Singh KB, Foley RC, Onate-Sanchez L (2002) Transcription factors in plant defense and stress responses. Curr Opin Plant Biol 5: 430–436.
Sothornvit S, Pitak N (2007) Oxygen permeability and mechanical properties of banana films. Food Res Int 40: 365–370.
Steeves RM, Todd TC, Essig JS, Trick HN (2006) Transgenic soybeans expressing siRNAs specific to a major sperm protein gene suppress *Heterodera glycines* reproduction. Funct Plant Biol 33: 991–999.
Sumner LW, Mendes P, Dixon RA (2003) Plant metabolomics: Large-scale phytochemistry in the functional genomics era. Phytochemistry 62: 817–836.
Till BJ, Jankowicz-Cieslak J, Sagi L, Huynh OA, Utsushi H, Swennen R, Terauchi R, Mba C (2010) Discovery of nucleotide polymorphisms in the *Musa* gene pool by Ecotilling. Theor Appl Genet 121: 1381–1389.
Umezawa T, Fujita M, Fujita Y, Yamaguchi-Shinozaki K, Shinozaki K (2006) Engineering drought tolerance in plants: discovering and tailoring genes to unlock the future. Curr Opin Biotechnol 17: 113–122.
Van den Berg H, Faulks R, Fernando Granado H, Hirschberg J, Olmedilla B, Sandmann G, Southon S, Stahl W (2000) The potential for the improvement of carotenoid levels in foods and the likely systemic effects. J Sci Food Agric 80: 880–912.
Vilarhinos AD (2004) Cartographie génétique et cytogénétique chez le bananier: caractérisation des translocations. Ecole Nationale Supérieure Agronomique, Montpellier, France.
Vinocur B, Altman A (2005) Recent advances in engineering plant tolerance to abiotic stress: achievements and limitations. Curr Opin Biotechnol 16: 123–132.
Wall MM (2006) Ascorbic acid, vitamin A, and mineral composition of banana (*Musa* sp.) and papaya (*Carica papaya*) cultivars grown in Hawaii. J Food Compos Anal 19: 434–445.
Willadino L, Gomes EWF, de F. e Silva EF, Martins LSS, Camara TR (2011) Effect of salt stress on banana tetraploid genotypes. Rev Bras Eng Agríc Ambient 15: 53–59.
Wasinger VC, Cordwell SJ, Cerpa-Poljak A, Yan JX, Gooley AA, Wilkins MR, Duncan MW, Harris R, Williams KL, Humphery-Smith I (1995) Progress with gene-product mapping of the Mollicutes: *Mycoplasma genitalium*. Electrophoresis 16: 1090–1094.
Yamaguchi-Shinozaki K, Shinozaki K (2005) Organization of cisacting regulatory elements in osmotic- and cold-stress-responsive promoters. Trends Plant Sci 10: 88–94.
Yano-Melo AM, Saggin OJ, Maia LC (2003) Tolerance of mycorrhized banana (*Musa* sp. cv. Pacovan) plantlets to saline stress. Agric Ecosyst Environ 95: 343–348.
Zivy M, de Vienne D (2000) Proteomics: A link between genomics, genetics and physiology. Plant Mol Biol 44: 575–580.

Index

Color Plate Section

Chapter 11

Figure 11-1 Screenshot of the GNPAnnot genome browser showing a *Musa* BAC sequence. Genome features are dispatched in different tracks according the kind of analyses performed (e.g., Eugene, Repet, BlastX). By clicking one of the features, a pop-up window provides links to a detailed report. Links to the Apollo and Artemis software are available on the track dedicated to manual curation.

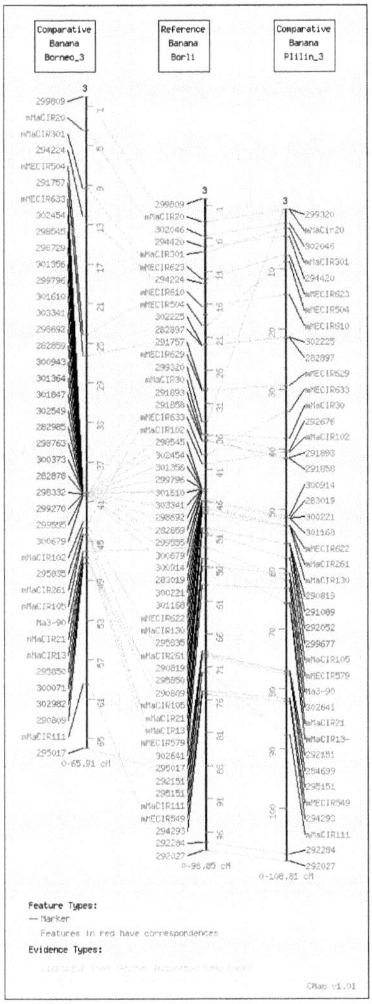

Figure 11-4 Graphical representation in CMap of correspondences between two banana linkage groups

Chapter 12

Figure 12-1 Black leaf streak symptoms in the variety BRS Maravilha (picture by Edson Perito Amorim).

www.ingramcontent.com/pod-product-compliance
Ingram Content Group UK Ltd.
Pitfield, Milton Keynes, MK11 3LW, UK
UKHW022215190926
475702UK00003B/371

9781578087884